CJ Lim + Ed

smartcities
resilient landscapes
+ eco-warriors

2nd edition

Ysaber
Keep
dreaming
2020

Routledge
Taylor & Francis Group

LONDON AND NEW YORK

Contents

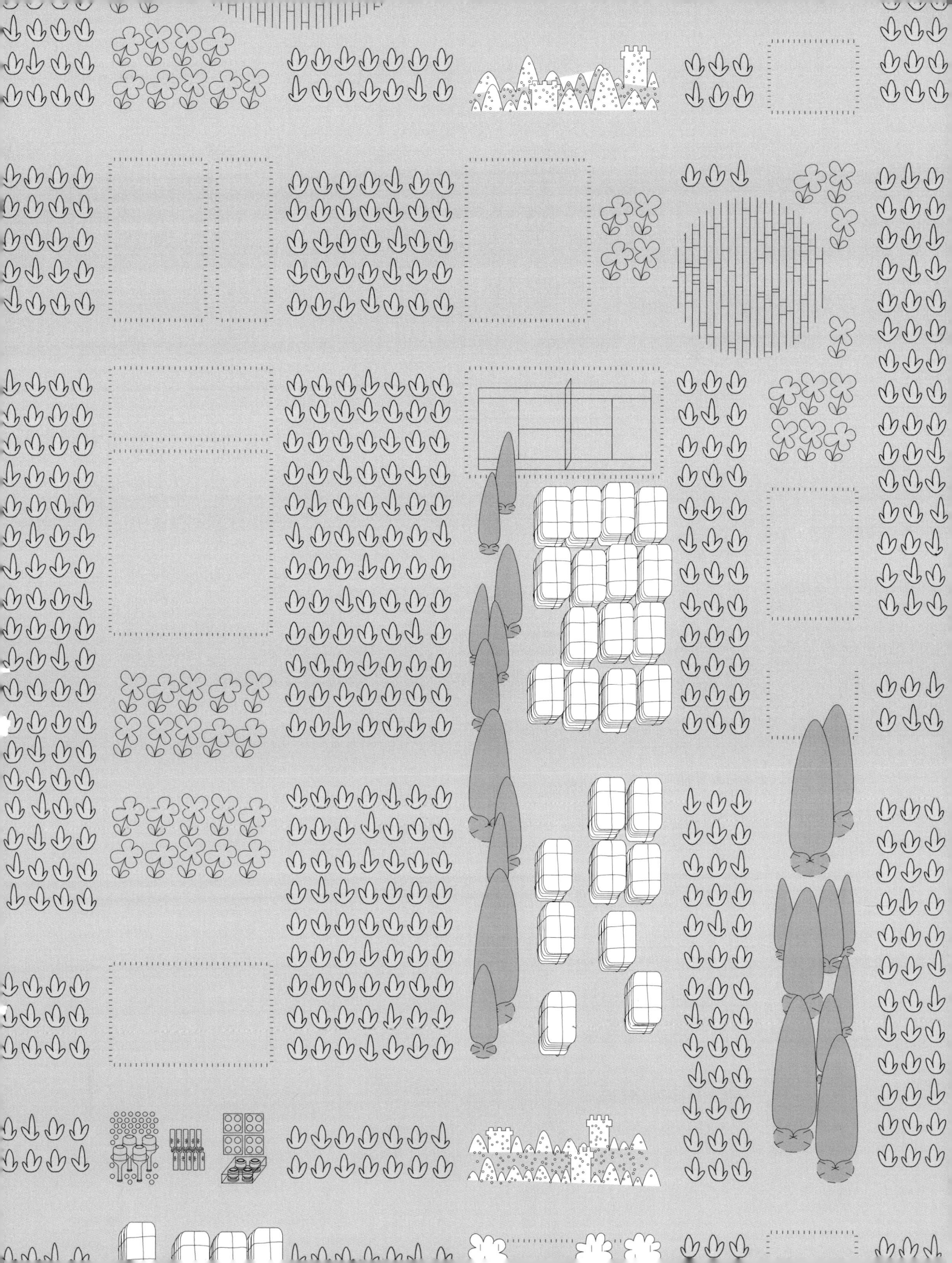

Smartcities, Resilient Landscapes + Eco-warriors

Dedicated to Matthew Wells, Andy Ford, and Colin Hayward

Second edition published 2019
by Routledge
2 Park Square, Milton Park, Abingdon, Oxon, OX14 4RN

and by Routledge
52 Vanderbilt Avenue, New York, NY 10017

Routledge is an imprint of the Taylor & Francis Group, an informa business

First edition published by Routledge 2010

British Library Cataloguing-in-Publication Data
A catalogue record for this book is available from the British Library

Library of Congress Cataloging-in-Publication Data
Names: Lim, C. J., author.
Title: Smartcities, resilient landscapes and eco-warriors / CJ Lim and Ed Liu.
Description: Second Edition. | New York : Routledge, 2019. | Previous editon: 2010. | Includes bibliographical references and index.
Identifiers: LCCN 2018056687| ISBN 9780815363248 (Hardback) | ISBN 9780815363255 (Paperback) | ISBN 9781351110037 (eBook)
Subjects: LCSH: Urbanization--Environmental aspects--Case studies.
Classification: LCC HT361 .L56 2019 | DDC 307.76--dc23
LC record available at https://lccn.loc.gov/2018056687

ISBN: 9780815363248 (hbk)
ISBN: 9780815363255 (pbk)
ISBN: 9781351110037 (ebk)

Typeset in DIN
by Studio 8 Architects

note: new chapters and case studies have green entries in the contents table

Preface

What is a Smartcity? 'Smartcity' is a vision. A vision of how the city of the 21st century might appear if we are serious about living sustainably and wish to be resilient. The Rockefeller Foundation '100 Resilient Cities' defines resilience as 'the capacity to bounce back from a crisis, learn from it, and achieve revitalisation. Communities need awareness, diversity, integration, the capacity for self-regulation, and adaptiveness to be resilient.' Cities are vulnerable and will increasingly be affected by anomalous climate change, natural catastrophe and urban stresses including chronic food and water shortages, pollution, a growing ageing population, and migration. Instead of a reactive approach to the manifold problems that contemporary life has thrown up, the Smartcity examines how we might live from first principles, taking the key component of any city – its people – as its starting point and raison d'être.

Following on from the success of the first edition, 'Smartcities + Eco-Warriors' (2010), this second edition reflects CJ Lim and Studio 8 Architects' latest research on resilience, ecology and urban sustainability, and has an additional nine case studies (these chapters have green backgrounds and green entries in the contents table). The explorations and critical thinking that began nearly two decades ago with proposals to cultivate awareness in the community landscapes of Chicago's DuSable Park and in Guangming Smartcity, have been adapted to achieve ecological self-regulation and environmental transformations in the Green Pension Plan in the UK, the City of a Thousand Lakes in Gaochun and Wanmu Orchard Wetland in Guangzhou.

The notion of the Smartcity is developed through a series of international case studies, some commissioned by government organisations, others speculative and polemic – visions of an urban future from a landscape perspective as opposed to a planning, environmental engineering or socio-economic one. A recurring feature of all the projects is the application of ecological sustainability, exploring potential opportunities to improve the ecological function of existing habitats or creating new landscapes which are considered beneficial to the local ecology and socio-economic values.

The central component of a Smartcity is the establishment of an ecological symbiosis between nature and built form to create diverse forms of resilient landscapes including and beyond urban agriculture. Reframing the way people think about the urban green revolution, the Smartcity explores the potential hybrid typologies of ecological programmes and landscape interventions that address the opportunities of the movement, and the role that nature and we as citizens play in the production of urban resilience. Trees, for example, have enabled communities to rehabilitate both their habitat and themselves from climatic and political catastrophe, while cultivating diverse and romantic imaginations of resilient landscapes.

Finally, the Smartcity is a manifesto and provocation. It should not be seen as an exercise in design monomania, but an invitation to planners, politicians, scientists, geographers and engineers to further a holistic dialogue and to stimulate activity in sustainability. The book therefore concludes with essays by hydro-environmental designer Anna Andronova, and food urbanist Carolyn Steel, representing diverging positions on the subject of resilient landscapes.

Urban Utopias + the Smartcity

UTOPIA

(noun) An imagined place or state of things in which everything is perfect. The word was first used in the book 'Utopia' (1516) by Sir Thomas More. The opposite of dystopia.[1]

At the time of writing, more than half of mankind, some 4.2 billion people, are living in urban areas. Asia is home to 54 per cent of the world's urban population, followed by Europe and Africa.[2] By 2050, the world's population is expected to grow to almost ten billion.[3] We are simultaneously experiencing a global food crisis resulting from low productivity, government policies diverting food crops to the creation of biofuels, climate change, and intensifying demands from an exponentially expanding population. 'The world is heading for a drop in agricultural production of 20–40 per cent, depending on the severity and length of the current global droughts. Food producing nations are imposing food export restrictions. Food prices will soar and, in poor countries with food deficits, millions will starve.'[4]

In November 1992, 1700 of the world's leading scientists issued a warning to humanity, urging a response to the unsustainably high consumption levels of finite energy resources, the reckless creation of deleterious effluent and the generation of greenhouse gases causing irreparable damage to vital planetary systems.[5] The 2005 Kyoto Protocol and the 2009 Copenhagen Summit have failed to deliver on their commitments to prevent detrimental anthropogenic effects on climate change. In the period from 1990 to 2008, the total carbon footprint savings by Europe was one percent but the developed world as a whole had its emissions rise by seven percent.[6] In 2015, at the Paris climate conference (COP21), 195 countries adopted a new global climate strategy – voluntary cooperation, self-regulation and political persuasion might achieve what previous quests for binding treaties failed to do. The Paris Agreement aimed to enhance the implementation of the United Nations Framework Convention on Climate Change (UNFCCC), which included 'increasing the ability to adapt to the adverse impacts of climate change and foster climate resilience and low greenhouse gas emissions development, in a manner that does not threaten food production'.[7] Despite the countless global initiatives, James Lovelock, author of 'The Revenge of Gaia' and 'The Vanishing Face of Gaia', foresees an unavoidable and radical climatic shift resulting in an environment less suitable for human habitation. In the absence of humanity mounting a massive 'sustainable retreat', he postulates 'a global decline into a chaotic world ruled by brutal warlords on a devastated earth'.[8]

At the same time, the world's economic order, premised on capital accumulation with scant regard to social wellbeing, employment and nature, is leading inexorably to extreme socio-economic differentiation

facing page: NASA Earth Observatory photograph of fields in Kansas: corn, sorghum and wheat crops using pivot irrigation.

1. 'Oxford Pocket Dictionary of Current English', Oxford University Press, USA, 2009

2. Population Division of the Department of Economics & Social Affairs (UN DESA), 'The 2018 Revision of the World Urbanization Prospects', The United Nations, New York 2018 [https://www.un.org/development/desa/publications/2018-revision-of-world-urbanization-prospects.html], retrieved 18 May 2018

3. Food and Agriculture Organisation (FAO), 'The Future of Food and Agriculture: Trends and challenges', The United Nations, Rome 2017

4. E deCarbonnel, 'Catastrophic Fall in 2009 Global Food Production', Global Research [www.globalresearch.ca/index.php?context=va&aid=12252], retrieved 10 May 2018

and a fractured society of the privileged and the dispossessed. The vast populations that make up modern cities result at best in weak social ties, at worst in mass control with concomitant violence and repression. As Lefebvre wrote in 'La Révolution Urbaine' in 1970, the big city sanctifies inequality, and is the most favourable milieu for the establishment of authoritarian power, pressing the countryside into servitude.[9] Long before the telematic assault of the virtual world, the Situationalists described the alienating nature of the city as a strange hybrid of crowd and solitude. The advent of the internet and online transactions has formed a society of now faceless, as well as nameless, strangers with an attendant diminution in social constraints motivated by anonymity.

Global famine. A poisoned earth. Societal collapse. Civilisation, it appears, is leading us down a path of ruin and steering us towards dystopia rather than utopia. Lovelock, in particular, portends an apocalyptic future one might expect from a science fiction author or religious prophet rather than a respected environmental scientist. In actuality, the complexity of weather systems and the factors that affect climate are still not well understood. However, deforestation and the burning of fossil fuels are leading to irreversible environmental damage, widely believed to be the greatest challenge facing mankind. Meanwhile, evidence of the current and escalating global deficiency in food security is uncontested but has only recently received media coverage or impetus to drive political action. The Paris Agreement and the 2030 Agenda for Sustainable Development have acknowledged the link between hunger and climate change. With the inevitable exponential growth of the urban environment, future cities incorporating mechanisms for food production, responsible energy use and social unity must be reassessed along with our visions for utopia.

By definition an unreachable destination, broadsides on utopia have been launched since its very inception. The word 'utopian' is more often than not used in the pejorative, pertaining to proposals featuring alternate realities rather than dealing with society's real and pressing ills. Such criticism misses the point and dismisses the potency of the utopic vision. Plato's 'Republic' (400 BC), Thomas More's 'Utopia' (1516) and Francis Bacon's 'New Atlantis' (1627) were intended as neither fantasies nor blueprints for reification, but reflections on the societies in which they were written. More significantly, they provided a stalking horse for the development and evolution of new communities that would improve on the status quo. Ebenezer Howard's garden city, for example, was inspired by the utopian tract, 'Looking Backward: 2000–1887', by the American lawyer Edward Bellamy. The third highest selling book of all time when published in 1888, Bellamy's novel immediately spawned a political mass movement and several communities living according to its ideals. Letchworth Garden City and Welwyn Garden City in the UK are founded on Howard's concentric plan of open space, parkland and radial boulevards. Housing, agriculture and industry are carefully integrated, and the developments remain two of the few recognised realisations of utopia in existence. There are valid concerns, however, that the tradition of utopian town planning as advocated by the Congress for the New Urbanism (CNU) and developments such as the Duchy of Cornwall-owned Poundbury, are elitist and non-inclusive. The cost of utopia is what lies outside utopia, the forgotten communities and infrastructure required to support it, a counterpoint that is sharply observed in the Peter Weir film 'The Truman Show', depicting the New Urbanist town of Seaside in Florida.

The 21st century has witnessed a phenomenal escalation in urban construction; entire cities are emerging fully formed in India and China rather than slowly evolving and accreting, made possible by the availability of affordable yet skilled labour, land and an uncompromising autocratic vision. Without invoking the term utopia, the aspiration and inspiration of nascent cities such as Dongtan in China and Masdar in the United Arab Emirates, both heralded as the first model eco-city, are both clear and vital. A model for how we should be living with improved modes of transportation, hydrological control systems, streamlined energy and supply programmes, and agencies for societal cohesion must surely be planned, albeit in forms protean enough to deal with the vicissitudes of urban living.

Sustainable design of the built environment has largely focused on discrete buildings, and we have become relatively adept at incorporating insulation, cooling, natural ventilation, solar control, greywater recycling, green roofs and renewable energy collection into architecture. Cities, though, are infinitely more complex than buildings, and the shift in scale to sustainable city design calls for a radically different approach to take advantage of the synergistic human-made and natural systems available. A conglomeration of buildings offers thermal efficiencies that are unachievable with smaller detached structures. The compact city, as championed by Richard Rogers, makes communal transport truly viable over the private car. During his address at the Reith Lectures of 1995, 'Cities for a Small Planet', Rogers presented a series of startling statistics demonstrating how the automobile has shaped the city. 'An efficient parking standard requires twenty square metres for a single car. Even supposing that only one in five inhabitants owns a car, then, a city of ten million (roughly that of London) needs an area about ten times the size of the City of London ('the square mile'), just to park cars.'[10] 'As transport by car becomes integral to city planning, the street corners and the shapes and surfaces of public spaces are all determined for the benefit of the motorist. Eventually the entire city, from its overall shape and spacing of new buildings to the design of its curbs, lamp posts and railings, is designed according to this one criterion.'[11] Now imagine a city with no cars – the possibilities are legion.

11

A sustainable high-density mixed-use city also allows waste products to be shared and recycled, land-use to be zoned vertically as well as horizontally, and the implementation of resilient landscapes that include urban agriculture and energy generation at a meaningful scale. In addition to environmental benefits, public and private space can be configured to promote social inclusion and economic growth. In short, the future of sustainable city design cannot be limited to sustainable buildings set within the outdated model of a European masterplan. Currently, the first phase of Masdar City, built by the Abu Dhabi Future Company and designed by Foster + Partners, has entirely replaced cars with an on-demand network of Personal Rapid Transit (PRT) systems.[12] The battery-powered and computer-navigated pods have covered a distance nearing 9000 kilometres on their transit route over six years, and have displaced CO_2 emissions equivalent to running 53 cars for a whole year. Significantly, the length of streets has been determined by wind fluid dynamics for urban cooling rather than vehicular traffic efficiency.

Traditionally, there has been a division of disciplines between architectural and urban planning. City design, a more inclusive term than urban planning, needs to embrace a number of disciplines that extend beyond land-use zoning and plot ratios – we need to engage agronomists, hydrologists, geographers, economists, transportation engineers, social scientists and politicians in addition to urban planners, landscape designers and architects. An urban infrastructure freed from the hegemony of the motorcar could and should manifest in a spatial manner radically different from the contemporary metropolis. Furthermore, existing car-based infrastructure – parking lots, motorways, service stations, driveways and garages – will require imaginative overhaul and programmatic adaptation.

5. 'World scientist's warning to humanity', authored by Henry Kendall, former chair of the Union of Concerned Scientists and endorsed by the majority of Nobel laureates in the sciences.

6. D Clark, 'Has the Kyoto Protocol Made Any Difference to Carbon Emissions?' The Guardian: Environment, 26 November 2012

7. The United Nations, 'Paris Agreement 2015', Article 2: 1b, p.3

8. J Lovelock, 'The Revenge of Gaia: Why the Earth is Fighting Back and How We Can Still Save Humanity', Allen Lane, London, 2006, p.154

9. H Lefebvre, 'La Révolution Urbaine', Gallimard, Paris, 1970

10+11. R Rogers, 'Cities for A Small Planet: Reith Lectures', Faber & Faber, London, 1997, p.36

12. 'Masdar City's PRT System Celebrates Milestone with 2 Millionth Passenger', www.masdar.ae [https://masdar.ae/en/media/detail/masdar-citys-prt-system-celebrates-milestone-with-2-millionth-passenger], retrieved 10 June 2018

'The world is sick. A readjustment has become necessary. Readjustment? No, that is too tame. It is the possibility of a great adventure that lies before mankind: the building of a whole new world ... because there is no time to be lost. And we must not waste time on those who laugh or smile, on those who give us ironical little answers and treat us as mystic madmen. We have to look ahead, at what must be built.'[13] Le Corbusier's commentary from 1967 might appear prescient, but could have been written at any time in the history of the city. The urban condition raises recurring as well as fresh challenges for every generation. In the past, architects have not been slow to offer their vision of utopia or ideal city, ranging from the polemic (Ron Herron's 'Walking City', 1964) to the serious (Le Corbusier's 'Radiant City', 1935), the futuristic (Paolo Soleri's arcologies) to the arcadian (Frank Lloyd Wright's 'Broadacre City', 1932). Tellingly, the architect's ideal city is frequently characterised by an immediately comprehensible visual order, whether as a grid or radial system. The meme of the concentric-ringed plan, for example, has been proposed by Filarete in the imaginary city of Sforzinda in 1465, John Claudius Loudon whose 1829 plan for London predated Howard's Garden City green belts by 69 years, and Claude Nicolas Ledoux in his proposal for the city of Chaux that centred around his half-completed Royal Saltworks at Arc-et-Senans. Konstantinos Doxiadis, on the other hand, is a celebrated exponent of the grid city, establishing a flexible plan for Islamabad that allows for gradual low-cost expansion. Other recurring motifs of the ideal city include a coalition between the countryside and the city, the orientation of buildings to a heliothermic axis to maximise daylight, and the liberation of the ground plane for public occupation.

Henri Lefebvre, the French sociologist and author of the seminal neo-Marxist works 'Critique of Everyday Life' and 'The Production of Space', argued that every society produces its own spatial practice and that without a distinctive space to mould it, a drive for societal change will never escape from its ideological beginnings. He ascribed the failure of the Soviet Constructivists of the 1920s and 1930s to them uncritically recycling the modern urban masterplan rather than inventing an appropriate new space to shape and be shaped by new social relations.[14] Planners, landscape designers and architects, then, have a crucial role as the agents of social change, and certainly the authors of the ideal city in their various incarnations saw themselves as such. At this critical juncture, the shape of the space that will help us contest climate change, social deprivation, and deficiencies in food, water and energy has to be re-imagined and re-produced.

As producers of space, architects represent a tiny minority of the rest of society who must in turn modify and refine the blueprint that has been mapped in front of them. They are well placed, however, to understand and design space that is of a human scale and more comprehensible to the general populace than large-scale zoning development maps. What will a city draped in a resilient landscape or an array of gasification plants look, feel, smell and sound like? As urban real estate becomes increasingly scarce, can we cross-programme public buildings and time-share streets with productive landscapes? These are questions that require a holistic understanding of socio-economic, political and ecological practice to answer. Fortunately, advances in visualisation graphics and computer rendering have made it far easier for designers to describe spatial propositions and as a consequence attract the necessary private and public sector investment to back them. Circumspection is necessary, however, to avoid the prevalence of visual information at the expense of less visceral information, and to ensure that the resulting built environment delivers more than a fleeting resemblance to its conceptual origins.

The central thesis of this book is the re-establishment of closed cyclical systems within urban and peri-urban areas and how they will manifest into resilient landscapes of a notional 'Smartcity'. The Smartcity differentiates itself from the 'Eco-city' by embracing new paradigms of sociological programme, environmental-driven form and ecological

12

interaction. The Smartcity's principal concern is not to overcome nature, nor to strive to preserve the natural environment in its original state, but to harmoniously integrate built form with nature. It is neither a fixed place or a singular approach but rather a manifesto for the production of resilient spaces relevant for the 21st century in the face of climate change.

The Smartcity is not a creation from a blank slate, but an evolution of long-standing sustainable principles that intertwine nature with contemporary desires for a healthier physical, mental and social existence in an increasingly alienating world. It aims to preserve and enhance natural and cultural resources, expand the range of eco-transportation, employment and housing choice and values long-term regional sustainability over short-term focus. The currency of an 'eco-' prefix has become devalued through overuse and abuse, and 'sustainability' is a blanket expression – clearly, some aspects of our lifestyle are worth sustaining and others are not. Deciding and acting on which category they fall into, however, is not as straightforward as it appears. Conservation of energy and the environment are key priorities, but so too is the conservation of heritage, tradition and human interaction. Each generation is the proprietor of its own values, and the current zeitgeist has reacted against the mass-produced and anodyne, whether in the guise of housing, jobs and clothing or fruit and vegetables. Without ignoring technological advances, the Smartcity embraces leanness and the low-tech by adopting an operating system that filters out excess and reboots our social space. Smartcity living does not ask for 'more' but determines how to use less in the creation of a healthier mental and physical existence.

At the forefront of the Smartcity manifesto and its diverse forms of resilient landscapes is urban agriculture. The hybridisation of agriculture and urban fabric can lead to an association that is symbiotic rather than parasitic, reducing carbon emissions and food shortages in addition to providing less tangible but equally significant environmental and social benefits. Food in most cultures is the glue that binds families and communities, and the restoration of the primal link between town-dwellers and their sustenance would constitute an important foundation to an increasingly ungrounded universe. The Smartcity manifesto reinstates food to the core of its governance.[15]

In this technologically advanced age we live in, there are shortages of food, shortages of basic living standards, and shortages of education and literacy. There should be no shortages of jobs. As a result of its resilient landscapes, the Smartcity programme comes with a host of fresh employment opportunities that are cross-sector and require a range of skills in the renewable energy, recycling, agriculture, and green construction industries. The business case for 'greening' the economy is robust. The potential for improving labour markets is greatest in developing nations, where over 40 per cent of the global workforce and their dependants are condemned to a life in poverty and insecurity.[16]

Smartcity strategies are inclusive, engaging all age groups, cultures and ethnicities. The Smartcity is an integrated holistic vision, not an appendix or a collection of unrelated ideas. The Smartcity calls for the renaissance of a manual universe in which we do things – grow food, play, travel, and design – from first principles again. It is a mindset questioning the way we live, driven by its inhabitants and prioritising human sustainability above all else. The rest will follow.

13

13. Le Corbusier, 'The Radiant City: Elements of a doctrine of urbanism to be used as the basis of our machine-age civilisation', Faber & Faber, London, 1967, p.92

14. H Lefebvre, 'The Production of Space', DN Smith (trans.), Blackwell Publishing, Oxford, 1991, p.59

15. CJ Lim, 'Food City', Routledge, New York, 2014, pp.185–187

16. M Renner, S Sweeney & J Kubit, 'Green jobs – towards decent work in a sustainable, low-carbon world', United Nations Environment Programme Report, September 2008, pp.5 + 73 [http://www.ilo.org/global/topics/green-jobs/publications/WCMS_158727/lang--en/index.htm], retrieved 18 June 2018

From Soil to Table

'And the LORD God made all kinds of trees grow out of the ground – trees that were pleasing to the eye and good for food.'   – Genesis 2:9

'And the LORD God commanded the man, "You are free to eat from any tree in the garden ..."'   – Genesis 2:16

Prior to their fall from grace for eating from the tree of knowledge, the fruit from the Garden of Eden provided Adam and Eve with food aplenty without having to toil for their sustenance. Since humanity's earliest days, we have yearned for immediate access to fresh healthy food. Refrigeration and rapid transport systems have, to a certain extent, made time and distance an irrelevance. In the United States, however, processing, packaging, transportation and storage of food account for ten percent of the total national energy budget,[1] with fresh produce travelling an average of 1500 miles from farmer to consumer.[2]

It is estimated that an acre of farmland is lost to urbanisation and highway production for every added person. It has been projected that by 2025, all food grown in the United States – the largest exporter of food worldwide – will be used for domestic purposes. Economically, this will result in an annual $40 billion loss of income.[3] When compounded by the reality that verdant fertile industrialised nations such as the UK have abandoned the objective of self-sufficiency and are hugely dependent on imported food, and that food security is indefensibly lacking for the billions, the need for increased food production and its equitable distribution is clear; a rapprochement needs to be reached between the praxis of urban living and food production.

The modern food industry epitomises the Marxist theory of alienation perhaps better than any other labour activity.[4] Unlike other trades and crafts that have followed esoteric and non-essential vectors, procurement of basic sustenance has always been a universal and innate occupation – foraging, hunting, husbandry and harvesting are straightforward exchanges of human capital in the form of energy expended and nutritional recompense. The abstraction of food from its origins through processing, portioning and packaging constructs a disassociation between the food producer and product, and between urban consumer and rural supplier. The consequence of this disassociation is that we, as consumers, are not seeing the clear effects of climate change and energy shortage on food production. Food supplied through the supermarket monopolies is still highly affordable despite the rising prices of the fuel essential for modern agriculture, but the cost of food we do not see takes into account the expense of environmental damage. Excess nitrogen runoff from fertilisers causes eutrophication of our

1. ME Webber, 'How to Make the Food System More Energy Efficient', Scientific American, 29 December 2011 [http://www.scientificamerican.com/article.cfm?id=more-food-less-energy], retrieved 20 May 2018

2. RS Pirog et al., 'Food, Fuel, and Freeways: An Iowa perspective on how far food travels, fuel usage, and greenhouse gas emissions', Leopold Center for Sustainable Agriculture Pubs & Papers, June 2001 [https://lib.dr.iastate.edu/leopold_pubspaper/3], retrieved 20 May 2018

3. D Pimentel & M Giampietro, 'Food, Land, Population and the US Economy', Carrying Capacity Network (CCN), 21 November 1994 [http://www.carryingcapacity.org/resources.html], retrieved 20 May 2018

4. K Marx, 'Economic and Philosophical Manuscripts of 1844' (The Paris Manuscripts), M Mulligan (trans.), Progress Publishers, Moscow, 1959

lakes and rivers, resulting in contaminated water and the destruction of aquatic ecosystems. Land-based ecologies are not spared either, despoiled by pesticides and herbicides. It is important to note that these are not simply costs to the environment, but fiscal costs to the general public in the form of subsidies, clean-up costs, and health treatment for poor nutrition, obesity, contaminated food and disease.

We, as city dwellers, need to re-engage with the roots of our sustenance in a way that does not involve abstract extruded vacuum-sealed meals if we are to alleviate the burdens of food production on the planet. The implementation of urban agriculture – the cultivation, processing and distribution of food within the city – would have the two-fold effect of making these processes transparent and offering a means for the re-establishment of food and its production as a social relationship rather than commodity. It would mean an end to a nonsensical boomerang trade that sees the UK importing 22 000 tonnes of potatoes from Egypt and exporting 27 000 tonnes in the other direction.[5] Resilient landscapes of urban agriculture would result in food immediacy within cities, providing nutrition and health benefits. It would create job opportunities, generate income for urban poverty groups and provide a social safety net. Urban organic waste would be turned into an agricultural resource. Social inclusion of disadvantaged groups and community development would be facilitated, and the city would benefit from urban greening and the maintenance of green open spaces. Bringing living food back to where we live would not re-establish the Garden of Eden, but there would be no second, third and fourth parties responsible for the commodification of produce, giving a new meaning to hand to mouth existence.

Urban agriculture is not a new phenomenon; its popularity and adoption has waxed and waned over the millennia, from the recycling of urban wastes and qanat tunnel irrigation networks in Ancient Persia for agriculture, to the stepped cities and farming terraces of Machu Picchu that can be considered as a precursor to hydroponics. In more recent times, victory gardens during the two world wars were employed to alleviate food shortages with rooftops, balconies, pontoons

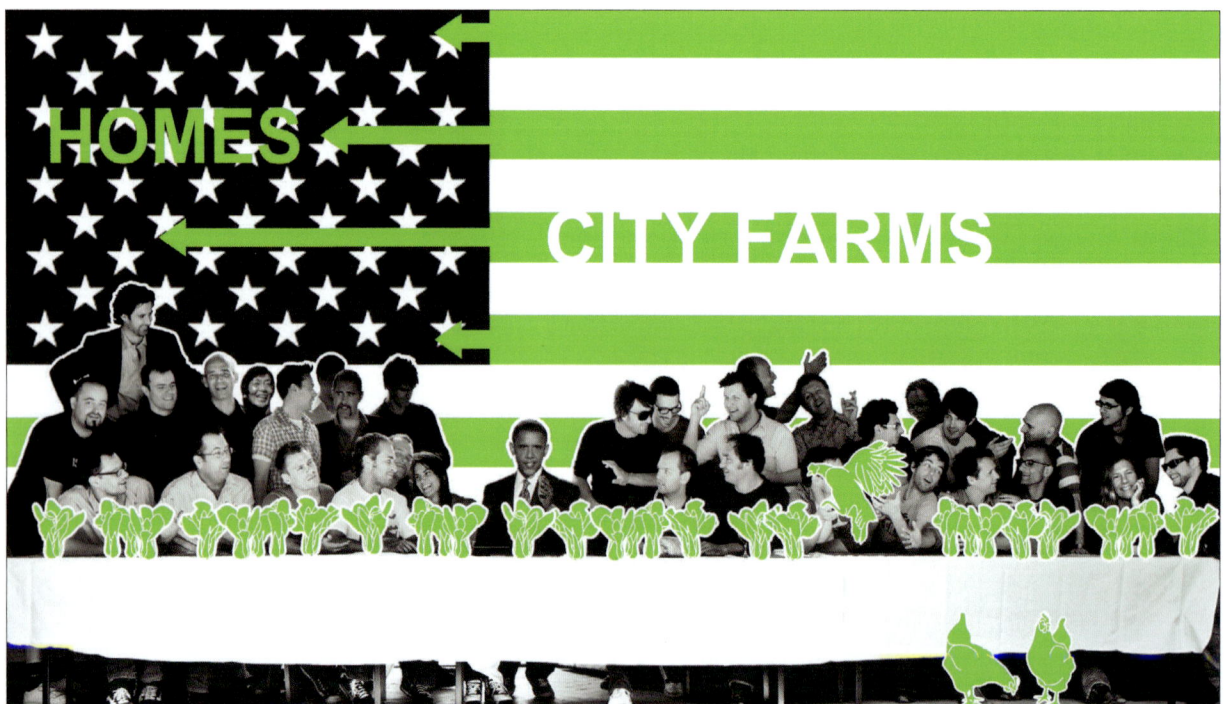

The Perpetual Motion Machine

'Mulberry trees are grown to feed silkworms and the silkworm waste is fed to the fish in ponds. The fish also feed on waste from other animals, such as pigs, poultry, and buffalo. The animals in turn are given crops that have been fertilised by mud from the ponds. This is a sophisticated system as a continuous cycle of water, waste and food ... with man built into the picture.'

– Jennifer Pepall, 'New Challenges for China's Urban Farms', IRDC, 1997

19

The Chinese mulberry dyke fishpond system, first introduced during the dying days of the Ming dynasty (16th century) in the northern part of the Pearl River Delta, is a striking model of a closed sustainable ecosystem deployed by mankind to provide food and clothing. The benefits of each link in the system had been known to the farmers of the area for many years, as reflected in the folk saying that 'the more luxuriant the mulberry trees, the stronger the silkworms and the fatter the fish; the richer the pond, the more fertile the dyke and the more numerous the cocoons'.[1]

Elegant forms of such closed systems in farming have fallen victim to new technologies, in particular the Haber Bosch Process to synthesise ammonium nitrate. The agricultural revolution enabled the evolution of an urban lifestyle, dramatically reducing the agricultural labour force and freeing the populace for other pursuits. Ironically, the unchecked growth of an urban lifestyle at the expense of agricultural land is now threatening the production of food that made the city's existence initially possible.

Urban agriculture provides an overdue mediation between the countryside and city, making possible a circular economy that has the same seductive clarity and well-tempered logic as the mulberry dyke fishpond system – the solid organic waste of city dwellers can be alchemically transformed via anaerobic digestion into gaseous energy and fertilising digestate; greywater and blackwater from showers, sinks and gutters can be treated and rechannelled to irrigate our crops provided they are in close enough proximity. With the added ingredient of sunlight, we have food from a living grocery store to propel another cycle of the human perpetual motion machine.

Since the 1920s, when Chinese exports of raw silk were at their peak, the mulberry dyke fishpond system has seen a sad decline after having evolved for over two millennia, the Chinese orthodoxy of a circular economy implacably usurped by urbanisation and industry. In recent years, however, cyclical systems founded on mulberry dyke farming have seen resurgence in academic circles as an alternative to unsustainable agriculture and have credible potential for real world application.

facing page: Dream Farm 2 Systems Diagram; Dr Mae-Wan Ho (ISIS).

1. Asia-Pacific Environmental Innovation Strategies (APEIS); Research on Innovative and Strategic Policy Options (RISPO).

Dr Mae-Wan Ho, geneticist and director of the Institute of Science in Society (ISIS), has been developing the 'Dream Farm 2', a model of an integrated, zero-emission, zero-waste farm that maximises the use of renewable energies and turns waste into food and energy resources. An implementation and extension of George Chan's Integrated Food and Waste Management System (IFWMS), Ho likens the farm to an organism, ready to grow and develop, to build up structures in a balanced way and perpetuate them. The closed cycle creates a stable, autonomous structure that is self-maintaining, self-renewing and self-sufficient.[2]

Key to the process is a zero-entropy or zero-waste directive that must be adhered to as far as possible. The human body tends towards this ideal, explaining why we age relatively slowly and do not spontaneously decompose. The Dream Farm becomes more productive as more life cycles are incorporated, with increasing amounts of energy and standing biomass stored within the system. Echoing the lessons of crop rotation, academic researchers have rediscovered that productivity and biodiversity are happy bedfellows in a sustainable system, with different life cycles reciprocally retaining and circulating energy for the whole system.

Circular economies are not restricted to agriculture. The reduction of energy demand through inter-seasonal heat transfer (IHT) is the perfect example of a Smartcity cyclical system. Excess heat during the summer can be collected in a thermal store and retained until wintertime, when it is redistributed through pipework to provide thermal comfort (usually with the assistance of heat pump technology), all the while building up a store of chilled water or chilled ground stores which will be used for cooling during the summer. At the scale of a city, the potential for reduction in energy demand is immense, making our current scattering of energy-autarkic houses pale into insignificance. Coupled with waste recycling and renewable offsets, a citywide net zero carbon lifestyle could actually be possible.

Harvard-based sustainability expert Nader Ardalan maintains that reduction of carbon emissions in the Middle East is possible purely by creating 'intelligent buildings that are inspired by the wisdom of the ancestors, who worked with nature to adapt to the demanding desert climate'.[3] The largest opportunity for cutting demand lies in the employment of holistic design and the configuration of buildings to passively heat and cool our environments for thermal comfort. As in the case of agriculture, antiquity provides us with lessons in our age of technological marvels. We need to look back to the Ancient Greeks, who reorientated entire city grids to increase southern exposure for passive solar heating in the winter months. We need to recycle the wisdom of cavemen who somehow knew that annualised thermal stasis is achieved at a depth of six metres below ground level and took advantage of the land's ability to mediate extremes of temperature. We need to duplicate the extensive underground labyrinths of the ancient Persians that were used to cool their buildings over 5000 years ago.

Shifting scale from seasons to days, temperature changes of the diurnal cycle can also be manipulated by using exposed heavyweight construction materials that retain daytime heat in their thermal mass to be released during the night. Cooling strategies in tropical climates utilise the water cycle, reducing temperature by releasing latent heat of evaporation. Here we can refer to the baud geer wind towers and cisterns of traditional oriental architecture, or the fountains and reflection pools of Moroccan courtyard riads.

The vast majority of discourse on energy conservation focuses on the shift from fossil fuels to renewable energies in the form of wind turbines, hydro-electric plants, combined heat and power (CHP), photovoltaics and ground source heat

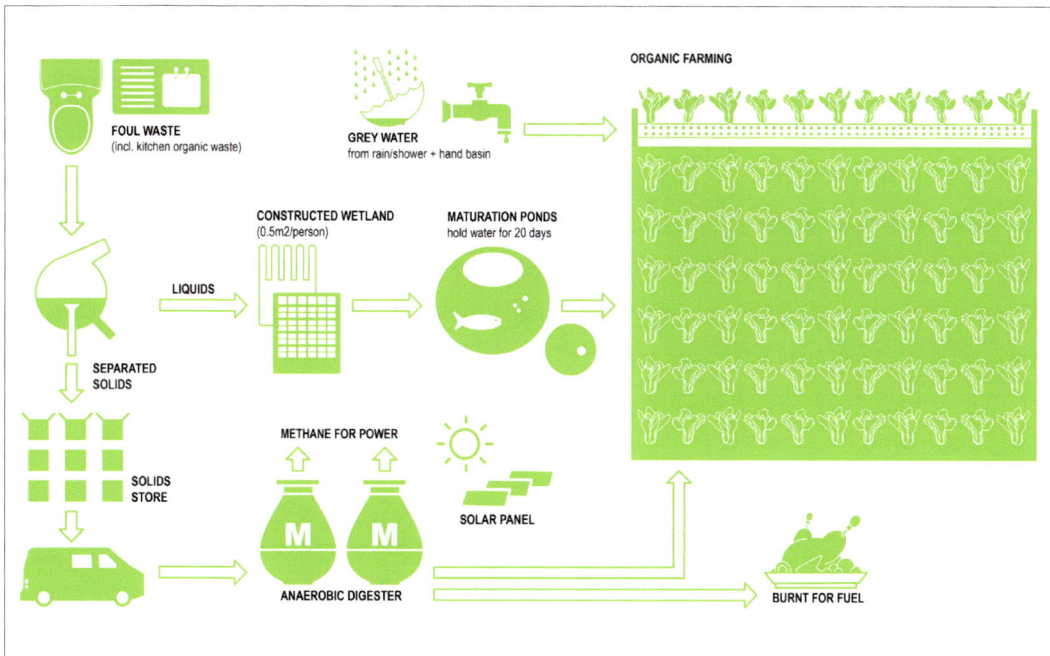

left: Imagining Recovery – The perpetual motion machine channels urban waste back into farming.

pumps. With ongoing concerns regarding the safety of nuclear power, such technologies serve a vital purpose, and in urban areas offer the only viable solutions. As an alternative to burning fossil fuels and releasing radioactive materials, heavy metals, volatile organic compounds, greenhouse gases and acids into the atmosphere, 'renewable' energy generation can only be welcomed. The idea that such energy is 'clean', however, is fundamentally flawed – biomass crops require food, water and energy for growth and transportation while photovoltaics and wind turbines require maintenance, replacement and significant energy resources in their production. In order to amplify the benefits of cleaner energy supplies, however, the reduction of energy consumption from the outset must be considered. Our first question should not be 'How do we generate more energy to feed our destructive lifestyles?' but rather 'How do we minimise our need?'

A second alternative to the generation of new energy is to share and recycle it. With hydro-electric, fossil fuel and wind plants situated in locations remote from cities, there are huge losses in efficiency and little potential for heat capture that could be used for district heating. Cogeneration fuel cells, now compact enough to be installed in an urban basement, provide one solution. The Industrial Symbiosis at Kalundborg in Denmark is a commercial-scale example of an energy-sharing cooperative often cited by industrial ecologists. Seventy-five miles west of Copenhagen on the coast of Denmark, this industrial eco-system is characterised by a network of trading companies in a closed cycle, where the waste or energy from one neighbour becomes a resource for another. 'From sharing one resource: waste water, in the 1970s, today eight companies are sharing 25 different resources, from fresh water and biogas to gypsium. This is reducing the CO2 emissions by over 635 000 tonnes per year and creating annual savings of DKK182 million for the enterprises cooperation and DKK106 million in socio-economic values.'⁴ The partnership was awarded the 'Win-Win Gothenburg Sustainability Award' in 2018.

2. MW Ho, 'Dream Farm 2 – Story so far', Science in Society Archive, 24 July 2006 [http://www.i-sis.org.uk/DreamFarm2.php], retrieved 16 March 2009

3. R Staley, 'The Architecture of Human Survival', yourmiddeast.com, 20 June 2013 [https://yourmiddleeast.com/2013/06/20/the-architecture-of-human-survival/], retrieved 20 October 2017

4. The Jury Motivation, 'Kalundborg Symbiosis', Win-Win Gothenburg Sustainability Award 2018, 19 June 2018 [http://winwingothenburgaward.com/item/2018-kalundborg-symbiosis/], retrieved 27 August 2018

5. C Weetman, 'A Circular Economy Hnadbook for Business and Supply Chains: Repair, remake, redesign, rethink', Kogan Page Limited, London, 2017, p.105

6. SE Haggar, 'Sustainable Industrial Design and Waste Management: Cradle-to-cradle', Elsevier Academic Press, Burlington, 2007, p.57

In the early 1970s the Statoil refinery agreed to provide waste gas as a fuel source to Gyproc that the latter was able to use as a low-cost fuel source. Treated wastewater was and is still sold to the nearby Asnæs fossil fuel power station which, losing 60 per cent of its energy through heat, began to temper thermal inefficiencies by providing heating to 3500 homes and selling process steam to the refinery and the pharmaceutical company, Novo Nordisk, for sterilisation purposes. The cooling water is also passed on to a fish farm, resulting in improved breeding conditions and growth in the warmer water. The 30 tonnes of annual ash by-product are recycled in the cement industry, and sulphur dioxide from flue gases are sold to Gyproc for gypsum production.[5] Wastes from all the symbiosis companies in the municipality are collected by Kara/Noveren I/S to produce electricity. Enzyme production at Novozymes A/S creates over 150 000 cubic metres of solid biomass as part of the fermentation process, which is exported as fertiliser, and yeast slurry from insulin production at Novo Nordisk, which is used in the pig-farming industry.[6] This circular economy, mirroring the symbiotic farming systems of the Dream Farm 2, has resulted in a considerable reduction in water, air and ground pollution whilst conserving natural resources. It is noteworthy that these environmental benefits are themselves by-products of profit-making commercial decisions.

Urban growth can no longer continue through synthetic linear processes that are colossally wasteful, discharging contaminants into the air, ground and water. Circular organic systems are in comparison regenerative, and it makes sense to ride the wave of these natural systems, harvesting the fruits of a harnessed ecology and feeding the process as necessary to reap disproportionate benefits from minimal investment. Natural systems of resilient landscapes are self-perpetuating and symbiotic. And it is high time that mankind rejoined these systems as a constructive rather than destructive force.

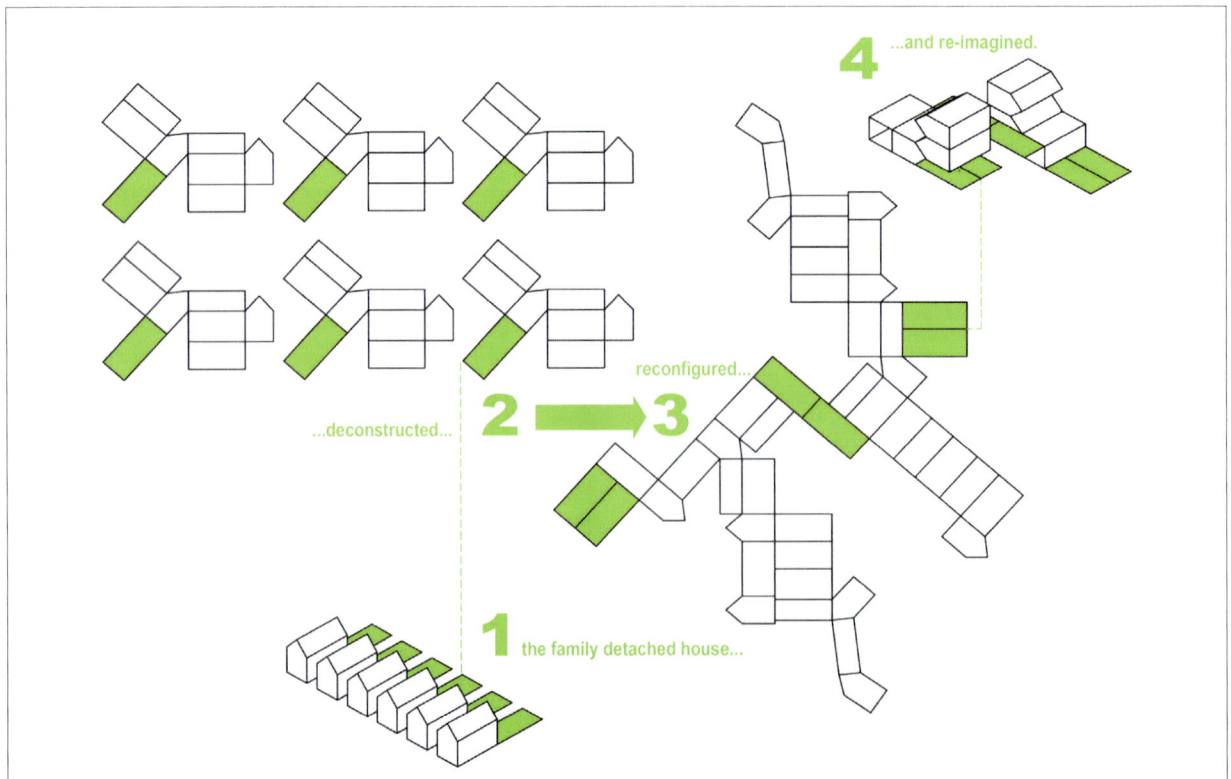

1 the family detached house...
...deconstructed...
2 → 3
reconfigured...
4 ...and re-imagined.

The American Dream Redux

The American Dream is 'that dream of a land in which life should be better and richer and fuller for everyone, with opportunity for each according to ability or achievement. It is a difficult dream for the European upper classes to interpret adequately, and too many of us ourselves have grown weary and mistrustful of it. It is not a dream of motor cars and high wages merely, but a dream of social order in which each man and each woman shall be able to attain to the fullest stature of which they are innately capable, and be recognised by others for what they are, regardless of the fortuitous circumstances of birth or position.'

– James Truslow Adams, 'The Epic of America', 1931

The Dream that James Truslow Adams captured for the American people in 'The Epic of America' was one based on social and ethical principles, reflecting a country putatively unencumbered by religious, class and racial boundaries with life prospects based on talent and determination rather than wealth and political connections. Successive generations have seen recalibrations of the Dream and somewhere down the line, it became synonymous with home and automobile ownership as a symbol of affluence, precisely what Adams declared the Dream was not.

facing page: Imagining Recovery – The deep-rooted American ideal of the single-family detached house and automobile requires re-evaluation in order to effect change.

Broadacre City, Frank Lloyd Wright's conception of the utopian city, reinforces this aspiration. A decentralised democracy, Broadacre City, or 'Free City' as Wright sometimes called it, was agrarian in nature, with a community built upon the transfer of a one-acre plot per citizen from federal land. In his own words, 'when every man, woman, and child may be born to put his feet on his own acres and every unborn child finds his acre waiting for him when he is born – then democracy will have been realised'.[1] Every family would own their own home in the form of the 'Usonian House' that would come in different-sized variants depending on need. The sizeable distances between individual dwellings and educational, religious and leisure establishments would lend primacy to the motorcar, with the larger Usonian houses incorporating five-car garages and pedestrian safety guaranteed only within the one-acre plots.

As a model for new social space arising out of the Great Depression and advancing technologies in telecommunications and the automobile industry, Wright's vision was every bit as revolutionary as, but divergent from, Le Corbusier's Radiant City, revealed three years later in 1935 and extolling the virtues of stacked high-density mixed-use living.[2]

1. MB Lapping, 'Toward A Social Theory of the Built Environment: Frank Lloyd Wright and Broadacre City', Environmental Review, vol.3, no.3, Spring 1979, Oxford University Press, p.14

In the light of today's world population growth and sustainability concerns, the Radiant City appears to be the more relevant model for city design. The ideal of the single-family detached house and

2. Le Corbusier, 'The Radiant City', Faber & Faber, London, 1967

automobile that has permeated the developing world is deep rooted but requires re-evaluation in order to effect change. The white picket-fenced suburban utopia of Wisteria Lane needs to be supplanted by a dream that is both leaner and more expansive, comforting but challenging.

The unattainable nature of utopia is less to do with an unreachable goal than the shifting of goalposts, what Gregg Easterbrook terms the Progress Paradox in his book of the same name. We have never been wealthier, lived longer and amidst less crime. The environment has also, with notable exceptions, become cleaner. However, there has been no commensurate increase in happiness, a contradiction that Easterbrook attributes to 'choice anxiety' and 'abundance denial'.[3] Similarly, the Easterlin paradox of 1974 posits that unlimited economic growth is not necessarily beneficial to contentment, correlating the 'happiness' index of countries at various levels of development; Easterlin showed that the inhabitants of low-income countries were not proportionally less happy than those of higher income nations.[4]

Nevertheless, the influence of the American Dream as a driving force for individual improvement in the 21st century cannot be underestimated; studies have shown that it is only following the 'Great Recession' of 2008 to 2009 with its home foreclosures, burgeoning unemployment and increasing energy costs that national attitudes to the Dream have soured. The global recession may be seen, though, as an opportunity as well as a catastrophe, enabling society at large to realign itself with a grounded value system that eschews rampant consumerism and exploitation.

Similarly, it is time for designers to reassess the values of their profession. In order to regain public and political confidence, design needs to offer intelligent solutions that focus on need and demonstrate added value. Beauty will not be judged purely through the lens aesthetic but through the elegance of efficient arrangements and systems. Modesty rather than narcissm will be the acceptable face of sustainable design. On the one hand, the currency of architectural design is severely devalued when it comes to economic renewal, even in the spheres of housing and commercial developments that it is associated with. Spaces can be designed to be functional, flexible, to have green credentials and, even to be beautiful. Design can improve quality of life and contribute towards the wider society; on the other hand, the contribution of construction professionals is for the most part guided or stymied by government policy and developing agencies. The real influence of the designer, whether of food packaging or a city master plan, lies in the visualisation of an alternative reality, a reality that is demonstrably better but conceivable only through the designer's shared vision. Too often, this alternate reality is seductive but bogus, used to market banal consumer products. As imagineers, however, designers are in a position to cajole the general public to embrace positive and profound change so that progress is not hampered by cultural bias and financial conservatism.

The compact city, which offers so many beneficial synergies, is at odds with the outdated American Dream. Change has to be gradual, and will be abetted by a modal shift in transportation such as that implemented in Curitiba, capital city of the Brazilian state Paraná, that will support physical interaction and societal cohesion. The resilient landscape of the Smartcity needs to demonstrate that shared experience and pooled resources can offer an improved and viable model to individual advancement. The simple notion that public space in the shape of a favourite table at a café, a park bench or a painting hanging in the permanent collection of a gallery can be sequestered into a shared but personal ownership is nonetheless a powerful one.

There are also encouraging signs that America's love affair with the automobile is in decline. 'Walkability' has become a buzzword amongst American estate agents, who have reported that housing values have shown a significant increase where schools and public transport facilities are within walking distance compared to the past. The expense of maintaining an additional dependency that provides only sporadic benefit coupled with vehicular congestion, competition for parking spaces, increasing fuel costs and pollution are slowly making the car a convenience that does not always justify the expense. There is also a growing sense that the vibrancy of dense mixed-use neighbourhoods is more appealing than the bland suburbia of Levittown shaped by streets and cars. The post-war mass-produced homes were built on a seven-square-mile tract of Long Island's potato and onion fields, and still stand as a harrowing indictment of America's 'general lust for conformity', and 'blind, desperate clinging to safety and security at any price'.[5]

Linked to the new American Dream must come a recalibration of the perception of beauty. The mowing of the front lawn in suburban America has become bizarrely ritualised, the beautiful manicured lawn a point of pride signalling conformity to a suburban code of conduct. The Canadian cultural critic and self-styled 'horticultural philosopher' Robert Fulford sees the lawn as an instrument for public shaming and social control: 'As the death of a canary announces the presence of gas in a mine, so a dandelion's appearance on a lawn indicates that Sloth has taken up residence in paradise and is about to spread evil in every direction. Pretty as they might look to some, dandelions demonstrate a weakness of the soul. They announce that the owner of the house refuses to respect the neighbourhood's right to peace, order, good government.'[6] Hyperbole aside, the manicured lawn is an unfathomable oddity to countries where the front yard is not central to their culture, rendered all the more contentious in the wider context of sustainability and food security. Scaled up, the great American lawn covers over 50 million acres of the country – more land area than is used for the growth of wheat or corn – and consumes vast water and energy resources for its maintenance. Perversely, it offers no spatial function; it is effectively wasted space despite the 'beauty' it proffers. Corralling this land resource for the production of food would result in a different kind of beauty that is neither skin-deep nor associated with the vanity of status. Resilient urbanism is characterised not only through the production of new social space, but also the creation of a new social aesthetic, and it is not inconceivable that this cultural anachronism, imported from England over a quarter of a millennium ago, will become a symbol of vulgar ostentation rather than good taste.

In addition to defining the American Dream, Adams is also remembered for his essay 'To "Be" or to "Do": A Note on American Education',[7] in which he declares 'there are obviously two educations. One should teach us how to make a living and the other how to live.' The relevance of his ideas remains undiminished.

3. G Easterbrook, 'The Progress Paradox: How life gets better while people feel worse', Introduction, Random House, New York, 2003

4. R Easterlin, 'Does Economic Growth Improve the Human Lot? Some Empirical Evidence', in 'Nations and Households in Economic Growth: Essays in honour of Moses Abramovitz', PA David & MW Reder (eds.), Academic Press, New York, 1974, pp.89–125

5. C Marshall, 'Levittown, The Prototypical American Suburb', The Guardian: Cities, 28 April 2015

6. R Fulford, 'The Lawn: North America's Magnificent Obsession', Azure magazine, July-August, 1998, pp.34–41

7. JT Adams, 'To "Be" or to "Do": A note on American education', Forum magazine, HG Leach (ed.), vol. LXXXI, no.6, New York, June 1929, pp.321–327

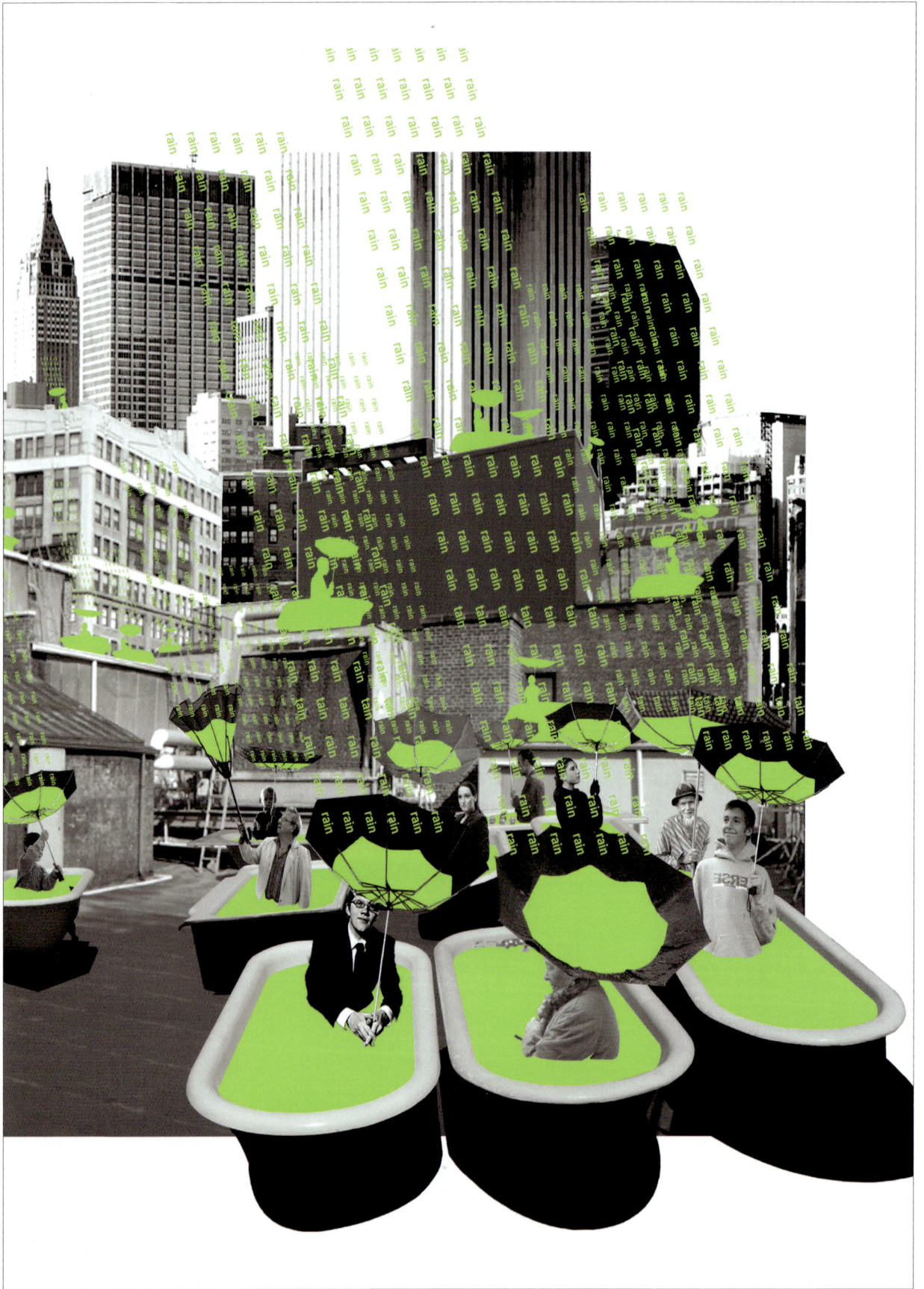

Rise of the Eco-warrior

William Sweet • Fernando Pereira • Chico Mendes • Mike Hill • Karel Van Noppen • Kenule Saro-Wiwa • Jill Phipps • David Chain • Bartolomeu Morais da Silva • Fernando Sarmiento • Celso Pojas • Danny Qualbar • Rolando Antolihao • Vicente Paglinawan • Isabelino Celing • Dorothy Stang • Sombath Somphone • Francisco Canayong • Jose Ribeiro da Silva • Maria do Espirito • Hernán Bedoya

27

In the developed world, the eco-warrior is a figure of ridicule, bringing to mind tree-huggers, hippies and holier-than-thou evangelists. The roll call of names above comprises just a few of the environmental activists who have been killed defending habitats and causes counter to the interests of powerful economic groups, sometimes with governmental ties. Many of those habitats are also home to, or represent the livelihoods of, impoverished indigenous communities. The atrocities included the murder in 1995 of Kenule Saro-Wiwa, who dared to speak out against the environmental damage to his homeland, Ogoniland, following decades of oil waste dumping by multi-national corporations, by the Nigerian military; Francisco Canayong, the president of a Philippine farmers' association, stabbed to death for rallying villagers to block a China-bound shipment of chromite ore from an illegal mine that was poisoning local water sources; Sombath Somphone's disappearance in 2012, which came after he spoke up for victims of a land-grab scheme that saw village rice fields bulldozed to make way for foreign-owned rubber plantations;[1] and Hernán Bedoya in Colombia, shot 14 times by a paramilitary group for protesting against palm oil and banana plantations on land stolen from his community in 2017. One of the most celebrated environmental martyrs is Chico Mendes, the rubber tapper whose murder in 1988 ignited international awareness of and support for the preservation of the Amazon rainforest against logging and ranching activities. In 2017 alone, the annual figures of international NGO Global Witness showed at least 207 land and environmental activists were killed across 22 countries, almost four a week.[2] The term 'warrior' is neither hyperbole nor ironic.

Today's eco-warriors include scientists, politicians, entrepreneurs, journalists, designers and farmers among their number, involving every age, ethnicity and gender. 'Women produce up to three-quarters of the food crops grown in West and Central Africa, and participate in agroforestry, tree nursery establishment and management, and community forestry action.'[3] Cécile Njebet, president of the African Women's Network for Community Management of Forests underlines that the empowerment of women farmers not only improves their lives but also the resilience of communities in the fight against poverty and climate change. The United Nations envisages if women had the same access to agricultural resources as men, they would be able to increase output on their farms by 20 to 30 per cent, raising total agricultural production by up to four per cent worldwide and reducing the number of hungry people

facing page: Imagining Recovery – greywater recycling and the experience of recovery.

1. Scott Wallace, 'Why Do Environmentalists Keep Getting Killed Around the World?', Smithsonian Magazine, February 2014, pp.64–69

2. B Kyte, B Leather & H Iqbal, 'Deadliest Year on Record for Land and Environmental Defenders, as Agribusiness is Shown to be the Industry Most Linked to Killings', Global Witness: Press Release, 24 July 2018 [https://www.globalwitness. org/en/press-releases/deadliest-year-record-land-and-environmental-defenders-agribusiness-shown-be-industry-most-linked-killings], retrieved 2 June 2018

3. D Ouya, 'Positive Action on Gender Supports Sustainable Development', World Agroforestry Centre, 12 September 2014 [http:// blog.worldagroforestry.org/ index.php/2014/09/12/action-on-gender-can-support-sustainable-development/], retrieved 3 June 2018

by up to 17 per cent.[4] Njebet is now leading the fight for the rights of women, the politically oppressed and the natural environment instigated by Wangari Maathai – the environmental and political activist, who founded the Green Belt Movement in Africa, and with the help of women, planted over 30 million 'trees of peace' to conserve the environment.[5]

In India, food sovereignty advocate Vandana Shiva describes the fight against agricultural biotechnology as 'a global war against a few giant seed companies on behalf of the billions of farmers who depend on what they themselves grow to survive'. After a lifetime rallying against food totalitarianism, she dismisses the American scientific organisations responsible for regulating genetically modified products, including the Food and Drug Administration, the Environmental Protection Agency, and the United States Department of Agriculture, as tools of the international seed conglomerates.[6] At the same time, Kehkashan Basu of the United Emirates, the 2016 Children's Peace Prize recipient, founded Green Hope UAE at the age of 12 to mobilise children and youth in the movement for a sustainable and green future. As the eco-warrior spirit has percolated down over the years from persecuted prophets to everyday members of society, there are two groups of unlikely players in the war to secure resilience for humanity and nature – storytellers and farmers.

There has been a tidal shift in public awareness of unsustainable environmental practices, and this can be partly attributed to storytellers in the mass media. Science has provided us with the statistical proof that humanity is both perpetrator and victim of widespread environmental damage, but facts and figures are poor vehicles for galvanising the electorate and grass roots action. At the end of the last decade, the film world saw a glut of movies with green agendas, most notably Davis Guggenheim's 'An Inconvenient Truth' (2006) and Robert Kenner's 'Food Inc.' (2009). The former documented Al Gore's commitment to exposing the 'planetary emergency' that global warming represents, the latter the detrimental effect the food industry is having on health, farmers' livelihoods, and the environment. Both documentaries were produced by Participant Productions, a company with a mission to raise awareness of world problems in the public consciousness through compelling narrative. Correspondingly, 'This Changes Everything' (2015) by Avi Lewis and Naomi Klein and 'Seed: The untold story' (2016) by Taggart Siegel and Jon Betz narrate similar David and Goliath battles to defend the future of our food. When even Bond movies tap into the zeitgeist – the villain in the 'Quantum of Solace' (2008), whose name is Greene, is not bent on world domination but on the monopoly of water supplies in Bolivia under the guise of a bogus environmental organisation – the critical mass necessary to effect change cannot be far away.

The second group of pivotal players are farmers, the individuals we have delegated to put food on our tables whilst we engage in more rarefied pursuits. In the Smartcity, with urban agriculture at the forefront, empowered farmers will take on a new instructional role, advising communities on how to best cultivate their crops. In a list of 50 people who could save the planet chosen by an expert panel assembled by the Guardian newspaper,[7] five are farmers or have had farming experience, including Al Gore who worked on his family's small holding as a boy. Also, recognised is Bija Devi who has been a farmer in the foothills of the Himalayas since the age of seven. She now spearheads an international movement to conserve cereal, pulse, fruit and vegetable strains that are at risk of extinction from modern agricultural practices. Having established an extensive bank of indigenous seeds, Devi travels around India disseminating endangered crops and the ancestral knowledge to cultivate them, simultaneously insuring against climate change, soil infertility and disease, and preserving a rapidly disappearing cultural tradition.

Outside of the Smartcity, the remit of farmers will become broader. Trained and reinstated as custodians of the land, a new breed of professional farmer will husband energy, natural ecosystems and forestry, arbitrating between the need for carbon sequestration, wildlife habitats, raw timber material and biomass. Provenance will enter the public lexicon in relation to energy and manufactured materials as well as food. Many farmers are working to create ecological resilience to sequester carbon dioxide and reduce agriculture's greenhouse gas emissions and their impact on the environment. Smallholder farmers in rural areas are vital to domestic food security. To reduce the risk of crop failure due to the impacts of climate change, farmers have made themselves more resilient by planting diverse traditional varieties, liberating themselves from commercial seed breeders and not using expensive modern hybrid seeds. A strong rural economy can reduce inequality, migration to urban areas, and sustain a balanced growth and flow of resources between rural and urban areas. Small farms are common in Japan, Norway and Switzerland, and account for 80 percent of the cultivated land in sub-Saharan Africa and South East Asia.[8] Farmers in Niger, over the last few decades, have managed the natural regrowth of native Faidherbia trees across five million hectares. According to the World Resource Institute, 'the Faidherbia fixes nitrogen in the soil, adapts the fields from wind and water erosion and contributes organic matter to soils with its fallen leaves. Compared to conventional farms, yields of maize in these agroforestry systems can be doubled and farmers in Ethiopia, Kenya, and Zambia are taking note.'[9] The Smartcity situates its sustainable vision for creating economic, environmental and social capital by restoring nature.

Smallholder farmers in low- and middle-income countries invest more than US$170 billion every year on their farms, making them collectively the single biggest investors in agriculture.[10] They are looking to strengthen their risk-coping capacities, secure tenure to the land they farm, and gain access to water and fair financial services to help build a resilient landscape. Sustainable development of food systems depends not only on public and external investments, but just as much on the policies and regulatory frameworks that uphold the autonomy and sovereignty of farmers and governments receiving aid. The resilience of farming communities is strengthened, above all, by sharing knowledge on developments of sustainable farming techniques and crop management. Bustani ya Tushikamane (ByT) is a farmer training centre for sustainable agriculture based in Morogoro, Tanzania managed by Sustainable Agriculture Tanzania and Janet Maro, an agronomist. In 2013 the initiative supported 2700 farmers through training, 46 percent of whom were female, and a further 833 farmers through the information centre.[11] Digital technology, for example mobile phone apps, which has improved access to quality data, including changing weather patterns and daily market conditions, is essential for empowering farmers, especially those in developing countries.

Sustainability must be accessible to everyone and resilience actions must be applicable to the practice of everyday life. Consumers respond poorly to browbeating activism and need fiscal incentives to use less, and to be given greater control over the energy they use. Lower-carbon products and services need to become desirable, which is where the aesthetic aspects of design need to be employed. We are often told that we are the last generation able to make any effective change in the future of our planet. The day that the eco-warrior dies is a day to look forward to, for it will mean that there are no more unconverted to preach to, that sustainable living has become normative rather than alternative. That day has not yet arrived. There are still battles to be fought and won.

29

4. Food & Argriculture Organisation of the United Nations, 'The State of Food and Agriculture, 2010-2011', www.fao.org, 2011, p.vi [http://www.fao.org/docrep/013/i2050e/i2050e.pdf], retrieved 2 June 2018

5. J Gettleman, 'Wangari Maathai, Nobel Peace Prize Laureate, Dies at 71', The New York Times: Africa, 26 September 2011

6. M Specter, 'Seeds of Doubt', The New Yorker: Annals of Science, 25 August 2014 [https://www.newyorker.com/magazine/2014/08/25/seeds-of-doubt], retrieved 1 June 2018

7. J Vidal, D Adam, A Ghosh et al., '50 people who could save the planet', The Guardian: Environment, 5 January 2008

8. KF Nwanze, 'Food Sustainability and the Role of Smallholder Farmers', The Economist: Intelligent Unit, 1 July 2014 [http://foodsustainability.eiu.com/food-sustainability-and-the-role-of-smallholder-farmers], retrieved 1 June 2018

9. H Gould, '10 Things You Need to Know About Sustainable Agriculture', The Guardian: Environment, 1 July 2014

10. I Fitzpatrick, 'From the Roots Up: How agroecology can feed Africa', Global Justice Now, February 2015, p.53

11. I Fitzpatrick, 'From the Roots Up: How agroecology can feed Africa', Global Justice Now, February 2015, p.38

Scenic Positions

'This whole mass of architecture which we had come upon so suddenly from amidst the pleasant fields was not only exquisitely beautiful in itself, but it bore upon it the expression of such generosity and abundance of life that I was exhilarated to a pitch that I had never yet reached.'

'"I don't understand," said he, "what kind of people you would expect to see; nor quite what you mean by 'country' people. These are the neighbours, and that like they run in the Thames valley."'

– William Morris, 'News from Nowhere', 1890

The sites for Smartcity systems vary in scale, geography, culture and time, consequently requiring adaptation according to context. The concept of terroir – the climate, topography, soil conditions and aspect of a piece of land for the production of wine – is equally applicable to the cultivation of edible produce, renewable energy, communities and resilient landscapes, with the additional region-specific variables of politics, land ownership, extant infrastructure and cultural bias.

The Marseille Unité d'Habitation, the purest example of Le Corbusier's vision for communal living, was given the unflattering moniker 'La Maison du Fada', French-Provençal for 'House of the Mad', when it was built. The systems buildings that followed in the wake of the Unité d'Habitation and corrupted its legacy are considered by many as the cause and symbol of deprivation and anti-social behaviour in the inner cities. On the other hand, while the very notion of the Unité would have been inconceivable in the United States, Le Corbusier's ideas would gain traction in South America and particularly Hong Kong, where the Parisian Plan Voisin of 1925 of giant cruciform towers has effectively been realised and is proliferating across Southern China. Acceptance of high-rise mixed-use living may be attributed to cultural factors as well as unusually high population densities.

At a macro level, the entire city becomes the site for a new urban land use – farming. Its location takes advantage of favourable adjacencies between agricultural processes and their raw materials, establishing a classic urban nutrient cycle. Urban solid waste and greywater can be used as fertiliser and irrigation; food transport and associated carbon emissions are removed from the equation. A growing number of communal farm initiatives in Madrid and Barcelona are being developed to include programmes in schools with the aim of providing basic agricultural skills. The culture of proximity consumption not only facilitates direct contact with the food chain, but also gives the children a special awareness of the value of water and avoiding its contamination.

facing page top to bottom: Imagining Recovery – Grocery shopping in New York's Central Park; Car boot allotments in Detroit.

following page: Imagining Recovery – Farming within the city takes advantage of programmatic and functional synergies.

1. B. Oakes, 'Sculpting with the Environment: A natural dialogue', Van Nostrand Reinhold, New York, 1995, p.168

2. W Morris, 'Useful Work Versus Useless Toil' in 'The Collected Works of William Morris', vol.XXIII, Cambridge University Press, New York, 2012, p.100

Due to the high premium of land in dense urban areas, urban agriculture is considered to be unfeasible within cities. The artist Agnes Denes eloquently articulated the disparity of land value in her installation 'Wheatfield: A Confrontation', in which she planted a golden field of wheat amongst the gleaming skyscrapers of downtown Manhattan. In the autumn of 1982, Denes harvested her crop that had a value of US$93 on land valued at US$4.5 billion. The piece was intended to 'call attention to our misplaced priorities and deteriorating human values'.[1] In the summer of 2009, the work was reproduced in the London borough of Hackney and the questions she raised regarding the true value of sustenance are more germane now than ever before.

The city farming model of Havana and the victory garden movement belie the assertion that the cost of land is prohibitive – brownfield sites, car parks, rooftops, window boxes, barges and riverbanks have all been appropriated to create a productive landscape in the past and can once again beautify our townscapes, all the while providing good nutrition and generating social capital. For overgrown sites and derelict rooftops, it is simply a case of replacing buddleia and nettles with berries, tomatoes and herbs.

William Morris' new world idyll in his utopian novel 'News from Nowhere' and Denes' visceral commentary owe their potency to the striking juxtaposition of the pastoral and urban, demonstrating how scale and context can effect beauty. A cabbage or wheat patch is not as conventionally beautiful as daffodils or tulips, but when scaled up, replicated a thousand fold, reconfigured into vertical surfaces or arrayed into pattern – in short curated – they can achieve the elegance of multi-sensory art and expand the limited palette of urban textures. In the Smartcity, buildings and roofscapes will transform in colour, volume and scent through the seasons. Morris' answer to the critics of his socialist vision, who argue that individuals will lack the incentive to work in a world where private property is abolished, is that work should be creative and pleasurable.[2] In European cities where waiting lists for allotments number up to 40 years, widespread cultivation of the metropolis offers a way to disintegrate orthodox distinctions between work, leisure and art.

Agricultural land in outlying areas for the growth of biofuels

City farming for food

Within dense urban areas, rooftops, windowsills, balconies and walls can be appropriated for the growth of edible crops, evoking the spirit of the Second World War victory garden. With the support of government policy, the public realm could be fully reclaimed – plazas, parks, waterfronts, boats, car parks and greyfield sites where appropriate sunlight levels are available are all viable locations. The metrics of progress will be clearly manifest, growing before our eyes while beautifying our environment.

A scattering of community growing programmes are already in action, and publicity surrounding Michelle Obama's White House Garden has helped raise public awareness of the potential productivity of our backyards. The 'Urban Farming Food Chain Project' birthed in Los Angeles has expanded its operations abroad to Jamaica, Canada and the UK. The vertical growing walls can be installed either as freestanding frameworks or cladding elements, which the architect, Robin Osler, logically argues do not consume valuable horizontal real estate. Their adoption therefore becomes more attractive to developers under economic pressures to maximise usable floor area. These initiatives may appear modest in scale but they play a vital role in stimulating the perceptual shift in how we think about and procure our meals. Walls of living food in the city can vitalise urban communities by bringing food provenance and human-scale interaction back to the table, consumer friendly but not pre-packaged. Edifices of spatial theatre, these verdant edible walls also raise the public's awareness of architectural possibilities in the city that are not limited to brick, concrete or glass.

Perhaps an even more intelligent use of space can be found in the Brick City urban farms of Newark in New Jersey, based on the Small Plot Intensive (SPIN) relay-farming model that was thought up by the Canadian farmers Wally Satzewich and Gail Vandersteen.[3] Using the simple device of a plastic crate or 'EarthBox', Brick City farmers are able to colonise disused sites.[4] The limited size of the units allows operations to decamp and re-root in other transient spaces as they become available. The deployment of EarthBoxes in Newark came out of necessity due to contaminated ground, but the containers have the added benefit of minimising water and fertiliser use.

Alternative materials in the city do not have to be organic. Renewable technologies in the form of photovoltaics and wind turbines are still inchoate and typically retrofitted to rooftops either as green marketing strategies or to satisfy government sustainability policy. There are, however, buildings in recent years that have broken from this mould, such as Toyo Ito's World Games Stadium in Kaohsiung, Taiwan, and Hamilton Associates, Strata building in London. The former, a sinuous river of blue that coils out to form a public plaza, is entirely covered in photovoltaic cells. A striking example of integrative interdisciplinary design, the stadium is classified as an independent power plant (IPP), potentially generating 1.14 gigawatt hours per year with surplus energy fed into the grid when the arena is not in use.[5]

Wind energy is more problematic, requiring strong laminar (unidirectional) wind and large swept areas to be efficient. The notion of buildings integrating turbines at their apex to power them is a seductive synergy of function and form. However, cities are typically situated in areas of low wind speed and the flow is turbulent by virtue of the buildings that constitute them. The top 20 metres of the Strata building

33

3. R Christensen, 'SPIN Farming: Improving revenues on sub-acre plots', UA Magazine, no.19, RUAF Foundation, December 2007, pp.25–26

4. J Weiss, 'Farm Fresh, in the City: Urban farming in Newark', www.nj.com, 24 October 2010 [https://www.nj.com/homegarden/garden/index.ssf/2008/10/farm_fresh_in_the_city.html], retrieved 18 March 2018

5. Guinness World Records, 'Largest Solar-powered Stadium', Guinness World Records 2016, Guinness World Records Limited, 2015, p.129

6. J Glancey, 'Spin City: London's Strata Tower', The Guardian: Architecture, 18 July 2010

7. F Estrada, WJW Botzen & RSJ Tol, 'A Global Economic Assessment of City Policies to Reduce Climate Change Impacts', Nature Climate Change 7, 29 May 2017, pp.403–406

8. D Despommier, 'The Vertical Farm: Feeding the world in the 21st Century', www.verticalfarm.com, 2008 [http://www.verticalfarm.com/?page_id=36], retrieved 14 June 2018

9. From Morris' lecture 'How We Live and How We Might Live' delivered to the Hammersmith Branch of the Socialist Democratic Federation (S.D.F.) at Kelmscott House, 30 November 1884. N Salmon, 'The William Morris Internet Archive: Works', Marxist Internet Archive

is a wind farm comprising three nine-metre diameter wind turbines. The angled elliptical concave surfaces into which the turbines are mounted create a venturi effect that channels the wind while minimising vibration and wind noise.[6] The projected energy returns are unremarkable, but the building does illustrate how energy farms and buildings can holistically coalesce.

The Smartcity thinks more in terms of inhabitable wind farms and photovoltaic parks rather than tower blocks with token micro-turbines and solar cells. The willful formalism of amorphous icon buildings favoured by marquee-name architects has shown that complex geometric forms are realisable using generative tooling software. There is no compelling reason why prodigious design skills and cutting-edge technology could not be channelled into developing clean energy morphology. A study published in Nature Climate Change claimed that by changing 20 per cent of the roofs of a city and half of its pavement to 'cool' forms, for example trees and vegetables, it would be possible to save up to 12 times the cost of reducing the temperature by 0.8 degrees.[7]

Shifting scale from innovative surfacing materials and interstitial spaces to buildings, the vertical farms championed by Dickson Despommier take the compact city argument for increasing plot ratio in advantageous areas to preserve remaining threatened ecosystems and apply it to agriculture. Despommier, professor of public health in environmental health sciences and microbiology at Colombia University, reasons that the success or failure of current crop production is wholly contingent on the vicissitudes of weather and disease, and any significant sustained deviation from an optimal range has catastrophic effects on yield. His solution lies in three-dimensional hermetic farms that ensure year-round high-yield crop production with minimal risk of infection from agents without the use of pesticides. The tower model also reduces the use of fossil fuels and takes advantage of energy-waste trades with other urban activities. He estimates that one vertical farm with a footprint of one square city block rising 30 stories would provide enough nutrition (2000 calories/day/person) to accommodate the needs of 10 000 people, employing technologies currently available.[8] The controlled conditions of the farms will permit analysis of the chemical composition of each plant, with gas chromatography employed to test flavenoid concentration guaranteeing flavoursome and ripe produce. Increased levels of artificial lighting necessary could be obtained by generating biogas from inedible plant waste on site.

In the case of the new city, the scope for more comprehensive Smartcity intervention is possible. New housing developments can be planned to integrate farming at the scale of landscape; buildings can be used to terraform the natural topography, be surfaced in growing media, orientated to receive or protect from sunlight, and integrate water conservation, inter-seasonal heat transfer and waste recycling mechanisms. As illustrated at the industrial symbiosis at Kalundborg and the theoretical Dream Farm, remarkable economies of scale are achievable at the magnitude of a city when symbiotic self-perpetuating cyclical systems are adopted. Morris' dream of an agrarian society is not compatible with contemporary urban existence or even universally desirable in today's age, but his insistence that 'the material surroundings of life should be pleasant, generous, and beautiful'[9] remains unassailable.

Cultivating Community

CULTIVATE

verb [trans.]

1 prepare and use (land) for crops or gardening.

• break up (soil) in preparation for sowing or planting.

• raise or grow (plants), esp. on a large scale for commercial purposes.

• grow or maintain (living cells or tissue) in culture.

2 try to acquire or develop (a quality, sentiment, or skill) : he cultivated an air of indifference.

• [usu. as adj.] (cultivated) apply oneself to improving or developing (one's mind or manners) : he was a remarkably cultivated and educated man.

> – 'The Oxford Pocket Dictionary of Current English', 2009

35

As evolving organic entities, communities grow, germinating from unlikely seeds and requiring careful nurture. They flourish when conditions are favourable, and when faced with a changing climate, they adapt to new environments or make way for better-suited alternatives; when faced with new arrivals, they are either crowded out or cross-pollinate, sharing resources and blending traits. Communities share much in common with agriculture, and one can show the way forward for the other.

Skid Row, in Los Angeles, home to one of the largest homeless populations in the United States, is one of the beneficiaries of Urban Farming, a Detroit-based non-profit organisation dedicated to eradicating hunger, founded by the singer Taja Sevelle. Together with the architects Elmslie Osler and Green Living Technologies, Urban Farming have installed a series of 30-foot-long-by-6-foot-high walls, each containing 4000 plants to supply the area's dispossessed with tomatoes, spinach, peppers, lettuce, leeks and herbs.[1] Just as significantly, the programme has drawn together diverse and disadvantaged members of the community of all ages and ethnicities as well as providing an opportunity to learn new skills and reducing local crime.

Food is a universal. It is cross-cultural, cross-gender, cross-class and cross-generational. As a key prerequisite for resilience, food is the great democratiser that defines our society and is an essential element of Smartcity living. Claude Lévi-Strauss, the French anthropologist, makes the point that culinary rites are not innate but learned;[2] the human digestive system is able to process almost any organic material, and the distinction of what is edible and what is not is a cultural convention. Food, as a social medium, communicates a veritable smorgasbord of meaning, from Eucharistic sacrament in the Christian church and religious separation in kosher law to etiquette in the formal dining room,

1. J Murdock, 'Vertical Food Gardens Sprout in LA's Skid Row', Architectural Record, 7 Oct 2008 [https://www.architecturalrecord.com/articles/4601-vertical-food-gardens-sprout-in-l-a-s-skid-row?v=preview], retrieved 28 Nov 2008

2. C Lévi-Strauss, 'Le Cru et Le Cuit', from 'Mythologiques I-IV', J Weightman & D Weightman (trans.), Plon, 1964

celebration on feast days, societal responsibility in soup kitchens and protest in hunger strikes. Living food in the city fulfils a yearning for the haptic and tangible as well as the digital, presenting a city framework that engages people rather than automata. The vegetable walls of the Urban Farming Project are an example of spatial phenomenology in the city, stimulating our eyes, ears, noses, minds and tongues – imagination made real, architecture that you can taste.

The deployment of agricultural and energy generation systems within urban environments is only part of the story. Much criticism has been levelled at planned communities, from Levittown in the United States to the three waves of post-war New Towns in the UK and the ongoing Thames Gateway project. The Thames Gateway is Europe's largest regeneration programme, and there are recurring concerns that the result will be a concentration of 'Stepford Suburbias' and 'Noddy Towns'.

Many of the perceived failures of Levittown and Milton Keynes can be attributed to the hegemony of the motorcar and the lack of opportunities for unskilled and lower income workers. These new towns were self-financing and whilst this rendered the large-scale development programme possible, any pioneering visions for an urban future were of necessity watered down to pander to the public's demand for private transportation. Vehicular–pedestrian separation remains vastly unpopular amongst Britons and Americans, stymieing the reification of truly sustainable urban environments.

The transcription of a new city from paper to lived reality usually ignores the genius loci, the distinctive atmosphere of a location. Established European metropolises such as London, with their rich, variegated histories, have become richly textured palimpsests, as extensively described by the urban chroniclers Peter Ackroyd and Ian Sinclair. Planned communities, however, do not have the luxury of having identities developed and nurtured over the passage of time. The same problems are faced by urban localities that suffer from social and economic deprivation, usually as a consequence of an anachronistic industrial heritage. In essence a mega-community, the city is a network of living systems that mutates or atrophies and dies.

Whether formed tabula rasa on an undeveloped site or integrated within an established metropolis, the Smartcity seeks to preserve the identity and the heritage of a place, ascribing as much importance to the past as the future. Traditionally, the character and industry of a settlement arose from the geographical uniqueness of the earth under and around it, whether it be from the geothermal springs of England's Bath Spa and onsen all over the volcanic region of Japan or the mining towns of Central Illinois, Southwestern Pennsylvania and West Virginia in the United States that resulted in railroad development across the continent. Similarly, viticulture in the Wine Country of Northern California that includes Sonoma County and Napa Valley came about from the unique variety of climate and soil conditions of the region, generating a form of employment, tourism and culture starkly different from, say, the Middle Eastern banking capital of Bahrain.

Financial institutions and agricultural terroir may appear unlikely partners, but in the wake of the 2009 global financial collapse, region-specific gourmet food has operated as a viable, if unlikely, form of currency. Credito Emiliano, a regional bank in Montecavolo, Italy, has been accepting Parmesan cheese as loan collateral since 1953. The bank owns two climate-controlled warehouses in which US$187.5 million worth of Parmigiano-Reggiano are stored.[3] The loans may account for less than one per cent of the bank's revenue, but are vital in preserving Montecavolo's culinary heritage and local economy, given that the cheese needs to age for five years. Other artisan food products such as prosciutto in San Daniele and brunello in Tuscany have also been considered as unconventional collateral; these commodities are highly site-specific, both in terms of raw material and local knowledge, and lend credence to Marx's labour theory of value: a commodity should be worth the amount of time and human labour invested in it.

The quest for urban identity must go beyond the current vogue for place-branding, an increasingly common exercise amongst municipal authorities in an attempt to regenerate inner city areas by attracting new residents and stimulating business investment. Places need to be made rather than hawked. Too often, the features that are marketed are generic, such as good transport links and green space. Too often, there is little coincidence between expectation and reality. Cities are not products. They are more complex and are not in direct competition with one another, operating in multiple sectors and at different scales. Environmentally too, there are compelling reasons for exploiting the genius loci. Certain social and environmental conditions are inherently better suited for certain processes. Food miles notwithstanding, it is still appropriate, for example, to rear and export lamb from New Zealand. Due to the abundance of land and moderate climate, the livestock are able to remain in the pastures all year round, without any food additives or growth hormones.

The arguments for localism as an answer to globalisation-linked identity loss extend beyond food to skills and construction materials. Built in the 16th century, the Mughal city of Fatehpur Sikri in the Indian state of Uttar Pradesh is almost entirely constructed in richly ornamented red sandstone, quarried from the same rocky landscape on which it stands. The 'Red City' is a world heritage site, as is the town of Bath, whose warm honeyed appearance comes from the extensive use of Bath stone. The use of aluminium in Reykjavik is a more contemporary instance of localism – as a by-product of Iceland's burgeoning aluminium industry that makes use of the country's natural geothermal resources, the city is characterised by brightly coloured corrugated aluminium facades, that, like growing vegetative walls, are recyclable.

Successful place-making and the fostering of community pride rely on differentiating one place from others and highlighting the character and activities of an area. Barcelona has used a combination of international events and architecture as the stimuli to reinvent itself since the Universal Exhibition of 1888 at Ciutadella, now the city's largest park. The Olympic legacy from the 1992 Games, including the creation of new beaches for public use, remains the standard for future Olympic cities. Significantly, the reconstruction of Barcelona in the wake of the Franco regime concentrated on improving the common urban fabric – schools, squares, museums, sewage treatment plants and community centres embellished by a relatively modest scattering of icon buildings by regional architects of international renown such as Santiago Calatrava and Enric Miralles. Taking a leaf out of the book of its Iberian neighbour, the port city of Bilbao has seen its fortunes completely transformed by Frank Gehry's Guggenheim Museum. Surveys have shown that 82 percent of tourists visit the city exclusively to see the museum.[4] More than a building, the Guggenheim effectively recentred the city, spearheading the economic reconstruction of the newly autonomous Basque country as a post-industrial service-based nation.

The cultivation of communities involves the generation of what Robert Putnam, author of 'Bowling Alone – The Collapse and Revival of American Community',[5] describes as social capital, the 'connections

3. A Migliaccio & F Rotondi, 'In Italy, Parmesan as Collateral for Bank Loans', The New York Times: Global Business, 13 August 2009 [https:// www.nytimes.com/2009/08/14/ business/global/14parma.html], retrieved 15 May 2018

4. M Bailey, 'The Bilbao Effect', www.forbes.com, 20 February 2002 [https://www.forbes. com/2002/02/20/0220conn. html#d30d76072036], retrieved 15 May 2018

among individuals – social networks and the norms of reciprocity and trustworthiness that arise from them'. Putnam emphasises the importance of bridging capital, the cross-connections between communities as crucial for the formation of a peaceful and multi-ethnic nation, and explores how public policy can facilitate or destroy social capital. The clearance of American slums in the 1950s and 1960s, for example, regenerated physical capital at the expense of arguably more valuable existing social bonding capital; similarly the consolidation of local post offices and small school districts in the name of efficiency has had unforeseen social costs. Gentrification, the influx of the affluent into inner city areas, has resulted in physical proximity with deprived neighbours but also as much hostile as positive social interaction; ethnic enclaves such as Chinatown and Little Italy in Manhattan are not ghettos and demonstrate the distinction between accommodating diversity and pushing for assimilation.

The Smartcity programme endeavours to support governments' priority to strengthen national and local food security by respecting land rights of the farming communities. Regrettably, land grabs are the common cause of violent clashes in the Chinese countryside in recent years, and have already created a 'floating' population the size of Spain or even Mexico with no access to land despite being officially classed as 'rural' under the country's 'hukou' or household registration system. In 2015, the Financial Times reported chilling concerns: 'Chinese history books teach that armies of landless people contributed to the chaos of the early 20th century, when local warlords battled for control of the country. Opponents of land reform within the ruling Communist party argue that privatising land would allow those armies to form again today, if rural families are left without fields to fall back on.'[6]

Where regeneration frameworks and environmental protocols must be global in scope, communities emerge at grass roots level, and crucial to the growth of Smartcities is a devolution of power to local representatives at the front line who are better placed than profligate high-level quangos to allocate resources and evaluate need. Social capital and resilience are formed at a human scale between individuals, and local structures need to be in place to enfranchise the disenfranchised and affect the disaffected.

Food is but one common ground between disparate communities. Housing, transport, water, heating, electricity, sewerage, waste treatment and data services are all or are all becoming basic necessities of urban existence, making them ideal sites for the creation of new social capital. Legislation requiring mixed housing tenures and a minimum provision of affordable housing in new developments is a step forward. The Smartcity takes this idea further through the banishment of the private motorcar and the widespread use of community-tailored and community-owned multi-utility service companies (MUSCOs). Where the provision of energy, water and waste treatment are not solely state controlled, MUSCO arrangements operating entirely transparently can return excess profits to the city in the form of lower consumer bills or capital injection into other community projects.

The panacea to a deleterious food industry is to replace damaging large-scale monoculture with larger numbers of smaller mutually supporting diverse permacultures. Communities share much in common with agriculture, and one can show the way forward for the other.

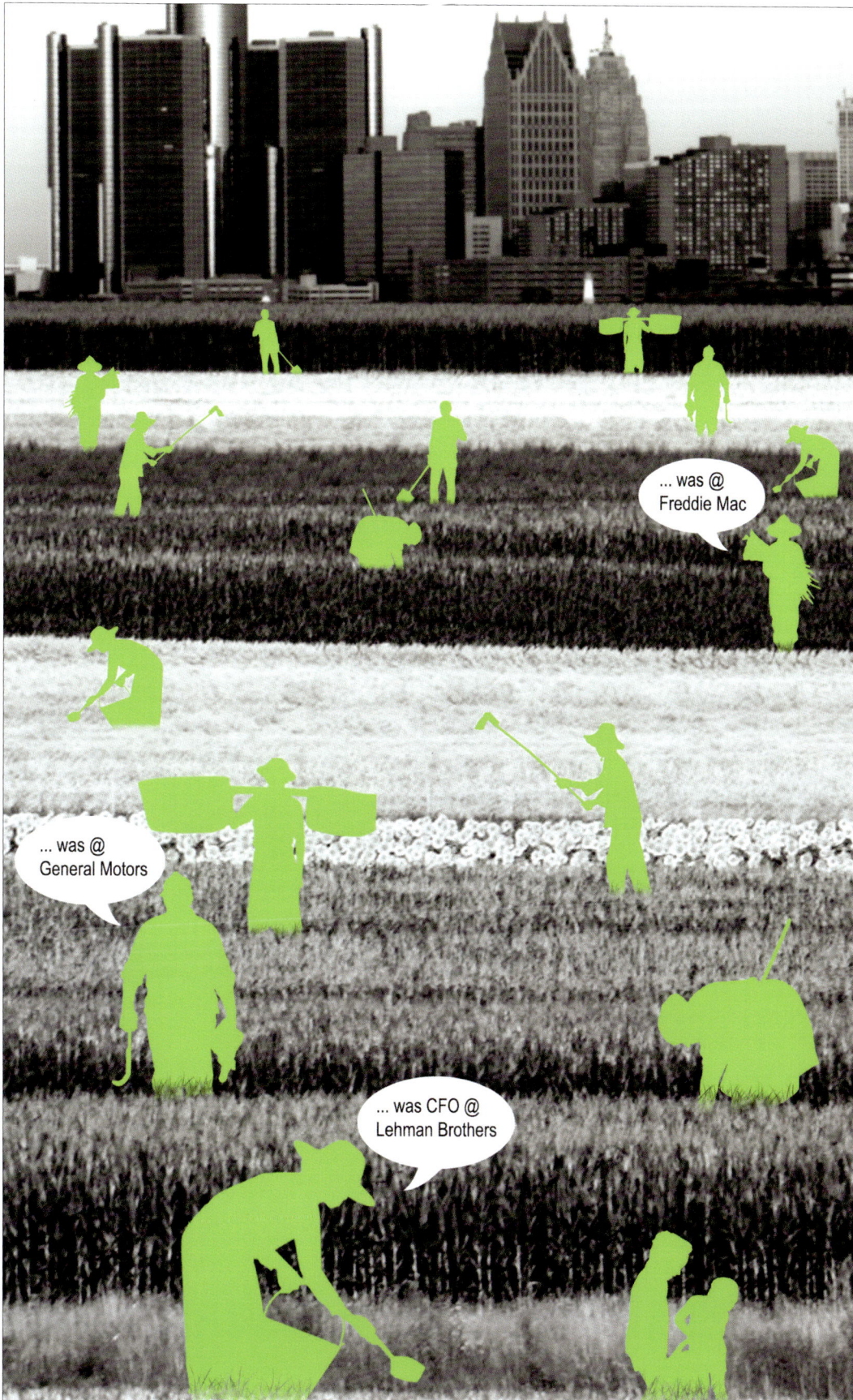

left: Imagining Recovery – Communal creation of a productive landscape in disused and neglected sites offers a means to cultivate employment and social responsibility.

5. RD Putnam, 'Bowling Alone – The Collapse and Revival of American Community', Simon & Schuster, New York, 2000

6. L Hornby, 'China Migration: Dying for land', Financial Times, 6 August 2015

a pictorial essay: Taipei
Resilient Landscapes

A Lexicon for the Smartcity

agritourism: The act of visiting an agricultural or horticultural operation for leisure, educational or active involvement purposes. Agritourism offers an alternative revenue stream for rural communities and an insight into food production for the general public.

agroforestry: A land use system in which the growth of herbaceous crops is combined with trees and shrubs to preserve and improve productivity. Trees draw water and nutrients from deeper soil and provide temperature moderation and mulch whilst herbaceous plants prevent soil erosion and the proliferation of weeds.

anaerobic digestion: The breakdown of biodegradable material into biogas and nutrient-rich digestate using microorganisms in an oxygen-free environment. Widely used in the treatment of organic wastewater, the biogas produced is used as a renewable energy fuel and the digestate pasteurised for use as fertiliser. Fibrous mass, a third product of anaerobic digestion, is either pasteurised and used to improve the structure and water retention capacity of soil or incinerated for electricity generation.

appropriate technology: Technological solutions that take into account environmental, social and cultural considerations and are sustainable within the communities in which they are employed. The term is usually used in reference to manual-based solutions in developing countries or socially and environmentally sensitive technologies in industrialised nations.

aquifer thermal energy storage (ATES): A type of low-temperature geothermal energy storage system using open loops in aquifers to store seasonal heat and cold when it is available and to retrieve it when required for space heating and cooling. ATES uses substantial quantities of water as a heat storage medium, although this water is constantly recycled.

artificial photosynthesis: The conversion of carbon dioxide and photon energy into carbon-based fuel molecules with minimal energy expenditure, mimicking natural photosynthetic processes. While the technology is still in its infancy, the use of solar energy to convert carbon dioxide released from burning fuels back into usable energy would establish a closed sustainable fuel cycle.

assimilative capacity: The capacity of an environment, usually aquatic-based, to receive waste or toxic material and convert it into harmless or useful material.

biochar: Charcoal produced from animal waste or plant residue biomass through pyrolysis.

biofuel: Fuel produced from renewable organic resources such as plant biomass, in contrast with fossil fuels that derive from long-dead biological material. The most common forms are bioethanol and biodiesel, which can be used either as additives or directly as transport fuel. Biofuel production is paradoxically energy intensive, and the appropriation of farmland for the growth of biofuels has led to significant increases in food prices as well as the destruction of natural habitats. However, agricultural waste can be used as a raw material, fast-growing crops such as switchgrass and hemp would minimise the impact on food production, and new technologies using cellulosic biomass conversion could reduce energy expenditure.

biogeochemical cycles: Natural cycles of the earth's atmosphere, hydrosphere, biosphere and lithosphere that involve biological, geological and chemical processes. Human activities can disrupt biogeochemical cycles by extracting material from reservoirs or depositing them into sinks, notably in the carbon cycle through the mining and combustion of hydrocarbons.

biomass: A renewable energy source derived from plant residue, vegetation or agricultural waste that is used in the production of heat or electricity. Fossil fuels are not considered to be biomass due to their long-established separation from the carbon cycle. Biomass crops tend to sequester more carbon than arable crops, but contribute in the same way as fossil fuels to greenhouse gas emissions when burnt.

bioregionalism: The establishment of social and environmental policies based on the ecological, geographical and cultural contexts of a region rather than political or economic boundaries.

blackwater: Water containing the waste of humans, animals or food. Separation from greywater permits the efficient processing of solid nutrient-rich waste into fertiliser or fuel and treatment of wastewater for reuse.

boardwalk: The main interconnected transport pathway of the Smartcity, modelled on the timber walkways found along beaches and wetlands. Although often located along an urban beach, the boardwalk combines the recreational qualities of a traditional boardwalk with urban functions. Capable of supporting emergency vehicles, the pathway is shared by pedestrians, cyclists and electric vehicles, expanding at key nodes into public plazas and transport hubs.

borehole thermal energy storage (BTES): An underground thermal storage system that uses closed loop boreholes to store solar heat collected in the summer for use in the winter. Boreholes are filled with a high thermal conductivity grouting material to ensure good thermal contact with the surrounding soil.

carbon capture + storage (CCS): The capture of carbon dioxide emissions from industrial sources and their injection into deep geological formations or ocean masses for permanent storage as a form of greenhouse gas remediation. The dehydration, compression and transportation of CO_2 require considerable energy and storage depends on the availability of empty spaces in the earth's crust with potential problems of leakage. Alternatives to storage include the use of the CO_2 to feed algae for biofuel production and artificial photosynthesis.

carbon offset deals: A trading mechanism designed to reduce greenhouse gas (GHG) emissions by funding renewable energy technologies in compensation for GHG-emitting transportation or electricity use. A single carbon offset or credit is equivalent to one metric tonne of greenhouse gases.

carbon sequestration: The removal of atmospheric carbon through biochemical or physical processes. Sequestration includes absorption into biomass such as crops, trees, soil and microorganisms as well as CCS permanent storage.

cash cropping: The cultivation of crops such as coffee, tobacco and flowers for sale, often specifically for export markets. The price of staple cash crops is set by global commodity markets, leading to vulnerability in times of excess supply. Intensive farming practices have also been responsible for soil depletion.

centre of excellence: A building or collection of buildings with a specialised programme offering resources in a specific focus area. Smartcity districts are sized and designed to be self-sufficient; they are provided with an identity in the form of a Centre of Excellence (CoE) that generates new avenues of communication between neighbouring communities.

circular economy: The optimisation of resource management and environmental efficiency by following the principles of reducing, reusing and recycling. Resource consumption and waste production is reduced, products are reused through repair and renovation, and waste products are recycled into resources to the fullest extent. 'Circular economy' is also the Chinese term for sustainability, a critical component of China's 18-year development plan for economic growth whilst mitigating negative ecological impacts.

closed system: A physical system that obeys the laws of conservation and does not interact with its external

51

environment. In reality, no system can be completely closed. The earth is a closed system with respect to matter, but an open system with respect to energy, receiving radiant solar energy that is, for all intents and purposes, inexhaustible.

cloud seeding: The dispersal of silver iodide particles into the atmosphere to act as nuclei for cloud formation. Marine stratocumulus clouds reflect sunlight back into space, reducing the amount of heat received by the earth's surface. The use of seeding to expand and distribute clouds to reflect solar radiation offers an alternative geoengineering approach to counteract global warming.

combined heat + power (CHP) + CCHP: A system that recovers waste heat from power generation to form useful energy such as steam or district heating. Combined cooling, heat and power (CCHP) additionally converts heat by-product into cooling energy by using absorption chillers and is also known as trigeneration. Multi-generation power plants are most efficient when waste heat or cooling can be used in proximity to its source.

Common Agricultural Policy (CAP): An agricultural programme enshrined in EU legislation designed to increase productivity, stabilise consumer prices, ensure a fair income to farmers, and preserve rural heritage. Due to the capacity of other nations to produce food at significantly lower prices, the EU has had little option but to heavily subsidise agriculture in its member nations while keeping prices high. In its initial formulation, the CAP paid scant regard to the environment, leading to intensive and detrimental farming practices that have been partially addressed by subsequent reforms.

concentrating solar power (CSP): The use of parabolic mirror arrays to focus large amounts of sunlight onto a small area for the generation of heat and/or electricity. Energy can be stored in phase-change materials to enable electricity supply through the night and during overcast conditions. CSP plants in desert environments

have the potential for generating vast amounts of energy, whilst waste heat can be used for desalinating water for crop irrigation. The crops would also benefit from shade beneath the mirror arrays, facilitating horticulture in an otherwise hostile environment. Transport of the energy to population centres still proves problematic, but CSP could pave the way for the development of urban centres in arid regions.

cool roof: A roofing system that reflects and emits solar energy, reducing the transfer of undesirable heat into a building. Cool roofs reduce cooling loads, reduce urban heat island effects and smog, and mitigate global warming. In addition to reflecting sunlight, cool roofs can emit infrared radiation, presenting a viable geoengineering technique.

cultivar: A plant species cultivated for its unique characteristics that are uniform, distinct, stable and retain those characteristics on reproduction.

cultural resources: Components of social capital including physical assets such as architecture and sculpture, but also more abstract and less quantifiable assets such as history, language, folklore and heritage.

cycle station: A vertical storage tower for rentable bicycles located at transport nodes to encourage a car-free environment in the Smartcity. Bicycles are stored via a lowerator mechanism to occupy a smaller footprint. Illuminated, the towers also act as navigational beacons.

domestic fuel cell: A small-scale electricity generator using replenishable fuel sources that can be located in urban environments and consequently at the point of consumption. Domestic fuel cells can be as small in size as a washing machine, and generate electricity from gas more efficiently than modern power stations. Using solid oxide technology, the CO_2 released is combined with water vapour, facilitating capture and storage.

EarthBox: A patented planting system comprising a recycled plastic container that houses a water reservoir and growing substrate. The EarthBox's self-contained nature allows plants to survive without constant tending, the cultivation of a productive landscape above contaminated ground and ease of transportation.

eco-warrior: An alternative term to environmental activist that also encompasses educationalists, farmers and businessmen.

ecotourism: Environmentally responsible travel to natural areas in order to enjoy and appreciate nature (and accompanying cultural features, both past and present). Ecotourism must promote conservation, have a low visitor impact and provide for beneficially active socio-economic involvement of local peoples [definition adopted by the World Conservation Union].

energy bonds: Also known as Clean Renewable Energy Bonds (CREBs), energy bonds are long-term secure investments in renewable energy projects. Significant investment is required to finance the conversion from a fossil fuel-based economy; necessary capital can be raised by governments, electric companies and clean renewable energy bond lenders who issue tax credit bonds on which interest is paid in tax credits rather than interest.

Engel's Law: A law in economics correlating the proportion of national income spent on food with welfare level. As income rises, proportional expenditure on food tends to fall. The Engel coefficient is often used as a metric for national living standards.

enhanced geothermal systems (EGS): A geothermal power technology that does not rely on natural rock porosity. Also known as hot dry rock geothermal, EGS involves the high-pressure injection of water into boreholes drilled several kilometres into the earth's crust to fracture and increase the surface area of hot rock. The water emerges as steam that can be used to generate electricity and heat, offering a viable clean energy source irrespective of geographical or geological location.

environmental remediation: The chemical, biological or physical removal of pollutants or contaminants from the environment. Remediation is usually monitored and based on regulatory controls.

eutrophication: A proliferation in plant growth, especially algae, stimulated by excess mineral and organic nutrients. Eutrophication occurs naturally but is usually triggered by human sewage and the leaching of chemical fertilisers from arable land. The sudden increase in vegetation has a detrimental impact on the aquatic ecosystem, depriving fish and other aerobic organisms of oxygen. Eutrophic conditions additionally affect drinking water treatment and the use of water bodies for recreation and leisure.

farming carpet: An expanse of land within the Smartcity used for the cultivation of arable crops. Colour, pattern and texture are considerations in plant selection.

Gaia hypothesis: The theory formulated by James Lovelock in the mid 1960s that the planet functions as a single self-equilibrating organism with the capacity for self-regulation.

gathering wells: Public spaces recessed into the landscape and used for a variety of recreational purposes from floral and sound gardens to swimming and fishing.

geo-engineering: The large-scale manipulation of natural systems to counteract climate change caused by human activities. Techniques include carbon sequestration, afforestation, solar shielding and stratospheric aerosols.

geothermal energy: Energy deriving from the heat below the earth's crust, extracted from steam, hot water or hot rock. The energy can be used for geothermal heat pumps, water heating or electricity generation using steam turbines.

53

greenhouse gases: Gases in the earth's atmosphere that absorb and emit radiation in the thermal infrared range. The gases prevent the release of heat into space creating a 'greenhouse effect' that makes the earth habitable. However, an increase in concentration of greenhouse gases [GHGs] partially caused by human activity has caused an imbalance in heat retention. Greenhouse gases include water vapour, methane, nitrous oxide, chlorofluorocarbons and carbon dioxide. The latter is the principal anthropogenic contributor to climate change, although water vapour accounts for the largest share of atmospheric GHGs.

greywater: Waste water from domestic washing.

groundwater: Water that has collected beneath the earth's surface in porous layers of rock called aquifers. A source of drinking water, the majority of groundwater has accumulated over the ages and is a finite resource, although rainwater adds to the store through percolation.

human capital: A form of capital stock consisting of accumulated skill and knowledge that contributes to productive power and economic value.

hydroponics: A technique for cultivating plants using a nutrient solution as a growing medium in lieu of soil. The productivity from hydroponic cultures is high due to the plants receiving a constant feed of nutrients with minimal risk of pest infestation. Growth can be further controlled in hermetic environments with artificial lighting and carbon dioxide flooding.

industrial ecology: An interdisciplinary field of research based on symbiosis between industrial and ecological systems. Industrial ecology involves the establishment of cyclical material and energy flows in which wastes from traditional waste-producing industries are used as resources in other processes while minimising the creation of harmful by-products.

integrated food + waste management system (IFWMS): The coordination of different types of production such that the waste output from one component feeds another, IFWMS is an agricultural permaculture system developed by environmental engineer George Chan.

inter-seasonal heat transfer (IHT): A means of reducing space heating and cooling demands by transferring extremes of temperature between seasons through the use of thermal stores. Pioneered by ICAX, IHT integrates solar thermal collection in summer with heat storage in thermal banks to double the efficiency and coefficient of performance of ground source heat pumps in winter.

kelp farming: Kelp, a fast-growing alga, can absorb carbon dioxide and be used as a sustainable biofuel as well as providing a habitat for marine life. A development of this carbon sequestration technique by PODenergy proposes the harvesting of kelp in large plastic 'stomachs' in which bacteria break down the kelp into carbon dioxide and methane. The former will be piped to the surface for energy generation, the latter stored in deep ocean sinks or used in ocean liming.

landfill: A site for the disposal of refuse through the compaction and containment of solid waste under soil to minimise its effect on the environment. As population pressures have increased, landfill sites have been redeveloped, requiring the employment of stabilisation and gas capping measures.

lawn pier: An elevated structure surfaced in grass for recreational purposes in the Smartcity, often extending over a productive arable landscape.

lifelines: The transport and communication pathways of the Smartcity.

marine turbines: Tidal energy has the advantage over wind and solar energies in being predictable. Regions with fast-flowing tidal streams have the potential to harvest colossal amounts of power with relatively small devices given the high energy density of tides.

monocropping: The cultivation of a single crop on arable land without the use of crop rotation. Whilst facilitating operational efficiencies, monocropping results in the diminution of soil fertility, overuse of chemical fertilisers and pesticides, and the erosion of biodiversity.

multi-utility service company (MUSCO): A community-owned or private–public delivery structure that provides a variety of utility services resulting in new operational synergies and a highly efficient customer interface.

ocean liming: The addition of calcium carbonate into the oceans. Carbon sequestration contributes to ocean acidity and threatens marine life. Limestone introduced into the water reacts with dissolved carbon dioxide to produce lime which neutralises the acid and increases the GHG absorption capacity of the oceans.

orchard hub: An urban agriculture component of the Smartcity in which fruit trees are planted to provide fresh produce, biomass, acoustic attenuation and carbon sequestration.

photovoltaics (PVs): Cell arrays, usually made from silicon with other trace elements, which convert solar radiation into electricity. PVs absorb photon energy to generate charge carriers that are attracted to a conductive contact, consequently transmitting electricity. Photovoltaic energy is particularly appropriate for use in remote locations where grid access is impossible and allows local use with minimal distribution losses. When connected to the grid, however, PV energy can reduce high-cost electricity at daytime peak demand, although it requires conversion from direct to alternating current.

planned obsolescence: The practice of artificially shortening a product's lifespan in order to increase replacement frequency and as a result, revenue. In the agricultural industry, many high-yield engineered seeds are infertile and become obsolete after each harvest, forcing farmers to purchase new seeds year on year. While switching to new strains protects from the possibility of disease-related crop failure, there is potential for the infertility gene to contaminate other fertile strains.

podcar: Also known as personal rapid transit vehicles, podcars are small automated vehicles that travel on guideways directly to the destinations of their individual passengers. Developments in battery technologies are enabling hybrid vehicles to travel large distances on a single charge. It is a little known fact that prior to the Ford Model T, a third of all cars were electric.

progress paradox: The proposition that despite great advancements in technology and improvements in key life quality indices, people are no happier than they have been in the past.

pyrolosis: The anaerobic decomposition of organic material to sequester carbon. The process is carbon negative, capturing up to 90 per cent of CO_2 that would otherwise be released through combustion into biochar, combustible gas and bio-oils. Used for centuries in the Amazon rainforests, widespread adoption of pyrolysis is considered to be a viable geoengineering solution that produces material for soil enrichment.

renewable energy: Energy generated from sources that can be easily replenished from natural processes or are practically infinite. Types include biomass, geothermal, solar and wind energies.

renewable energy sources act (EEG): A German parliamentary act granting priority to renewable energy and establishing minimum prices for its generation over the next 20 years. The structure of the EEG ensures high investment security and low credit interest rates, accounting for a 250 per cent increase in clean energy generation between 2000 and 2004. The scaling back

55

of payments for installations commissioned at a later date has averted operators waiting for the technology to become cheaper.

shallow ecology: An anthropocentric approach to ecology focusing on pollution and the management of natural resources in contrast with the more holistic approach of deep ecology that places equal value on human and non-human life.

sky garden: A recurring motif of the Smartcity ranging from rooftop kitchen gardens to vegetative walkways that line streets in the sky. The elevated nature of the gardens explores and fulfils the need for new connections and territories at multiple levels to render vertical living socially sustainable.

slow food movement: An initiative formulated by Carlo Petrini espousing high-quality small-scale farming and regional cuisine as a response to fast food culture. Slow food is the precursor to and part of the broader 'slow movement' that aims to resist the homogenisation and globalisation of towns and cities while seeking to improve the quality and enjoyment of living.

smart grids: The application of digital technologies within electricity distribution networks enabling improved stability and efficiencies in transmission, monitoring and demand management. Investment in smart meter devices will permit the establishment of variable tariffs in relation to the time of day, levelling out demand fluctuations. With a reduction in maximum generation capacity requirements, the load on power plants is reduced; decentralised and diversified power generation will allow clients to choose renewable energy sources and to supply as well as consume energy; improved transparency will encourage responsible energy use while online management will streamline the customer interface.

social capital: A concept employed in the fields of sociology and economics describing the reciprocal relationships of trust that enable the advancement and cohesion of communities. Usually regarded as a resource in the battle against societal problems, it has been acknowledged that social capital can improve the wellbeing of individuals at the detriment of society at large, exemplified by old boys' networks and criminal fraternities.

solar chimney: A tubular device used to amplify natural stack ventilation by using solar energy to heat air at the top of the chimney, causing an updraft and suction at the base of the chimney.

sound garden: A sunken circular structure offering aural stimulus and diversion. Sound gardens vary in scale from amphitheatres for outdoor musical performance to small contemplation gardens focusing the mind on ambient sound.

sustainability: Forms of progress that meet the needs of the present without compromising the ability of future generations to meet their needs (definition from the World Commission on Environment and Development, 1987). Sustainable development is differentiated from green development in its inclusion of cultural and economic as well as environmental factors.

urban + peri-urban agriculture: The production of food in or, in the case of peri-urban agriculture, around cities and its integration within urban economies and ecologies.

urban beach: A recreational feature of the Smartcity adjacent to natural or artificial water bodies. More than an area for aquatic play, the urban beach is a multi-use space catering to all ages and challenging the formal codes and territories of traditional urban space.

victory garden: A kitchen garden planted during the world wars to relieve pressure on food production deriving from the war effort. In addition to backyards, victory or war gardens took over the rooftops, sidewalks, vacant lots and public parks of major European and American cities.

56

volatile organic compound (VOC): An organic compound that vaporises under normal conditions and engages in atmospheric photochemical reactions. VOCs can be carcinogenic and can contribute to the formation of ozone and smog.

waste-to-energy (WtE): The creation of usable heat and electrical energy from waste sources, usually through incineration. Incineration plants incorporate material and energy-recovery programmes along with emissions monitoring. Incineration decreases the volume of compacted waste by approximately 95 per cent, and is therefore adopted in preference to landfill in countries of high population density. However, fly ash by-product requires disposal in toxic landfill sites. Other WtE technologies include gasification, pyrolysis, fermentation and anaerobic digestion.

wind power: The conversion of kinetic wind energy into usable mechanical and electrical energy through the use of turbines. Wind energy derives from differential solar radiation that causes change in atmospheric pressure. Due to its intermittent and non-dispatchable nature, wind power needs to be supplemented by good inter-regional transmission lines or energy storage infrastructure.

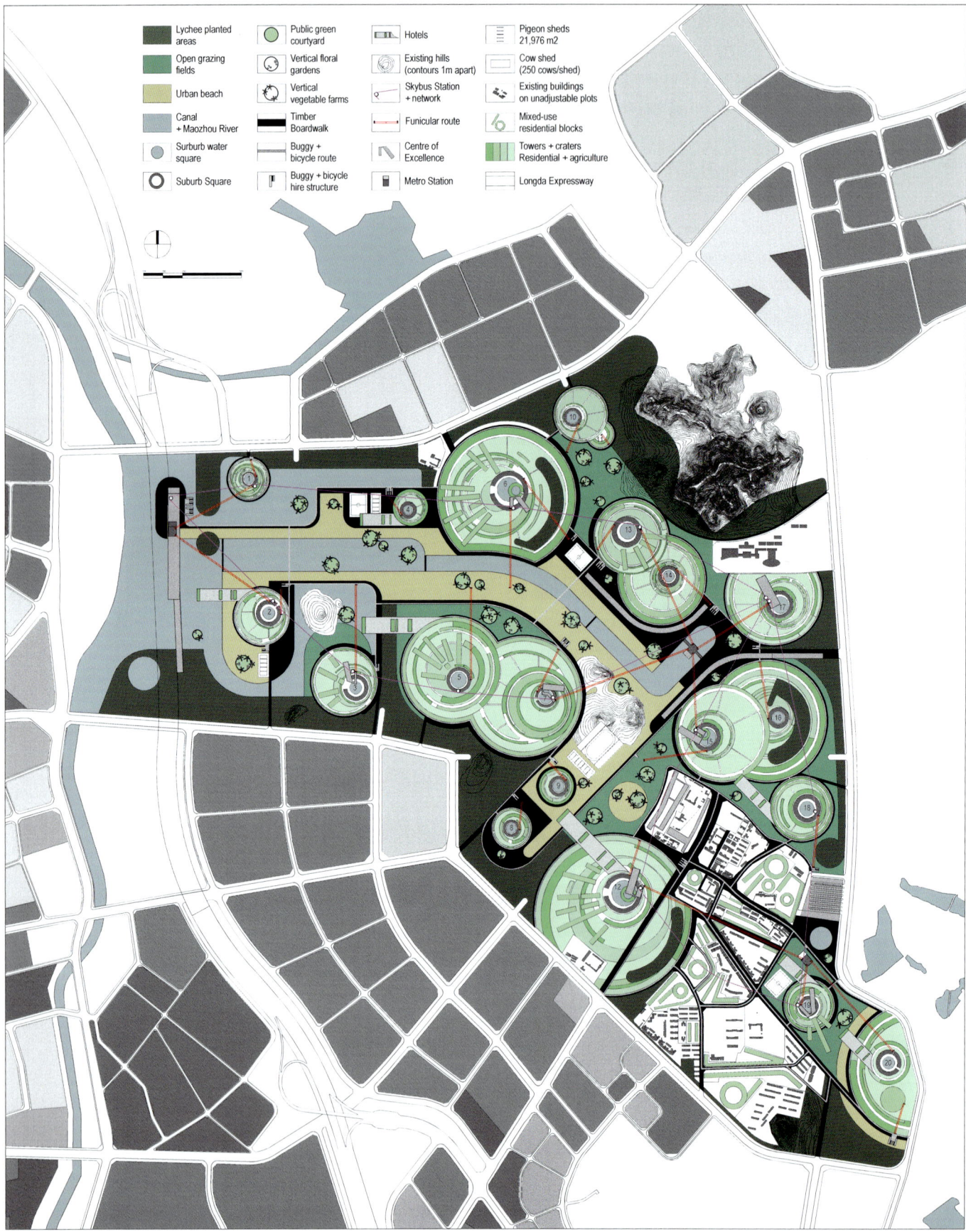

■	Lychee planted areas	◉ Public green courtyard	▭▭ Hotels	▤ Pigeon sheds 21,976 m2
■	Open grazing fields	◉ Vertical floral gardens	▭▭ Existing hills (contours 1m apart)	▭ Cow shed (250 cows/shed)
■	Urban beach	◉ Vertical vegetable farms	Skybus Station + network	Existing buildings on unadjustable plots
■	Canal + Maozhou River	▬ Timber Boardwalk	Funicular route	Mixed-use residential blocks
◉	Surburb water square	▭ Buggy + bicycle route	Centre of Excellence	Towers + craters Residential + agriculture
◎	Suburb Square	Ⅰ Buggy + bicycle hire structure	▭ Metro Station	▭ Longda Expressway

Guangming Smartcity China

'Every utopia – let's just stick with the literary ones – faces the same problem: What do you do with the people who don't fit in?'

       – Margaret Atwood, 2013

Along the roadside, groups of men hunker down, waiting. Some of them idle away the hours gambling with small counters; others smoke out of boredom. Their trade is farming, but they no longer have farms on which to ply this trade. After the initial novelty of being moved from traditional rural houses to modern 'aspirational' high-rise developments, some of China's new urban residents have found a living running small businesses or hiring themselves out as construction labourers. Many others are surviving off a sudden windfall in cash compensation that is as double-edged as it is finite.

facing page: Guangming Smartcity masterplan.

A 2011 report by the Chinese Academy of Social Sciences found that between 40 and 50 million Chinese farmers, from a total of 250 million, have been divested of their livelihoods to make way for industrialisation and urbanisation – it is anticipated that three million people would lose their land every year.[1] All land in China is state-owned with plots granted to farmers on long leases, which has smoothed the way for land expropriation. Despite growing concerns regarding food self-sufficiency and a rural reform plan designed to secure farming land rights, local authorities have largely been able to ignore directives from the central government.

Legislation to preserve farmland is increasingly critical as China supports one fifth of the world's population with only ten percent of the world's cultivated land.[2] Positively, more land has reverted to farming than appropriated for construction in the past few years, and the relocation of rural dwellers into urban environments has freed up land for cultivation. At the same time, the rural migration has displaced important social bonds in the form of native-place networks; villagers retain a strong allegiance to their place of origin that is reflected in their attitudes to upbringing, life rituals and employment.

Over the past four decades, China's cities have been developing at an overwhelming rate; some of them are even bigger than in many industrial nations. One third of China's workforce moved out of agriculture between 1990 and 2015 and will have moved from the countryside to the cities, triggered by the desire of rural inhabitants to take part in China's economic boom.[3] The nation has eclipsed the United States in the consumption of basic agricultural and industrial goods, and is now the world's largest consumer of grain, meat, coal and steel. With such huge industrial, agricultural and economic shifts come major demands on resources, as well as environmental issues.

1. L Hornby, 'China Migration: Dying for land', Financial Times, 6 August 2015

2. T McMillan, 'How China Plans to Feed 1.4 Billion Growing Appetites', National Geographic, February 2018

3. J Manyika et al., 'Jobs Lost, Jobs Gained: Workforce transitions in a time of automation', McKinsey Global Institute, December 2017, p.4

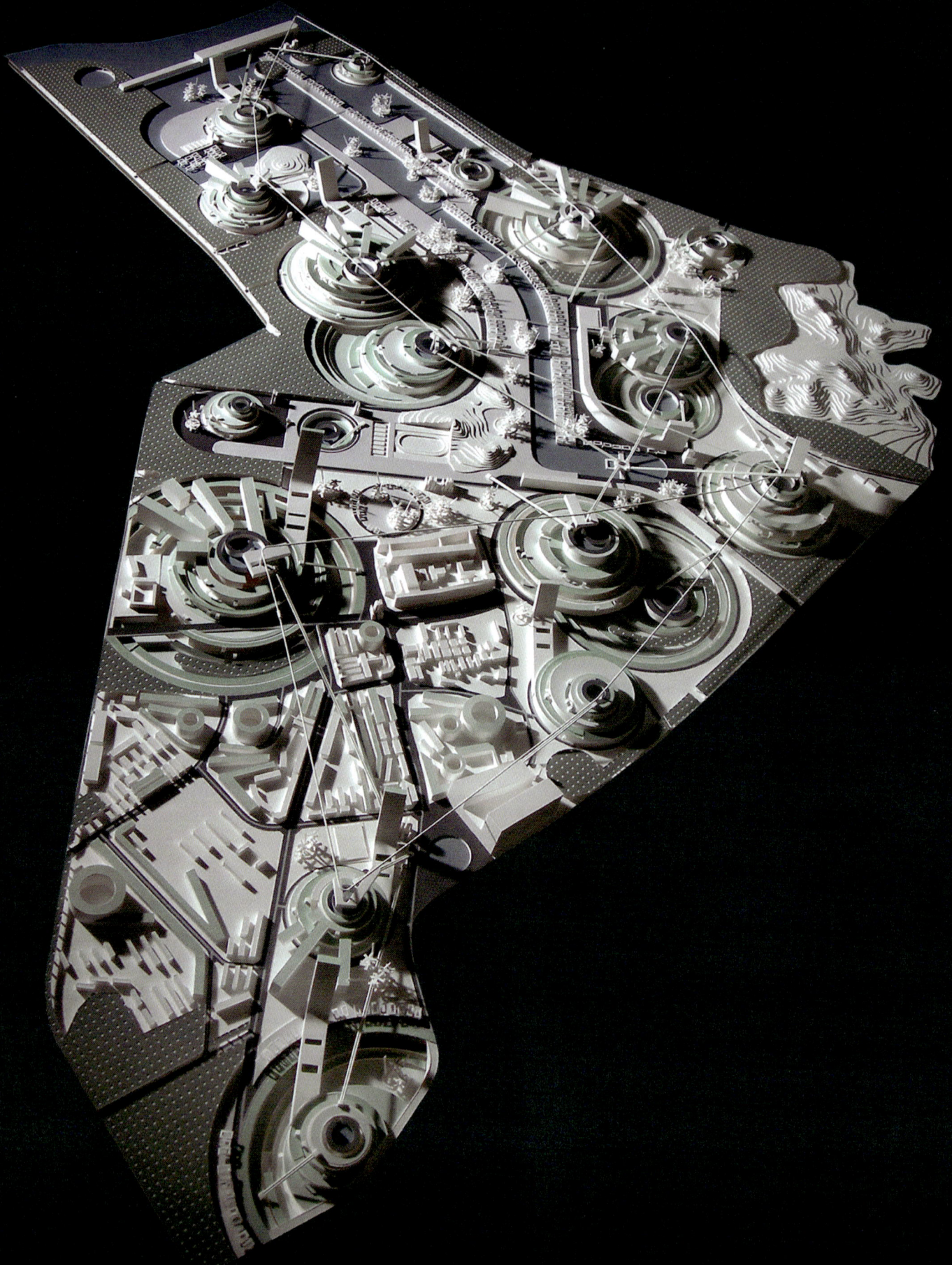

A new town for the 200 000 residents of Guangming in Shenzhen presents the opportunity to develop a city paradigm reconciling the contrasting needs of urban growth and rural preservation – in essence a Smartcity. China has a history of having built the largest and most spectacular cities before the modern era, with Beijing reaching some two million people as long ago as the 17th century AD. However, it has also continued to be a land of villages and farmers. Guangming will continue this agricultural heritage, creating a hybrid city at the vanguard of eco-sustainability and pioneering a new way of urban living. Guangming Smartcity covers an area of 7.97km2, northwest of Shenzhen in Guangdong Province. The site is surrounded by the region's agricultural land, with the Longda Expressway, Maozhou River and Gongming Industrial Cluster to the west, and Guangming Industrial Park to the south. It is 18km to Shenzhen International Airport, 40km to Hong Kong and an hour's drive from the city centre of Shenzhen.

61

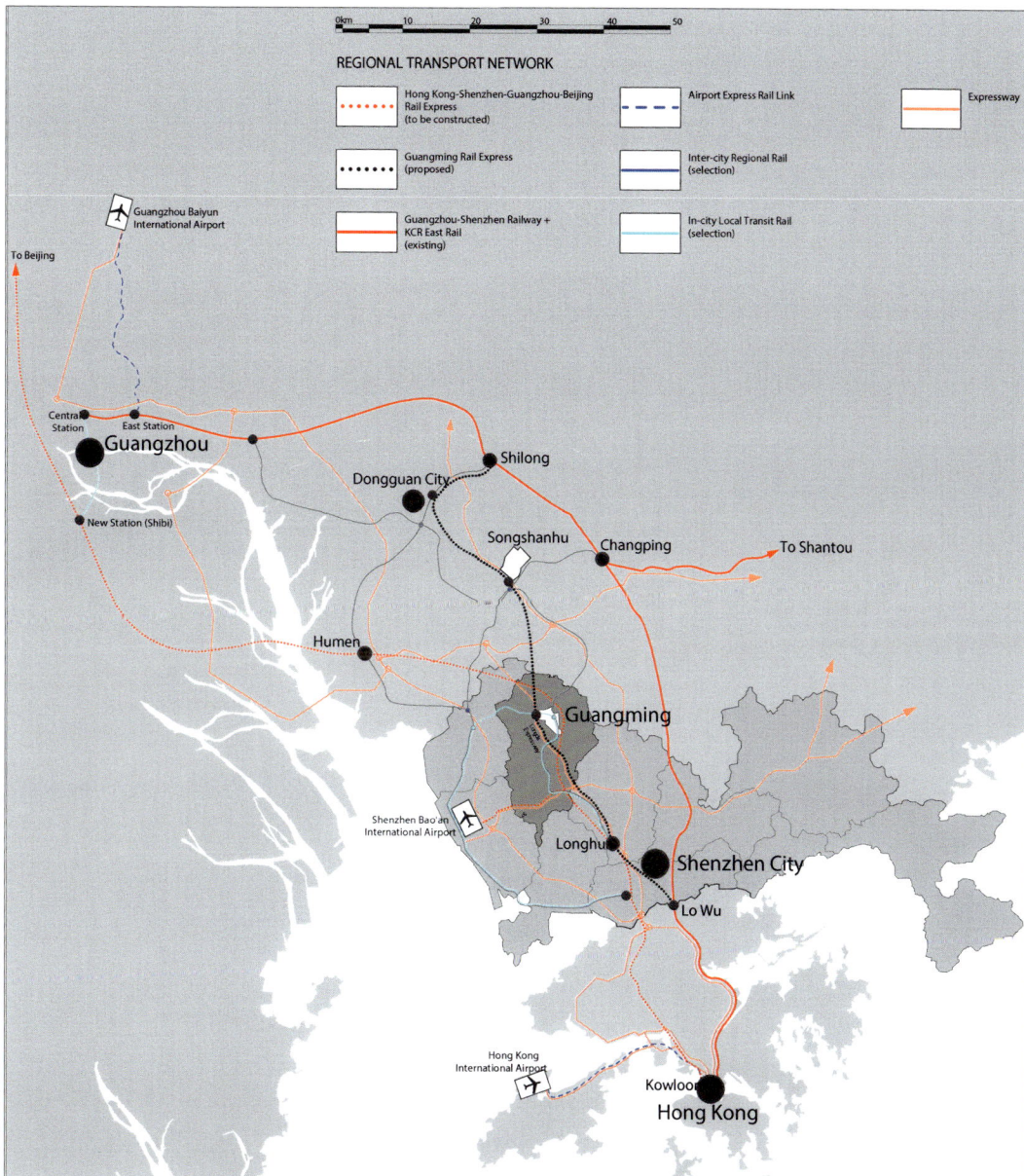

facing + following page: Model of Guangming Smartcity.

left: Location plan of Guangming Smartcity and the regional transport network.

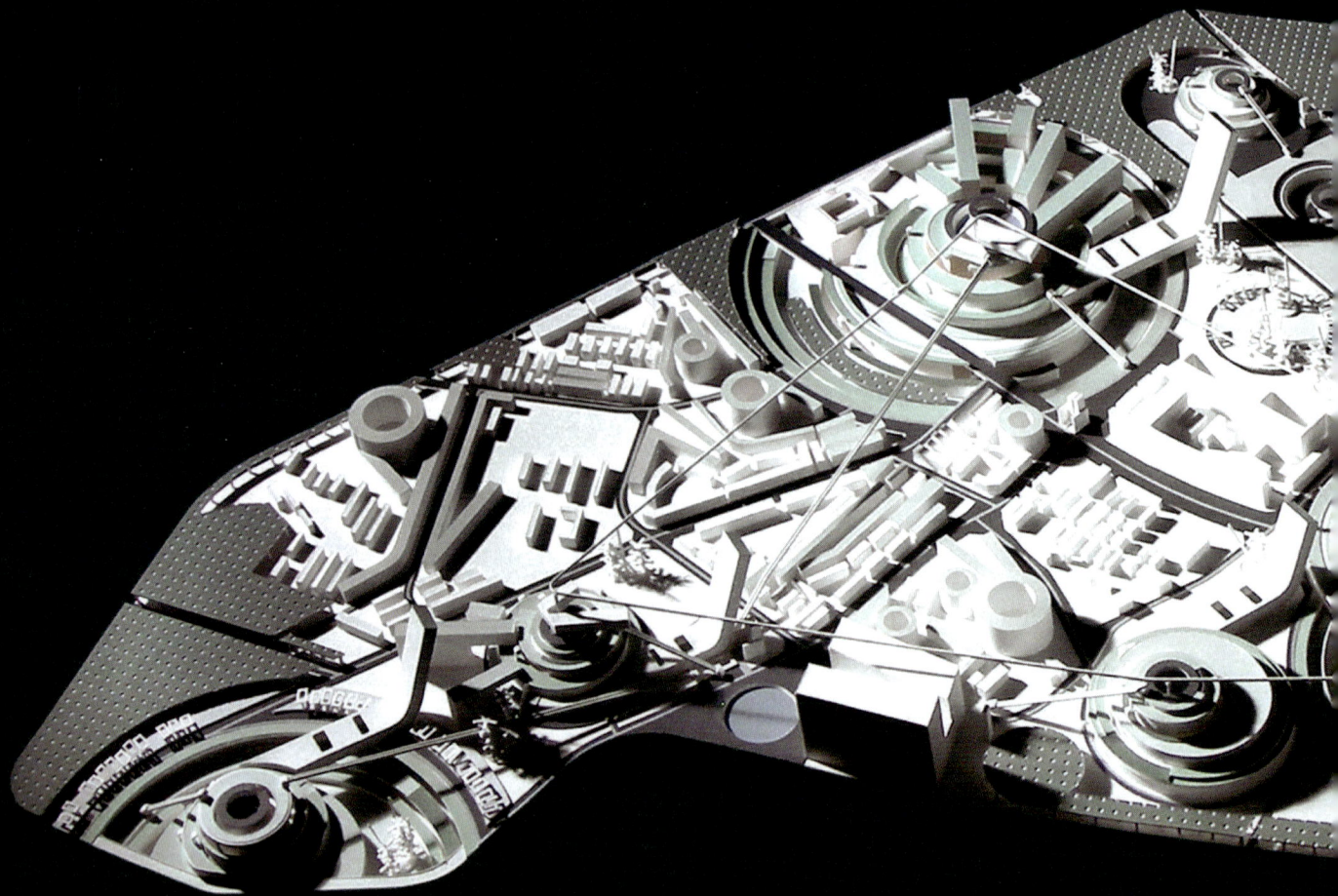

Regional Development + New Programmes for the Smartcity

The programmes for Guangming Smartcity have been established with the development of the region clearly in mind. Fundamentally, the site cannot become an isolated island city – it must support, complement and act as a generative seed for the region and beyond. The proposals for Guangming Smartcity, fully supported by an effective transport infrastructure, integrate with the surrounding Bao'an district at various levels, providing civic, commercial, recreational, agricultural, cultural and tourist facilities. Overlaid onto this basic function of a town centre, the programmatic backbone of the Smartcity will comprise organic urban agriculture, agritourism and eco-gastronomy.

As well as servicing its own residents, Guangming Smartcity will act as a civic and commercial centre for Gongming and Songshanhu. Besides providing the usual services and necessities of daily life, the commercial centre will specifically promote eco-products, organic foodstuffs and holistic living in keeping with the sustainable ethos of the development. In this way, the Smartcity will not duplicate the retail and service industries in Shenzhen City or Dongguan, whilst simultaneously reinforcing the brand image of Guangming as a model of sustainable living.

Agricultural Context

China is currently struggling with the problem of increasing the efficiency of its agricultural production whilst finding employment for a vast number of rural migrants in cities already suffering from high unemployment rates. Within the next three decades, it intends to reduce its farm employment to ten per cent of the labour force. In the long term, the influx of new urban dwellers will create a new market for goods and services in the city, boosting employment and the country's GDP. In the short term, the erstwhile farmers lack the skills required for working in an urban environment, have been cut off from their social infrastructure, and are discriminated against by established city dwellers. The hybridisation of city and arable land in Guangming Smartcity offers resilience by allowing farmers to retain their land, and by extension their social insurance that they do not receive if they work in cities, whilst offering opportunities to train in new employment sectors. Part-time farming becomes viable, and the city's diversification into high-end and high-yield agriculture presents an alternative career route, maintaining both a connection with the soil and financial parity with other vocations.

The local practice of dairy, vegetable, fruit and pigeon farming will be retained and modernised with advanced farming techniques. Guangming Smartcity will continue to be the principal supplier of milk and vegetables to Hong Kong, but will use aquaculture and hydroponics to increase crop yields. Disease prevention will be improved, and urban nutrient waste recycling introduced to establish a circular economy. Together with the innovative vertical farms and floral gardens, arable laboratories and institutions specialising in nutrition and food science, the Smartcity will be ideally placed to be a testing ground and partner for the South China Agricultural University in Guangzhou. Traditionally, education amongst villagers is seen as a means to migrate to urban employment; graduate level rural education could reverse this trend, allowing farmers to adopt new techniques and correct age-old urban prejudices.

The existing farming community, together with new agricultural schools, will play an important role in sustaining a skilled workforce in the local area that are able to run farms as viable businesses. Local food production will establish a strong sense of community and substantially help reduce energy and fuel consumption from food transportation. Guangming Smartcity locates people where the food grows instead of moving food to the people.

Livestock and crops will be allowed to grow at their natural pace without pesticides or preservatives and animals will be reared without the use of growth hormones. Livestock manure and human organic waste will be recycled for nutrients in anaerobic digesters, exemplifying the efficiencies of permaculture.

At present, most organic food in China is grown for the export market. However, with the standard of living and purchasing power of discerning Chinese consumers on the rise, greater demand is expected from the domestic market.

65

Ecogastronomy + Agritourism

Ecogastronomy promotes healthy eating along with protecting the environment. With its accessibility to the finest local produce and livestock and commitment to organic farming techniques, visitors from the region and abroad will be able to enjoy first-rate cuisine and understand the provenance of their food. The Smartcity is well placed to become an international venue for symposia concerning the production, understanding, development and sharing of food.

Despite the high population density of Guangming Smartcity, layered planning leaves generous areas for public relaxation and leisure in the form of an urban beach and canal. The grazing fields and aquaculture terraces will also offer scenic beauty and green space in stark contrast with the urban fabric of industrial parks outside the city. The beach will be a closer and more convenient leisure destination than the coastline with luxury hotel and villa accommodation facilities, making the city an ideal weekend resort for Gongming and Shenzhen. Consequently, the Smartcity can act as a tourist base for other recreational locales such as Songshanhu that offer complementary leisure pursuits such as fishing and boating.

left: Local organic produce.

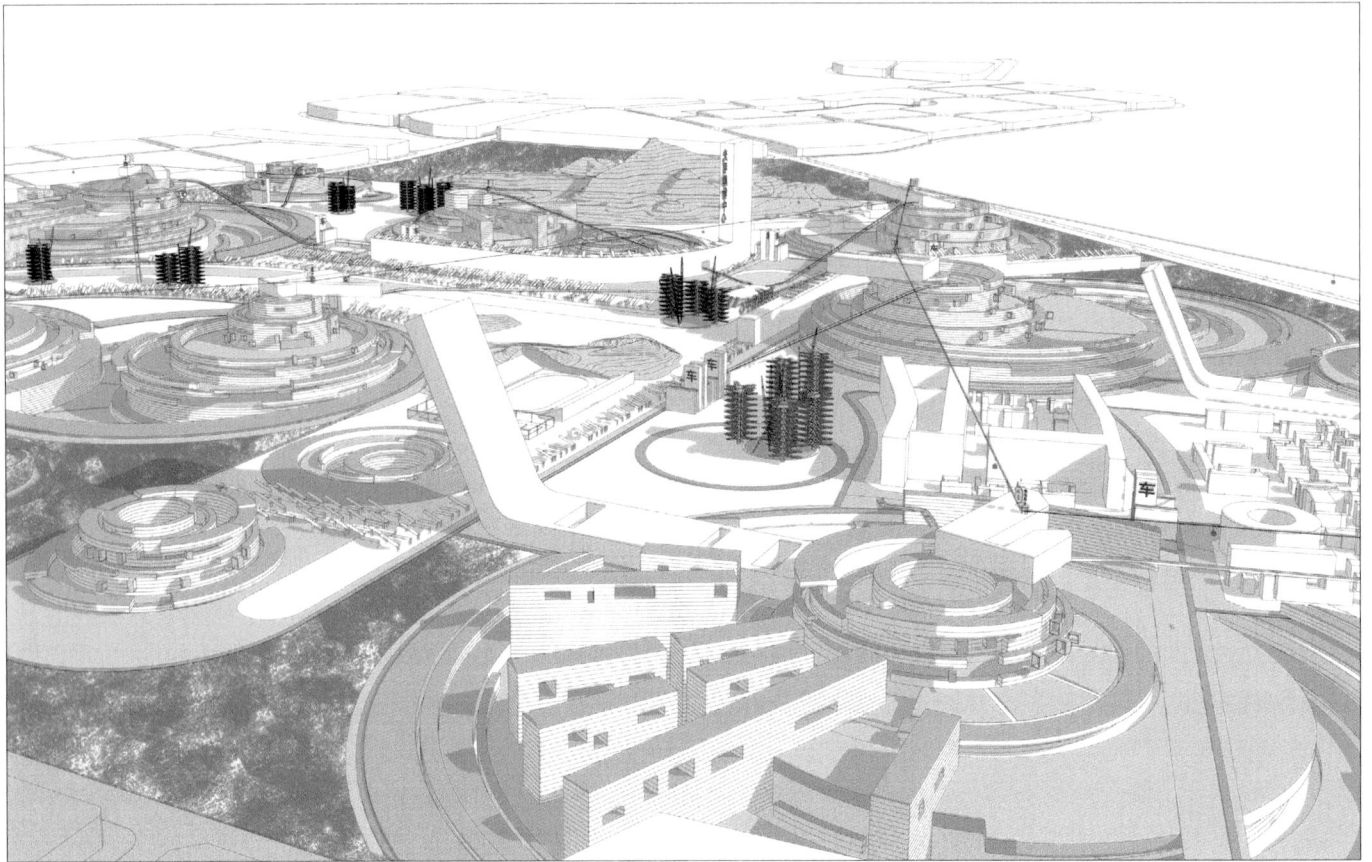
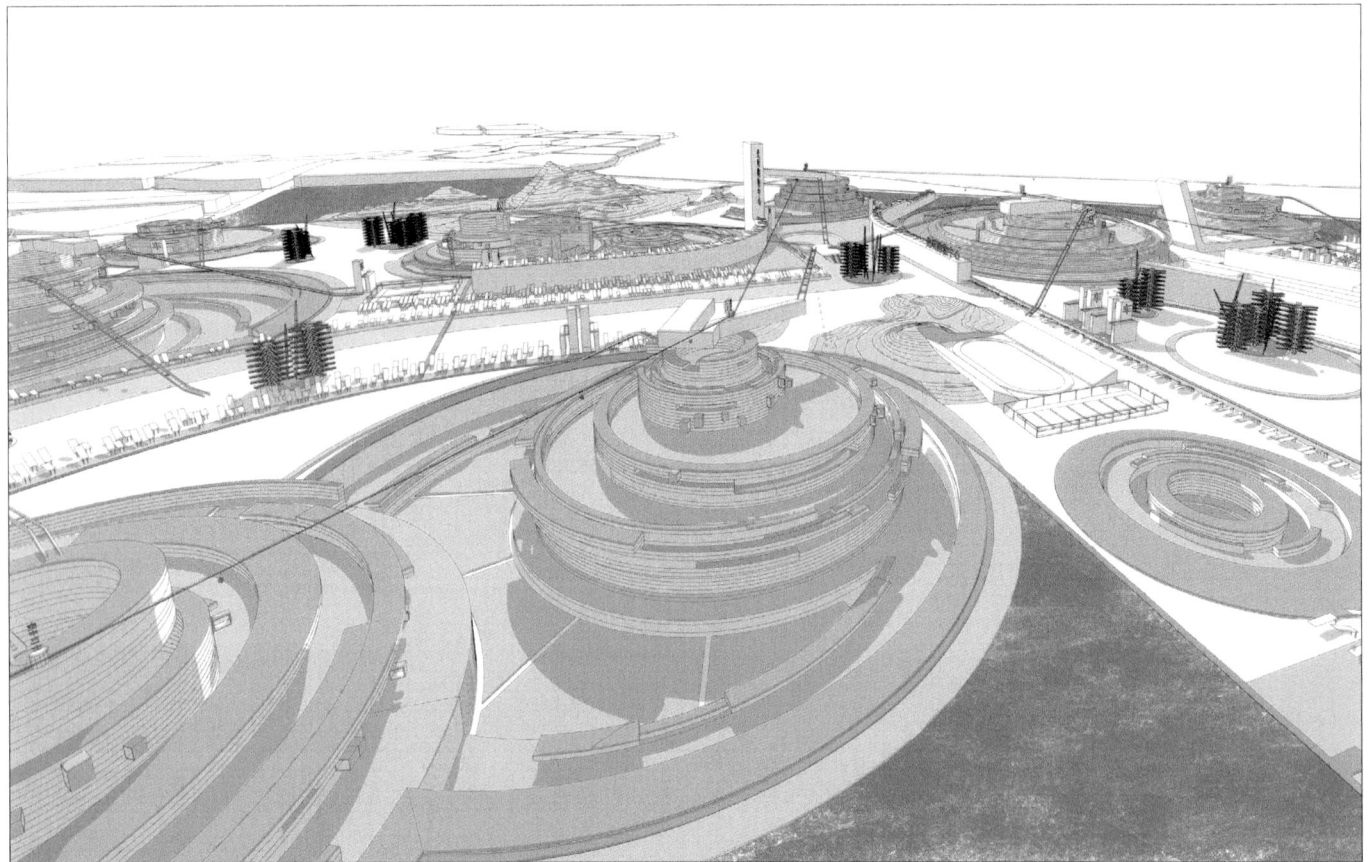

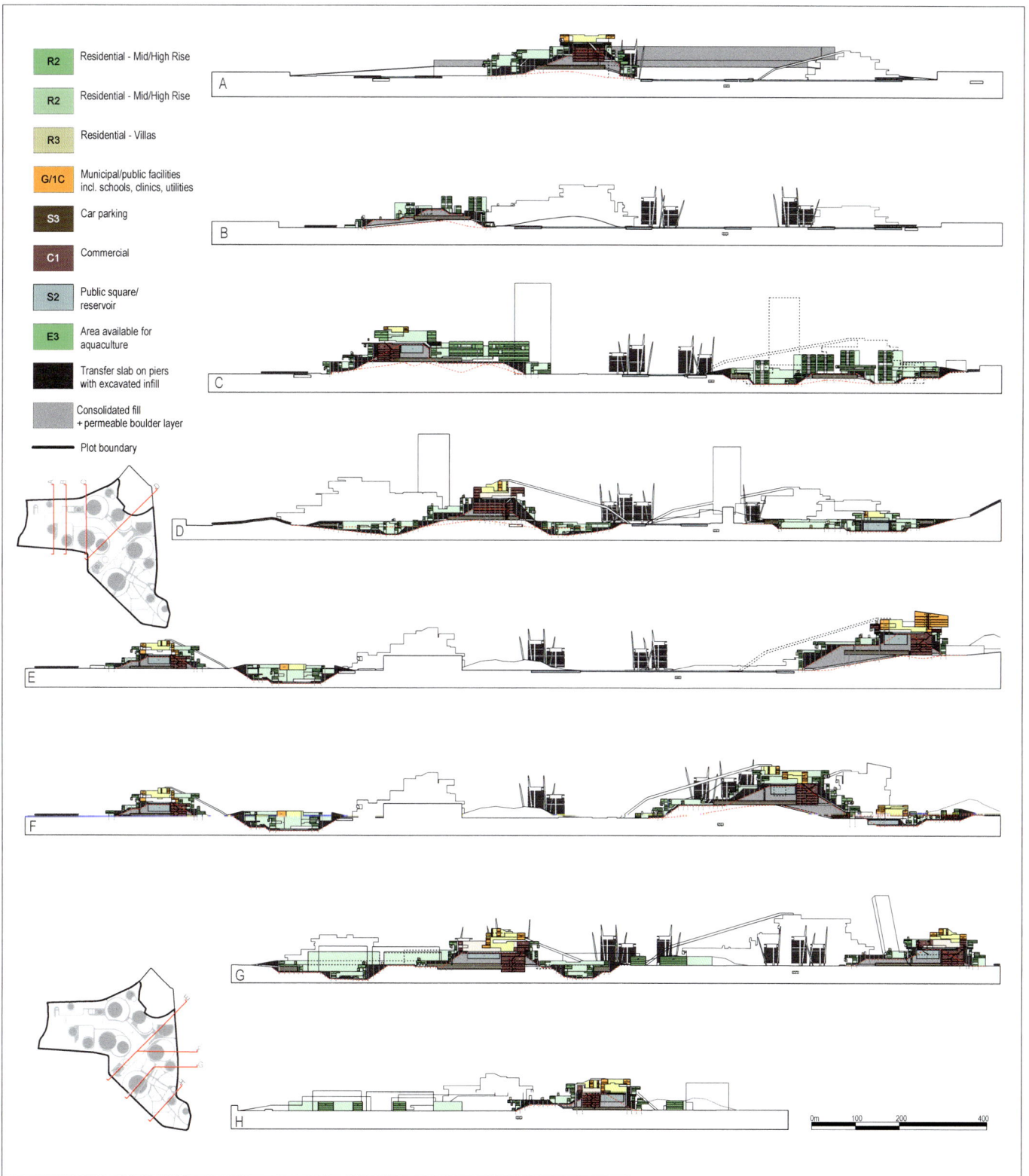

R2	Residential - Mid/High Rise
R2	Residential - Mid/High Rise
R3	Residential - Villas
G/1C	Municipal/public facilities incl. schools, clinics, utilities
S3	Car parking
C1	Commercial
S2	Public square/ reservoir
E3	Area available for aquaculture
	Transfer slab on piers with excavated infill
	Consolidated fill + permeable boulder layer
———	Plot boundary

above: Sections showing land use distribution in the Smartcity.

facing page: Views of suburb towers and craters.

City Framework

Guangming enjoys rich mountainous and river resources that are exploited in the city's morphology. The Smartcity is arranged into optimally sized clusters of housing and farming suburbs that manifest as towers and craters, shaped by and augmenting the existing undulating topography. The towers and craters borrow from the technocratic formalism of the Japanese Metabolists and the utopian trope of concentric ringed streets and buildings, but introduce a third vertical dimension calibrated to storey-height terraces. This stepped arrangement improves the solar angle for natural lighting within the apartment buildings and offices; natural cross ventilation is possible and the distances between buildings can be reduced to increase housing density without adverse overshadowing. Most significantly, the terracing creates level rooftop surfaces that can be used for farming without fear of erosion and slippage. The result is an unprecedented spatial connection between mass housing and arable land. Where necessary, the terrain is reshaped by redistributing excavated material from the craters to the towers and the addition of inert but non-biodegradable landfill that has been accumulating around China's cities at an alarming rate.

The density of the tower and crater suburbs prevents urban sprawl and advocates compact land use patterns that assist in limiting the carbon footprint of the town centre's residents. Each tower or crater is self-sufficient with its own high street, suburb square and individual community identity.

In the centre of the development area is an artificial beach and canal leading into the revitalised Maozhou River where a reed bed water filtration system is introduced. The beach is a recreational oasis that contrasts with the ubiquitous open space which, although scenic, is fully productive. A boardwalk encircles the beach and connects the tower and crater communities providing a place to meet and socialise. Traversed by bicycle or electric buggy, the boardwalk also plays host to the practice of tai chi, jogging and constitutional walks.

Over 80 vertical kitchen garden farms can be found scattered throughout the central city, and collectively they form the city's arable research institute. The city's vertical floral gardens sit alongside the vertical kitchen garden farms around which people can sit, mingle and enjoy the scent of local flora. The interstitial landscape is used to graze livestock.

The Smartcity is a car-free zone and individual suburbs are designed with their own local municipal facilities to ensure day-to-day activities are within walkable distances. The primary transport system between neighbourhoods will be via light rail (MTR). Electric or biogas sky-buses run between the urban plazas at the top of each tower and crater community that are individually accessed by funiculars, lifts and escalators. The City Hall lies on an axis along the Maozhou River with the Central Train Station and overlooks the entire city from the top of the largest suburb tower. Six other Centres of Excellence sitting astride the remaining suburb towers complete an urban sky court in a reinterpretation of the traditional city square.

Lychee and longan orchards, the fruit of which are renowned in the region, border the site and act as a filter for the clean inner heart of the city. The orchard belt is periodically interrupted by peripheral car parks and recycling centres serving the community as well as incoming visitors who can also deposit waste for local energy generation. The bank of trees extends along the embankments of the Maozhou River, metaphorically reaching out and embracing the Smartcity's neighbouring towns. Guangming's new centre is self-sufficient but not inward looking, its environmental programme offering a resource for the region to the mutual benefit of both parties.

top: Suburb plaza water reservoir.

bottom: Lychee orchards as natural pollution filters.

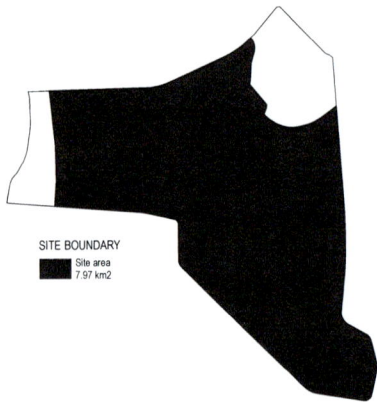

SITE BOUNDARY
■ Site area
7.97 km2

EXISTING BUILDINGS
■ Building footprints

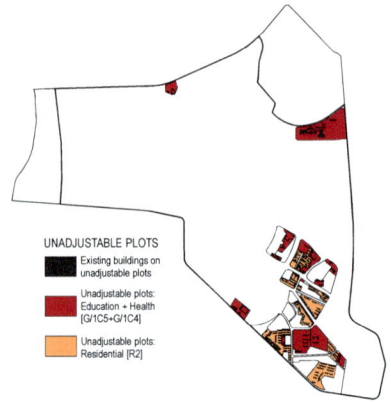

UNADJUSTABLE PLOTS
■ Existing buildings on unadjustable plots
■ Unadjustable plots: Education + Health [G/1C5+G/1C4]
■ Unadjustable plots: Residential [R2]

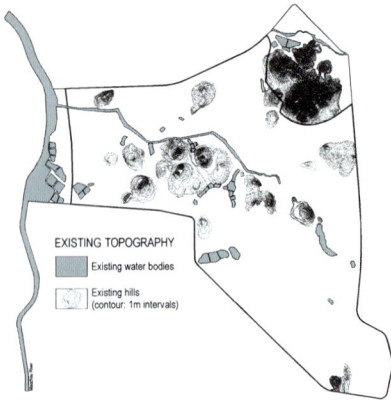

EXISTING TOPOGRAPHY
■ Existing water bodies
Existing hills (contour: 1m intervals)

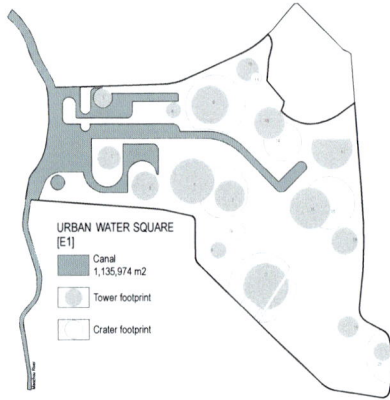

URBAN WATER SQUARE [E1]
■ Canal
1,135,974 m2
□ Tower footprint
□ Crater footprint

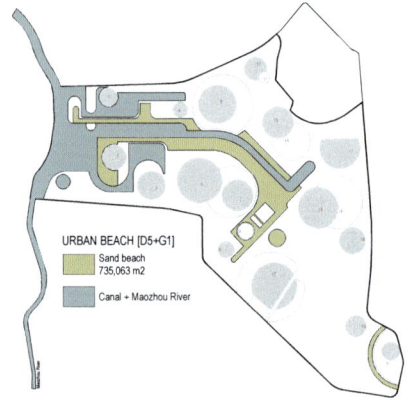

URBAN BEACH [D5+G1]
■ Sand beach
735,063 m2
■ Canal + Maozhou River

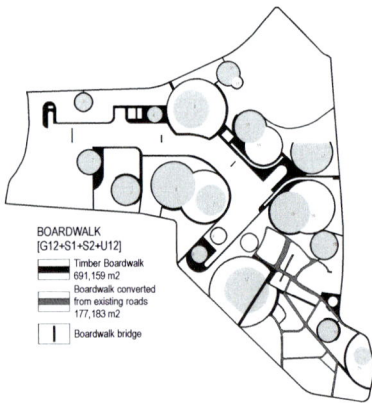

BOARDWALK [G12+S1+S2+U12]
■ Timber Boardwalk
691,159 m2
Boardwalk converted from existing roads
177,183 m2
□ Boardwalk bridge

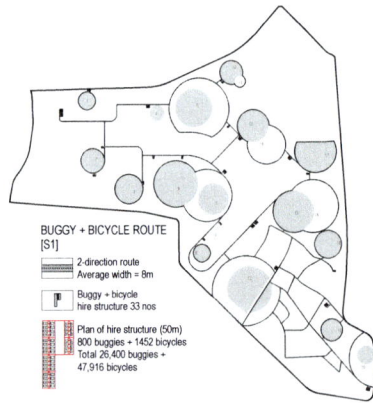

BUGGY + BICYCLE ROUTE [S1]
2-direction route
Average width = 8m
Buggy + bicycle hire structure 33 nos
Plan of hire structure (50m)
800 buggies + 1452 bicycles
Total 26,400 buggies + 47,916 bicycles

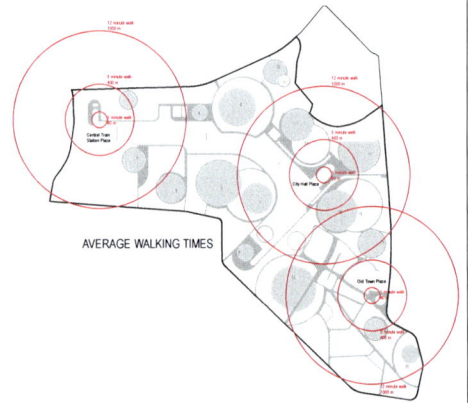

AVERAGE WALKING TIMES

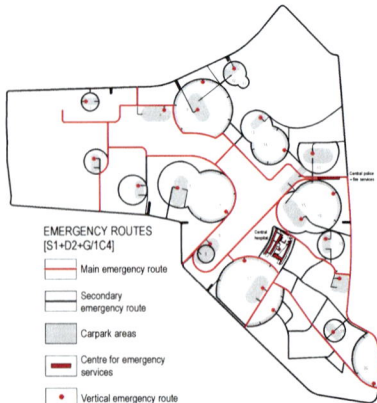

EMERGENCY ROUTES [S1+D2+G/1C4]
— Main emergency route
Secondary emergency route
■ Carpark areas
■ Centre for emergency services
• Vertical emergency route

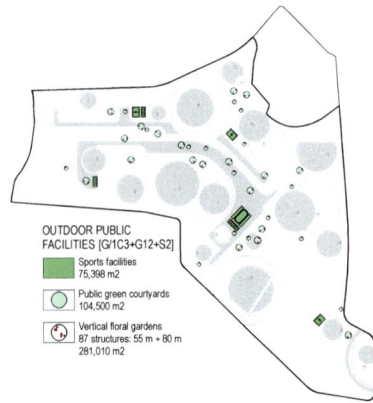

OUTDOOR PUBLIC FACILITIES [G/1C3+G12+S2]
■ Sports facilities
75,398 m2
○ Public green courtyards
104,500 m2
○ Vertical floral gardens
87 structures: 55 m + 80 m
281,010 m2

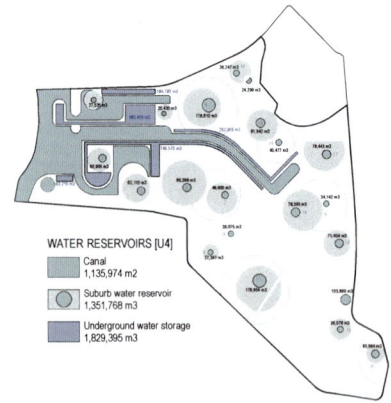

WATER RESERVOIRS [U4]
■ Canal
1,135,974 m2
○ Suburb water reservoir
1,351,768 m3
■ Underground water storage
1,829,395 m3

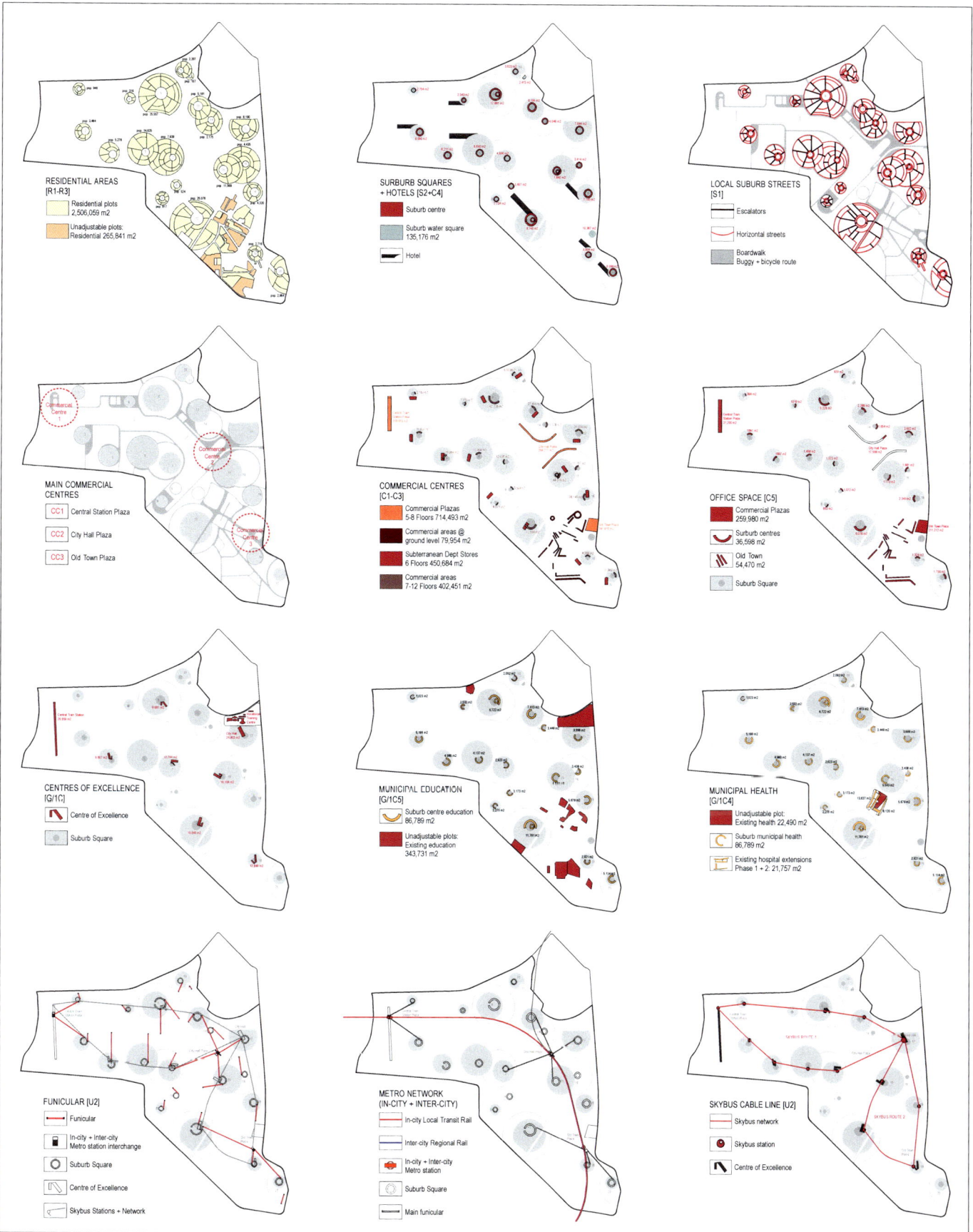

RESIDENTIAL AREAS
[R1-R3]
- Residential plots
  2,506,059 m2
- Unadjustable plots:
  Residential 265,841 m2

SURBURB SQUARES
+ HOTELS [S2+C4]
- Suburb centre
- Suburb water square
  135,176 m2
- Hotel

LOCAL SUBURB STREETS
[S1]
- Escalators
- Horizontal streets
- Boardwalk
- Buggy + bicycle route

MAIN COMMERCIAL
CENTRES
- CC1 Central Station Plaza
- CC2 City Hall Plaza
- CC3 Old Town Plaza

COMMERCIAL CENTRES
[C1-C3]
- Commercial Plazas
  5-8 Floors 714,493 m2
- Commercial areas @
  ground level 79,954 m2
- Subterranean Dept Stores
  6 Floors 450,684 m2
- Commercial areas
  7-12 Floors 402,451 m2

OFFICE SPACE [C5]
- Commercial Plazas
  259,980 m2
- Surburb centres
  36,598 m2
- Old Town
  54,470 m2
- Suburb Square

CENTRES OF EXCELLENCE
[G/1C]
- Centre of Excellence
- Suburb Square

MUNICIPAL EDUCATION
[G/1C5]
- Suburb centre education
  86,789 m2
- Unadjustable plots:
  Existing education
  343,731 m2

MUNICIPAL HEALTH
[G/1C4]
- Unadjustable plot:
  Existing health 22,490 m2
- Suburb municipal health
  86,789 m2
- Existing hospital extensions
  Phase 1 + 2: 21,757 m2

FUNICULAR [U2]
- Funicular
- In-city + Inter-city
  Metro station interchange
- Suburb Square
- Centre of Excellence
- Skybus Stations + Network

METRO NETWORK
(IN-CITY + INTER-CITY)
- In-city Local Transit Rail
- Inter-city Regional Rail
- In-city + Inter-city
  Metro station
- Suburb Square
- Main funicular

SKYBUS CABLE LINE [U2]
- Skybus network
- Skybus station
- Centre of Excellence

Tower + Crater Suburbs

The sizes of the tower and crater communities are carefully designed both to provide a variety of environments and to take advantage of shared resources. Environmentally, there are synergistic benefits of common walls, reduced energy consumption and improved structural integrity. Socially, the planned populations of each community are optimised to be autarkic in terms of education, health, commerce and recreation, catering for residents of all ages and from every stratum of society. A significant proportion of its housing stock will be affordable units, and all buildings will meet the accessibility needs of an increasingly ageing population. The radially stacked arrangement of the housing addresses the generational domestic politics of the family, especially those households comprising more than two generations. A broad range of typologies including houses, apartments, villas, studios and care homes cater for the community spectrum, all within a stunning agrarian landscape.

73

Each suburb has a main street at its apex with local shops and services to meet quotidian needs: a tailor, a grocer, a health clinic with medics, dentists, opticians and a natural healing centre, a cinema, a post office, banks, schools, religious centres and office buildings. The suburb square operates as a farmers' market, community gathering space and outdoor arena for concerts and festivals. A reservoir is located in the centre of the square that stores rainwater, contributes to the summer cooling strategy, and provides an attractive backdrop for community events. Although the tower and crater suburbs possess similar basic amenities, each has its own individual character and specialisation in the eight Centres of Excellence which comprise a mediatheque, the International Food Festival convention centre, the Museum of Eco-Gastronomy, the Agricultural University, the International Food Forum, the International Culinary and Catering College, Guangming Smartcity Central Station and Guangming City Hall. This urban arrangement encourages social communication between the different tower and crater communities.

Tower Formation

The proposal amplifies the natural topology of the area by excavating and filling certain areas. Existing hills are built up with material taken from adjacent earth cuttings. The base of each tower is naturally formed. Onto this, large boulders taken from nearby bedrock are piled. Stone columns, local compactions of the rock, are installed using standard methods. A car parking basement in reinforced concrete, base slab, columns and cover slab is placed as the building-up work proceeds. Over the final hill profile, geo-textile layer and sand is applied. This seals the boulder layer through which cooling air is to pass towards the buildings above.

Reinforced concrete pad foundations are set on the stone column locations and an under-croft built up. Suspended slabs support the public spaces at their chosen level. Areas of solid build-up support roadways and agricultural areas. Piers for the large annular accommodation structures bear on the stone columns and pass up through the ground plane to support pre-stressed in-situ placed reinforced concrete transfer slabs. These slabs are stiffened by concrete cross-walls. Simple plates of concrete support a variety of accommodation completed in reinforced concrete, block-work, timber and lightweight steel framing.

facing page: Residential modules located in-between layers of urban agriculture in suburb towers and craters.

BLOCK 15: LAND-USE + PLOT DIVISION SECTION   Scale 1:2500

BLOCK 12: LAND-USE + PLOT DIVISION SECTION   Scale 1:2500

Legend:
- Residential - Mid/High Rise
- Residential - Villas
- Municipal/ public facilities incl. schools, clinics, utilities
- Tourist accommodation
- Commercial
- Public square/ reservoir

0 25 50 75 100m

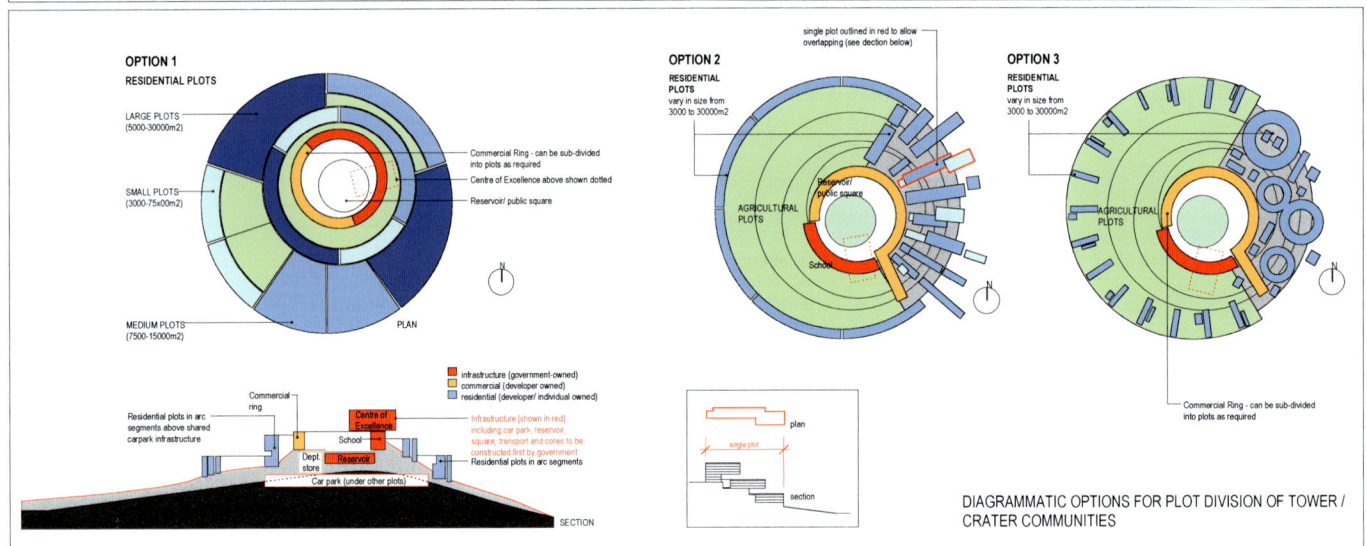

OPTION 1
RESIDENTIAL PLOTS

LARGE PLOTS (5000-30000m2)

SMALL PLOTS (3000-75x00m2)

MEDIUM PLOTS (7500-15000m2)

PLAN

Commercial Ring - can be sub-divided into plots as required
Centre of Excellence above shown dotted
Reservoir/ public square

OPTION 2
RESIDENTIAL PLOTS vary in size from 3000 to 30000m2

single plot outlined in red to allow overlapping (see section below)

AGRICULTURAL PLOTS

Reservoir/ public square

School

OPTION 3
RESIDENTIAL PLOTS vary in size from 3000 to 30000m2

AGRICULTURAL PLOTS

Commercial Ring - can be sub-divided into plots as required

- infrastructure (government-owned)
- commercial (developer owned)
- residential (developer/ individual owned)

Residential plots in arc segments above shared carpark infrastructure
Commercial ring
Centre of Excellence
School
Dept store
Reservoir
Car park (under other plots)

Infrastructure [shown in red] including car park, reservoir, square, transport and cores to be constructed first by government
Residential plots in arc segments

SECTION

plan
single plot
section

DIAGRAMMATIC OPTIONS FOR PLOT DIVISION OF TOWER / CRATER COMMUNITIES

A Day in the Life in Guangming Smartcity

Mrs Lam – Elder

1. Mrs Lam wakes up early and sweeps the apartment.
2. She then ascends to the suburban sky square.
3. Where she practises tai chi on the reservoir.
4. Mrs Lam meets several of her friends for dim sum at the food court and picks up some groceries from the market a short walk away.
5. She then goes to see the herbal doctor at the health clinic in the municipal quarter of the square.
6. Mrs Lam returns home and has some rice and steamed vegetables for lunch.
7. Mrs Lam has an afternoon nap.
8. In the early afternoon, she goes to visit her friend who lives on the outer residential ring.
9. Mrs Lam returns home to prepare dinner for herself and her son who will soon return from working in the lychee groves.

Zhang Siu Ming – Teenager

1. After breakfast, Siu Ming meets his schoolmate Siu Fun who lives a few doors away.
2. Siu Ming and Siu Fun walk to school which is in the municipal district.
3. Siu Ming spends the day at school. For lunch he can eat in the school dining room, on the rooftop playground or go to a café in the nearby suburb square.
4. After school, Siu Ming ascends to the Skybus platform outside the Centre of Excellence.
5. He takes the cable car to the local stadium where he plays football.
6. After the game, he takes the Skybus back to the tower community where he lives.
7. Siu Ming returns home via the suburb square.
8. After having a quick meal, Siu Ming finishes his homework and goes to bed.

Mr Jiang – Farmer

1. Mr Jiang lives at the base of the main inner ring with his elderly mother, Mrs Lam. He rises early.
2. Mr Jiang spends most of the morning tending his vegetables in the aquaculture field outside his home.
3. He is able to return home for lunch as the farmland is so close.
4. In the afternoon, Mr Jiang takes the escalator down to the base of the tower and makes his way to his lychee orchard.
5. He spends several hours pruning his trees.
6. Mr Jiang returns home via the escalator.
7. At home, he has dinner with his mother.

Mr Zhang – Industrial Park Worker

1. Mr Zhang lives on an inner ring with his wife, son and baby daughter.
2. After breakfast, he descends to the underground car park by lift.
3. He leaves Guangming Smartcity by car via one of the two peripheral link roads and drives to the Innovation and Hi-tech Industrial Park where he works.
4. Mr Zhang returns to the tower community after work.
5. He enters the department store from car park level and picks up a few household items.
6. He then ascends to the food court in the commercial ring at the top of the tower and meets his wife and daughter for dinner.
7. After enjoying a leisurely meal, they take a short walk around the shops before taking the lift down to their floor.
8. Mr Zhang and his family return home and retire for the evening.

The new structures are arranged so that movement joints fall on radial divisions of the site. This allows for the segmental development of each area. The central square and reservoir of the tower rests on the built up fill on a flexible pre-stressed concrete base. Vertical ducts in the perimeter walls bring cooling air upwards from the rock layer.

Plot Division + Flexibility

Despite the strong form of the circular towers and craters, there is inherent flexibility in the design proposal at a range of scales. Each tower and crater can be considered to be a series of linear streets, wrapped into concentric rings. Each ring is subdivided into plot segments separated by radial circulation paths, ensuring that each plot is easily accessible. Plot sizes range from 3000m2 to 350 000m2. The ratio and sizes of these plots can be easily redistributed to suit land conveyance demands.

Additionally, the master plan incorporates hybrid tower and crater communities in which the circular forms are opened up to incorporate more conventional orthogonal urban typologies that radiate from the same centre. This has the benefit of introducing a wider range of places and characteristics, integrating with the existing settlement, and allowing for complexities of land attornment.

At a community scale, the spatial and structural design of the towers and craters also invite future change. Based on standard depths of six or eight metres for natural ventilation and lighting, apartments are arranged in 40m2 modules. Residences can be 40m2, 60m2, 80m2, 120m2 or 200m2 in size; the ratio will again depend on the mix of tenure and demand, but a column and beam system will allow walls to be reconfigured so that smaller apartments can be expanded into larger ones, or vice versa.

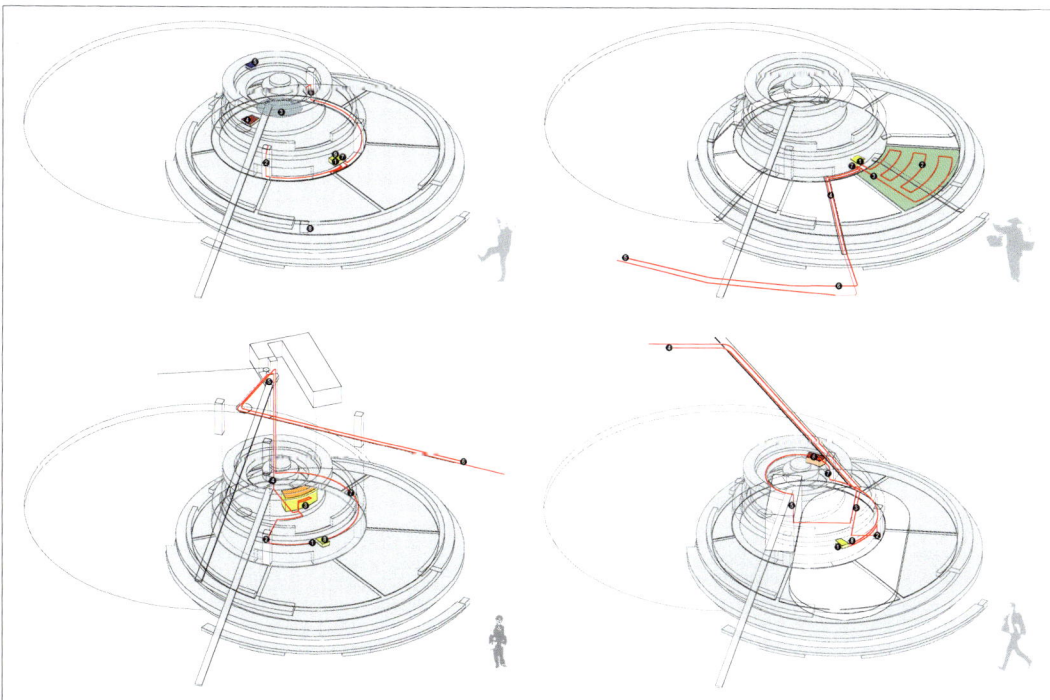

facing page top: Section showing land-use and plot division.

facing page bottom: Diagrammatic options for plot division in the tower/crater communities.

left (clockwise from top left): A day in the life of Guangming Smartcity: Mrs Lam – elder; Mr Jiang – farmer; Mr Zhang – industrial park worker; Zhang Siu Ming – teenager.

following page left: Executive recommendations and measures.

following page right: Generic plan of suburb tower.

Table 1 Technical and Economic Index Table for the Main Urban Construction Land

序号 No.	用地性质 Land Use	用地面积（公顷）Area [Ha]	建筑面积（万平米）Floor Area [10 000m2]	建筑密度 Building Density	建筑密度 Plot Ratio	绿化率 Green Ratio	人口容量 Population Capacity
1	居住用地 Residential Land	277.80	495.80	42.79%	1.78	0.57	189 831
2	商业+服务设施用地 Commercial + Service Facility Land	42.80	276.79	85.42%	6.47	0.15	18 072
3	政府和社团用地 Government + Community Land	66.70	149.72	36.69%	2.24	0.63	N/A
4	绿地 Green Space (incl. beach + agricultural)	296.11	0.00	0.00%	0.00	1.00	N/A
5	其它用地 Miscellaneous (incl. Roads + Transport)	113.69	6.64	5.84%	0.06	0.00	N/A
	合计 Total/ Mean	797.10	928.95	23.40%	1.17	0.77	N/A

Table 2 Technical and Economic Index for Block 15 (Typical)

地块编号 Block No.	用地性质 Land Use	用地面积(公顷) Area [Ha]	面积（万平米）Floor Area [10 000m2]	建筑密度 Building Density	容积率 Plot Ratio	绿化率 Green Ratio	人口容量 Population Capacity
15	居住用地 Residential Land	14.17	31.17	54.74%	2.20	0.45	11 568
	商业+服务设施用地 Commercial + Service Facility Land	1.04	4.33	94.96%	4.17	0.05	N/A
	政府和社团用地 Government + Community Land	0.35	1.93	76.72%	5.52	0.23	N/A
	绿地（包括农业用地）Green Space (incl. Agricultural Land)	1.10	0.00	0.00%	0.00	100.00	N/A
	其它用地 Miscellaneous	0.00	N/A	N/A	N/A	N/A	N/A
	合计 Total	16.66	37.43	54.09%	2.25	0.46	11 568

Table 3 Technical and Economic Index Table for Residential Land

类别 Item	编号 No.	名称 Name		单位 Unit	数量 Quantity	百分比 %	平方米/人 m2/ Person	备注 Remarks
用地规模 Land-use Scale	1	居住用地 Residential Land		ha	152.15	19.09%	8.06	% is residential building footprint area to site area. Floor area per person is 26.26m2
	2	其中 Including	二类居住用地 R2 Land	ha	150.90	18.90%	8.03	
			二类居住用地+商业用地 R2 + C1 (commercial) land	ha	193.70	24.30%	N/A	% is R@ + C1 building footprint area to site area. Floor area per person is 40.97m2
	3	居住户数 Number of Households		Households	56 591	N/A	N/A	
	4	平均每户人数 Average no. of Persons per Household		Persons	3.34	N/A	N/A	
	5	居住人口 Resident Population		In 10 000 persons	18.88	N/A	N/A	
	6	居住用地总建筑面积 Total Building Area in theResidential Land		m2 (in 10 000)	922.31	N/A	48.84	% figure to total floor area on site
	7	其中 Including	住宅建筑面积 Residential Floor Area	m2 (in 10 000)	495.52	53.75%	26.25	
			商住建筑商业面积 Commercial Area in the R2/C1 Complex	m2 (in 10 000)	276.79	30.01%	14.66	
			配套公建面积 Public Service Area	m2 (in 10 000)	149.72	16.23%	7.93	Area includes education, health + government admin facilities
	8	平居户居住建筑面积 Average Floor Area/ Household		m2	162.98	N/A	N/A	Figures for residential, commercial and public service area per household.
	9	人口毛密度 Residential Density		persons/ha	236.90	N/A	N/A	
	10	居住用地总建筑密度 Total Building Density of the Residential Land		%	N/A	44.39%	N/A	Figures for building density of all land excluding lychee groves, beach, canal + broadwalk + grazing fields.
	11	居住区容积率 Plot Ratio for the Residential District			N/A	2.28	N/A	Figures for building density of all land excluding lychee groves, beach, canal + broadwalk + grazing fields.

Table 4 Main Public Service Facilities Planning List

序号 No.	类别 Category	项目 Item	数量 Quantity			备注 Remarks
			现状 Current Status Reservation	规划增加 Planned Increase	合计 Total	
1	教育设施 Educational Facility	幼儿园 Kindergarten	5 406	43 855	49 261	Existing figures estimated
		小学 Primary	22 084	142 295	164 379	Existing figures estimated
		初中 Middle	25 266	85 377	110 643	Existing figures estimated
		高中 High	21 516	94 864	116 380	Existing figures estimated
		职业训练 Vocational Training	101 286	12 744	114 030	Existing figures estimated
2	医疗卫生设施 Medical + Healthcare Facilities	医院 Hospitals	15 560	174 056	189 616	Existing figures estimated
		诊所 Clinics	0	28 050	28 050	
3	文娱体育设施 Sports + Recreational Facilities	海滩 Beach	0	693 123	693 123	
		体育馆 Stadium	0	60 295	60 295	Sports playing fields currently on site, area unknown.
		网球/篮球 Tennis/ Basketball	0	19 320	19 320	Existing figures unknown
4	行政管理与社区服务设施 Admin + Community Service Facilities	市政厅 City Hall	0	24 903	24 903	
		地方行政单位 Local Municipal Admin	31 536	540 965	572 501	Figures include police, fire + postal services.
5	对外交通设施 Intercity Transport Facilities	火车站 Railway Station	0	26 659	26 659	
		巴士站(包括长程) Bus Station (incl. long distance)	30 202	35 202	65 404	
6	道路交通设施 Urban Traffic Facilities	塔堡/环山进入道路 Residential Access Roads	0	131 956	131 956	Existing access roads uncalculated
		步道 Boardwalk	0	868 342	868 342	
		轻轨车站 Light Rail (MTR) Stations	0	8 325	8 325	Figures for above ground entrances only

awareness: Guangming Smartcity

LAND USE + SPATIAL CONTROL

1 Vehicular Access Route

2 Vertical Circulation + Services Core

3 Escalator

4 Services/Goods Delivery route +
Loading Bay

5 Sewage Recycling (Anaerobic Digester)
+ Trigeneration (CCHP Plant)

6 Funicular

7 Car Park Access Ramp

8 Car Park

9 R2 Residential (Mid/High Rise Housing)

10 Municipal Administration

11 Buggy + Cycle Store

12 Convenience Store/ Retail

13 R1 Residential (Villa/Townhouse)

14 Function Hall/Multi-purpose Community
Space

15 Aquaculture Terrace

16 Ventilation

17 Household Waste Recycling Collection

18 Pedestrian Access to Car Park

19 Department Store

20 Car Park Ramp to other Levels

21 Suburb Square (Commercial + Municipal
Centres)

22 Funicular Station

23 Plaza

24 Reservoir

25 Centre of Excellence

26 Photovoltaic Array

27 Funicular/Skybus Interchange

28 Skybus Line

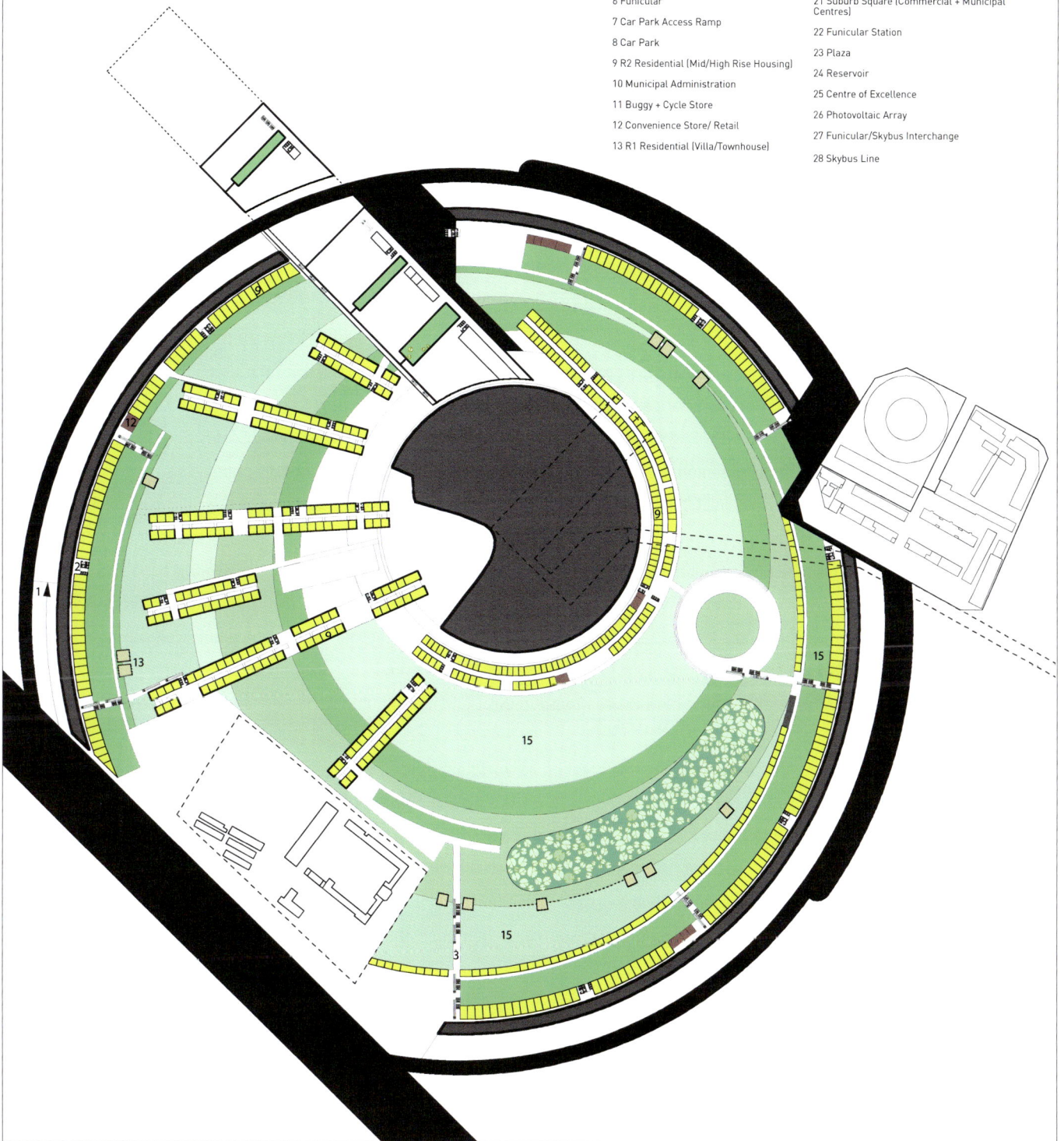

Environmental Sustainability of the Smartcity

The environmental implications traditionally associated with accommodating 200 000 people in a new high-density urban environment are substantial. An orthodox approach to waste and water processing, energy consumption and transport would be neither appropriate nor desirable when a holistic approach can be adopted that integrates architectural, landscape, services, agricultural and civic systems.

In Western industrialised countries, lighting and comfort heating or cooling within buildings can be responsible for over half of greenhouse gas (GHG) emissions. Many of the cities in these countries were designed and built in a time of perceived infinite resources, but even as we have come to the realisation that resources are severely depleted and that our actions have caused profound detrimental effects on the planet, change has been slow. The response has generally been characterised by ineffective approaches relying on 'end of pipe' solutions – buildings are designed and built to poor standards and greenhouse gas emissions are 'offset' or negated through green technologies such as wind turbines, solar thermal heating or solar PV panels. This methodology does not address environmental sustainability in the long term, and energy production quantities and performance standards often fall short of demand. Reluctance to adopt different, and often low-tech, approaches to construction has ironically left industrialised nations relying on emerging science; a fear of change leading to faith in ever more alien solutions.

It has been calculated that the earth has around 11.3 billion hectares of productive land and sea space and 6.1 billion people, equating to 1.85 hectares per person if shared equitably and ignoring other species.[4] This figure has been used to calculate the 'ecological footprint' of different countries; the average citizen in the United States uses over nine hectares to sustain their way of life, meaning five planets would be required to support the world's population were this lifestyle to be adopted as a benchmark. In 2004, Chinese citizens used on average under one global hectare per person according to a WWF Living Planet Report, a figure that had increased to 2.1 hectares by 2008 due to China's rapid transition from rural landscape to cityscape. The relationship between rural and urban ecological footprints, however, is not a straightforward one and varies according to territory. In wealthy countries, the lifestyles of rural dwellers can be effectively urban in nature with consumption exacerbated by distance, resulting in increased per capita emissions compared to their city counterparts. Moreover, modern agricultural practice and deforestation contribute heavily to carbon emissions while compact city arrangements can minimise consumption through shared public transport and efficient space heating.

The city framework for Guangming Smartcity takes a hybrid approach that capitalises on rural and urban advantages to help China maintain economic growth and increase the wellbeing of its people without exerting undue pressure on the rest of the world's natural resources. The scale of the Smartcity means that almost all sustainable technologies are economically viable, allowing the design process to focus on the relative merits of different approaches in terms of social and environmental impact; equally, the opportunity for integration is almost unprecedented as 'waste' metamorphoses from material to be disposed of to a potential resource, as demonstrated in the industrial symbiosis at Kalundborg.

As well as reusing traditional 'waste' on site for energy generation and fertiliser, Guangming Smartcity demonstrates a positive influence on its neighbours by incorporating waste streams from neighbouring areas, exemplifying how the Smartcity can avoid the traditional pitfalls of Western development. Where Santa Monica (8.3 square miles) has an ecological footprint 350 times its size (2914 square miles), Guangming Smartcity will become a net importer of 'waste' and a net exporter of energy while promoting tourism, education and food production to the benefit of the wider region.

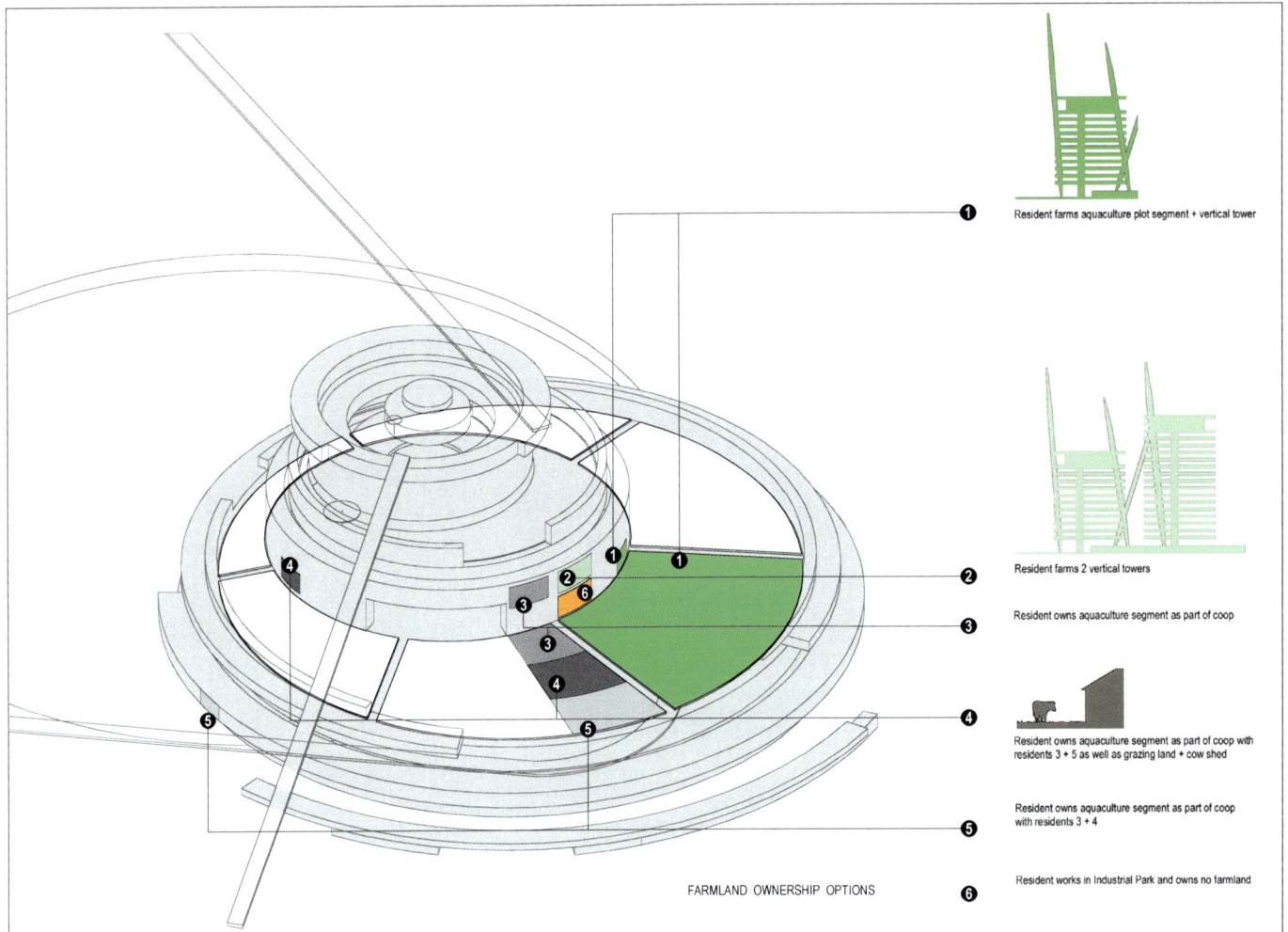

- ① Resident farms aquaculture plot segment + vertical tower

- ② Resident farms 2 vertical towers

- ③ Resident owns aquaculture segment as part of coop

- ④ Resident owns aquaculture segment as part of coop with residents 3 + 5 as well as grazing land + cow shed

- ⑤ Resident owns aquaculture segment as part of coop with residents 3 + 4

FARMLAND OWNERSHIP OPTIONS

- ⑥ Resident works in Industrial Park and owns no farmland

Crop 农作物	Field Area/m2 农地区／平方公尺	Growing area/m2 种植区／平方公尺	Plants/m2 植物／平方公尺	Expected yield per annum* 每年期望产量
Pak Choi 青菜	2 509 886	2 007 909	30-35	61 015 335 pcs
Tomato** 蕃茄			4	18 774 tonnes
Lettuce 萵苣			12-20	30 038 319 pcs

* Figures shown for monoculture over entire available growing area. It has been assumed that 5-8% of seedlings will not take up.
** 2-3kg yield per tomato plant.

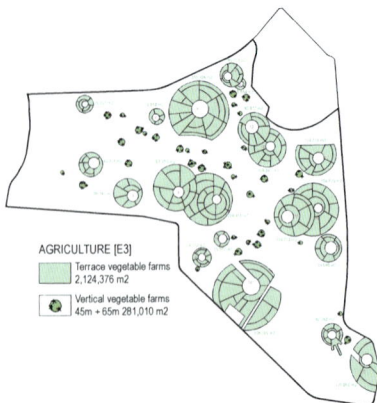

AGRICULTURE [E3]

- Terrace vegetable farms 2,124,376 m2
- Vertical vegetable farms 45m + 65m 281,010 m2

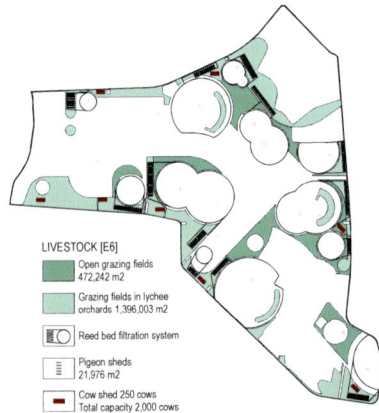

LIVESTOCK [E6]

- Open grazing fields 472,242 m2
- Grazing fields in lychee orchards 1,396,003 m2
- Reed bed filtration system
- Pigeon sheds 21,976 m2
- Cow shed 250 cows Total capacity 2,000 cows

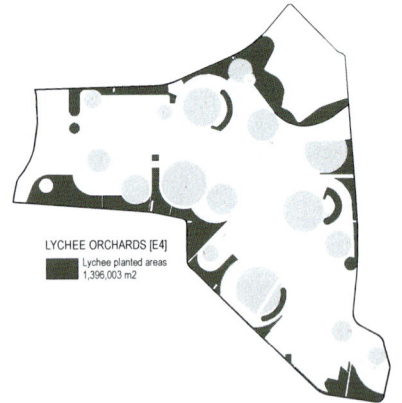

LYCHEE ORCHARDS [E4]

- Lychee planted areas 1,396,003 m2

Organic Food Production

Roofs in most cities serve only one function – to shelter – often creating bleak, uninspiring surfaces. In keeping with the philosophy of 'nothing is waste', every square metre of Guangming Smartcity is used, in this case as a hydroponic membrane capable of growing significant amounts of food. Beneath the growing surface, a gravel substrate is used to clean household water. The city consequently integrates the three functions of shelter, water purification and crop cultivation into the same space in addition to improving thermal insulation and surface water retention.

The hydroponic system uses an absorbent medium such as vermiculite or mineral wool instead of soil; a nutrient-rich solution is passed through the medium allowing plant roots to absorb the required minerals. Hydroponic farming is one of the most efficient cultivation methods in terms of crop spacing and access to sunlight, providing crops with more nutrients and using less energy in the process. Furthermore, hydroponic crops stay fresh for longer once harvested as they can be harvested without killing the plant; there is no need for soil disposal and sterilisation, and soil-borne diseases are virtually eliminated.

Conservative estimates suggest that the available roof space in Guangming Smartcity can produce a substantial quantity of produce. A combined growing area of 450 hectares over the site could produce either 18 800 tonnes of tomatoes or 61 million pak choi annually.

Vertical kitchen garden farms and laboratories can be found scattered throughout the central city with accompanying vertical floral gardens. The facilities located in each vertical farm contribute to research projects, forming a major centre for agronomic and nutritional science. Each tower is constructed as an array of growing trays projecting from a cylindrical circulation spine. The trays are configured in pairs with a central gantry and staggered to maximise photosynthetic reaction. The preservation of endangered crop species is encouraged through research in arable laboratories located at the top of each tower.

As part of the city's urban agricultural programme, livestock is farmed as well as vegetables, fruit and flowers. The flat pockets of land in-between the craters and towers are used as grazing fields for livestock.

Located in each suburb is a farm shop run by the city's utilities management organisation where produce is marketed at a price specified by the farmer. Unsold vegetables will be broken down in the anaerobic digesters, contributing to methane production used in electricity generation. The system will reach a natural equilibrium as the shop cannot exert pressure on the farmer to sell crops at a reduced profit, but if prices are set too high, produce is recycled. The outlets will act as a hub to connect everyone within the community. It is by such interconnections on a local scale that effective sustainable urban master plans can be implemented.

81

previous page: Farmers – the new eco-warriors.

facing page top: Farmland ownership options; Expected crop yield.

facing page bottom: Plan diagrams of terrace + vertical farms (left), and open grazing fields.

4. WWF Global, 'WWF Update on Alarming State of the World', 21 October 2004 [http://wwf.panda. org/?15995%2FWWF-update-on-alarming-state-of-the-world], retrieved 16 April 2018

Waste Treatment

Traditional sewage treatment is very energy intensive, using 65 000 gigajoules a day in the UK. At Guangming, waste is dealt with as efficiently as possible through a combination of natural low energy processes, but is also used to produce methane powering the city, and fertiliser for the city's farms and flower gardens.

Traditional methods of sewage treatment deal with all waste flows together, which includes vast quantities of relatively clean greywater from showers, baths and washing machines; moreover, the water used to flush toilets is often fully treated to drinking quality. This has the effect of diluting the actual sewage and making it much harder to remove and process. In the Smartcity, blackwater is kept separate, enabling the city's sewage to produce rather than consume energy. Greywater will be cleaned in gravel-bed hydroponics (GBH) systems in the roof farms, dramatically reducing the quantity of water requiring full treatment. Once processed, the water is clean enough to discharge into local watercourses, or is 'polished' and recycled to potable standards.

Kitchen sinks will be installed with waste macerators so that organic waste can be combined with foul drainage from toilets. This blackwater passes through an aquatron device that separates liquid from the solids. The former continues to natural reed beds or further GBH systems for purification while the latter are transported to an anaerobic digester. Here, the solids are broken down naturally into methane, which is captured and used to generate electricity, a nutrient-rich liquor called digestate that can be used as fertiliser once pasteurised, and fibrous mass which can be applied to soil to improve retention capacity or be incinerated for electricity.

Internal Climate Control

Air conditioning is generally regarded as a necessity in the region, but the landscape and component infrastructure of the Smartcity allow natural low energy design to counter this convention. Calculations suggest that no mechanical space cooling will be required in residential dwellings and commercial buildings will require only 50 per cent of their usual demand. This is achieved through reducing direct solar access during the summer, the thermal mass of exposed heavyweight construction materials, and a labyrinth cooling system located in the base of the tower structures that transfers inter-seasonal heat. By drawing air underground where the temperature is a constant 22°C throughout the year, hot summer air is cooled using the labyrinth walls as a heat sink; in winter, cold air is preheated by the warmer conditions underground. To reduce the embodied energy required for its construction, the labyrinth is built from excavated rock rather than concrete. The interstitial space between the crushed rocks creates a convoluted path to increase the contact period for heat exchange, functionally replicating a conventional concrete labyrinth.

The Smartcity is a car-free zone but provides underground parking for residents who require personal vehicles for travelling beyond Guangming's centre. The car parks will be ventilated through 'solar chimneys' at the top of the towers. Air in the chimney rises as it is warmed by the sun, drawing fresh air through the car park while phase change material (PCM) is used to store the excess heat effectively so the system can continue to run through the night. In the same way that labyrinth cooling uses constant underground temperatures to cool the air, the air being drawn through the car park is cooled as it spends time in contact with the thermally massive walls underground. This cool air is not suitable for ventilation, but is made use of by circulating it around the external envelope of a subterranean shopping mall to reduce cooling loads.

RESIDENTIAL COOLING

Warm air

Cool air

1   Warm air intake from above
2   Air pre-cooled in rock labyrinth
3   Pre-cooled in circulation corridors
4   Pre-cooled air drawn through apartments

DEPARTMENT STORE
COOLING

Warm air

Cool air

1   Warm air intake from car park entrances
2   Air cooled in underground carpark
3   Cooled air passes through envelope around
    department store
4   Glass top to envelope creates solar chimney
    effect and air flow to ventilate carpark and
    cool department store
5   Temperature in department store also
    moderated by evaporative cooling using
    water reservoir above

Target Supply
Air Temperature
to Flats

45 °C
40 °C
35 °C
30 °C
25 °C
20 °C
15 °C
10 °C
5 °C
0 °C

Period 1   Period 2

Jan Feb Mar Apr May Jun Jul Aug Sep Oct Nov Dec

External Air Temperature
Labyrinth Output Temperature

Rock Labyrinth Performance: Summer

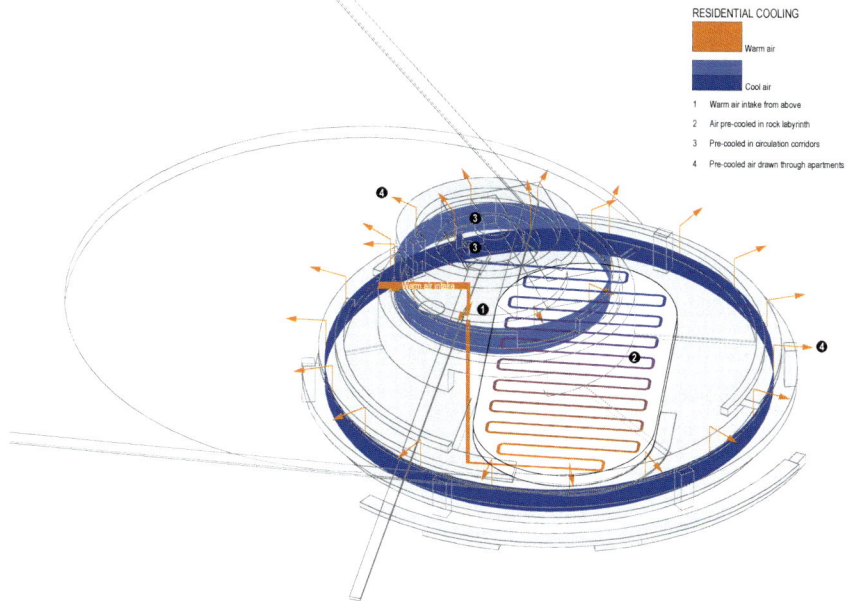

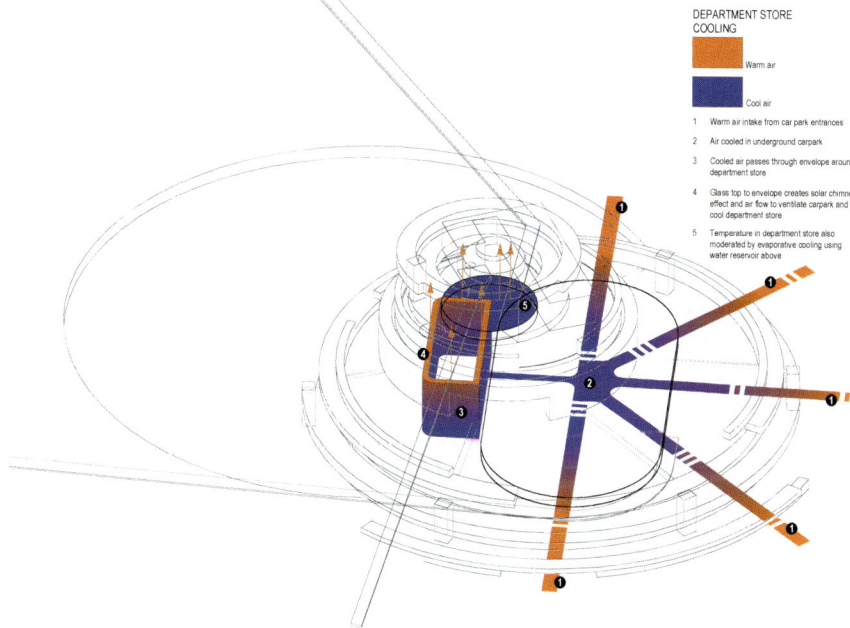

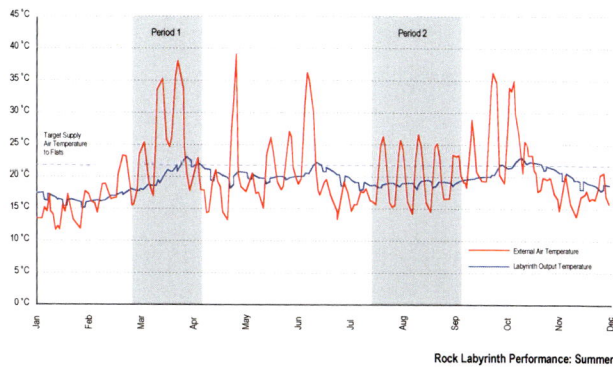

top: Cooling strategy for residential
units.

middle: Cooling strategy for
the underground commercial
development.

bottom: Rock Labyrinth Cooling
System – seasonal performance.

Low energy displacement cooling is employed in all the larger internal city spaces. Fresh air is supplied at low level from wall or floor mounted diffusers at around three degrees Celsius below the desired room temperature, forming a reservoir of cool air across the floor. On contact with a source of heat gain, the air warms and rises through natural buoyancy to settle in a ceiling reservoir leaving the occupied zone fresh and cool. The depth of the stale air reservoir is controlled by the careful distribution of high level extracts.

Energy Demand + Generation

Every strategy employed at Guangming has been chosen for the benefits it confers in terms of energy demand reduction, efficient resource use, and improved wellbeing. Often the benefits of a strategy are multi-faceted, offering gains in more than one area. This means that the energy demand for the new city is the lowest it can be, requiring less energy to be produced, no matter which route is taken. Using benchmarks for developments in the surrounding region, a development of this size using standard construction methods to meet modern living standards would require 321 750MWh/year; Guangming's demand is estimated at 127 110MWh/year, while still achieving the highest standards of modern living.

To complement the most effective methods of reducing energy demand, the Smartcity's energy supply strategy adopts the same fundamental principles of economy, practicality, synergy and anti-wastefulness.

Anaerobic digestion has been used with great success throughout China, generating biogas energy equivalent to 280 million tonnes of coal every year and meeting almost 14 percent of the nation's energy demand. The quota generated at Guangming will depend on which waste streams are harnessed. If limited to human and domestic organic waste, the methane produced could constitute around half the city's energy supply; with the addition of organic matter from farms, shops and restaurants, the city would become a net energy exporter, generating a new revenue stream.

The Smartcity will also employ waste-to-energy plants (WtE) where incinerators will burn non-recyclable municipal solid waste (MSW) in high-efficiency furnaces to produce steam or electricity. Modern air pollution control systems will be fitted and emissions continuously monitored. The intention is to expand the scale of operations to assimilate waste from neighbouring Gongming, increasing electricity production and reducing waste build-up in the region. Waste heat from electricity generation, normally lost to the atmosphere, will be used to provide 'free' hot water for the city and channelled into absorption chillers for additional cooling.

China is already a major supplier of photovoltaic panels with future demand both in China and abroad expected to increase dramatically in the coming decade. Integrated PV panels along the riverside boardwalk will provide shading as well as contributing to the Smartcity's electricity output. The scale of Guangming would permit the sanctioning of an entire PV factory, benefiting the local and national economy, and driving down unit costs. Guangming's PV farm represents a guaranteed investment, confirming a colossal order of panels years in advance and enabling manufacturers to bypass 'start-up' and 'growth' phases to full-scale production, furthering China's standing as a world leader in PV production.

Typical Energy Demand vs Guangming Energy Demand

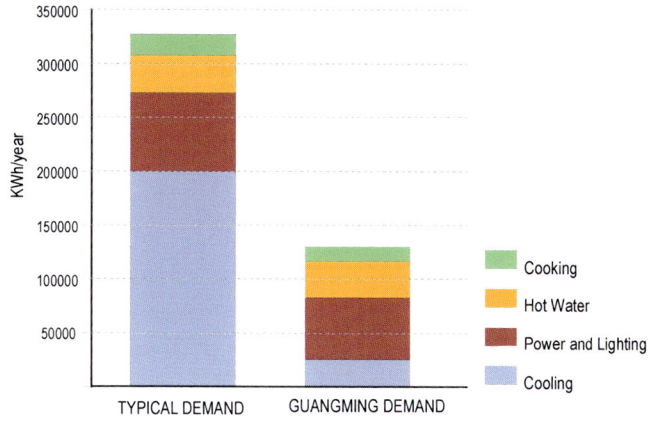

KWh/year

350000
300000
250000
200000
150000
100000
50000

TYPICAL DEMAND   GUANGMING DEMAND

- Cooking
- Hot Water
- Power and Lighting
- Cooling

Electricity Demand + Potential Renewable Supply

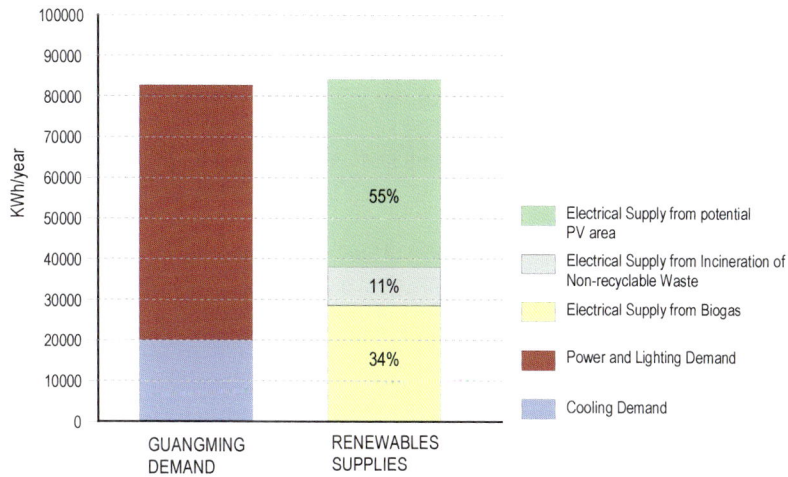

KWh/year

100000
90000
80000
70000
60000
50000
40000
30000
20000
10000
0

GUANGMING DEMAND   RENEWABLES SUPPLIES

55%
11%
34%

- Electrical Supply from potential PV area
- Electrical Supply from Incineration of Non-recyclable Waste
- Electrical Supply from Biogas
- Power and Lighting Demand
- Cooling Demand

top: Guangming Smartcity's energy demand.

middle: Renewable supply potential.

bottom: Plan diagrams showing distribution of anaerobic digestors + reed beds (left), photovoltaic fields (middle), and recycling centres.

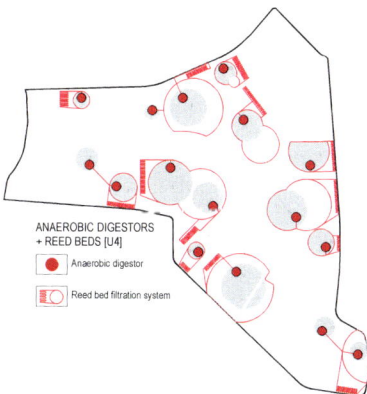

ANAEROBIC DIGESTORS + REED BEDS [U4]
- Anaerobic digestor
- Reed bed filtration system

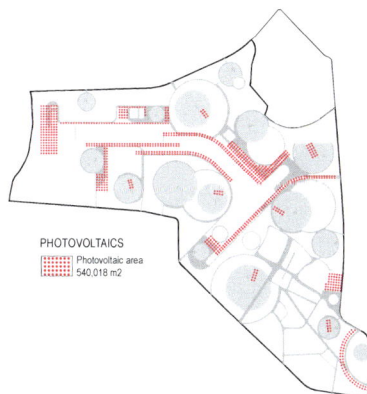

PHOTOVOLTAICS
- Photovoltaic area 540,018 m2

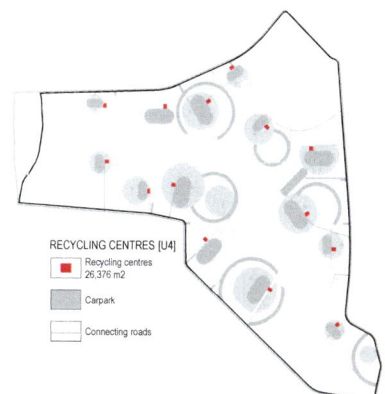

RECYCLING CENTRES [U4]
- Recycling centres 26,376 m2
- Carpark
- Connecting roads

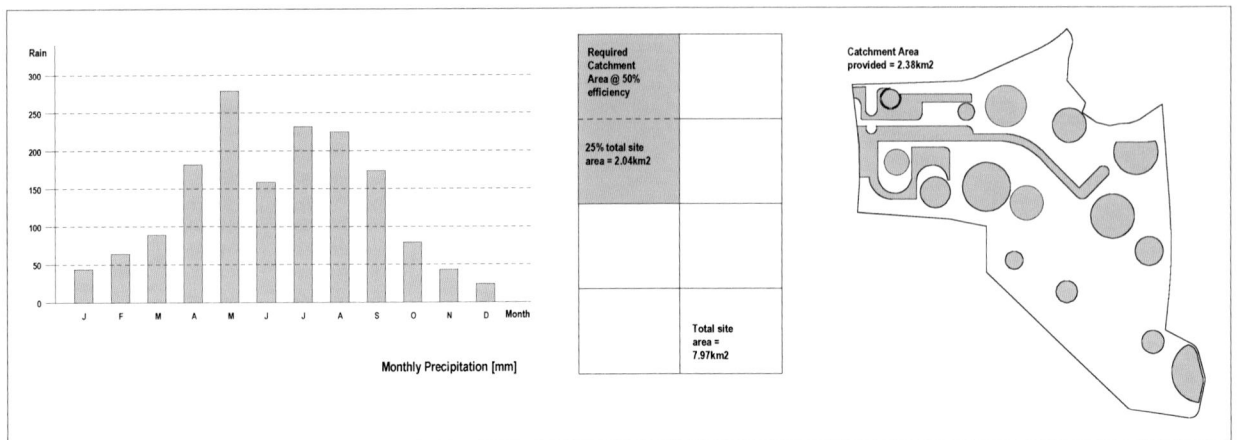

Hydrology

The varying topography of the land with its hills and valleys is naturally suited to the formation of large water bodies that benefits a multitude of uses. Working with the existing surface water network, the natural tributary that eventually merges with the Maozhou River to the east of the site is sculpted to form a central canal. The disposition of form in the amplified towers and craters results in a cascade of water bodies, stepping down from one to the next in the circular aquaculture terraces.

The water bodies will act as a medium for transport, agriculture and recreation with all uses mutually reinforcing. They will provide the main water storage for the whole development and be linked to the reservoirs at the apex of each tower. In addition to the rainfall directly collected, the water storage bodies adjacent to the canal will gather surface runoff and water drained from the hills, increasing the catchment area. There is approximately 1575mm of rainfall in Shenzhen per year, peaking between May and September; relatively little water is available between October and March. Storage will be required during this dry period, although greywater and blackwater recycling will mitigate the shortage.

Transport

Guangming Smartcity is a car-free city designed to ensure that every resident is able to carry out all their needs within their local area – to live, to play, to learn and to work – while still having the freedom for inter-local and regional mobility if they wish. The transport infrastructure will provide the foundation for a dense diverse urban community with an environment that encourages the use of public space. Walking and cycling in a humane environment will be given the highest priority, leading to a human scale urban community that places high value on personal contact.

The public transport system has a clearly defined hierarchy fundamental to the design of an efficient infrastructure that interfaces well with private modes of transport. This is especially important at Guangming where public transport systems must deal with vertical as well as horizontal movement.

Residents of the Smartcity are served at regional inter-city scale by car, rail and bus. The Longda Expressway on the west boundary of the site provides the main regional road traffic connection with Guangming. The expressway is elevated to accommodate the extension of the city boundary to the edge of the Maozhou River, removing the physical barrier to

Gongming village and its industrial cluster whilst forming a striking gateway to the development and offering an elevated panorama over Shenzhen's green haven. The gateway incorporates Guangming Central Train Station, enabling easy access and interchange with road traffic and a visitor car park. At the base of the structure, the underground mass transit rail can be accessed, together with buggy and cycle hire stations for local travel. To put Guangming on the map as a vanguard city and tourist hub, and to create an attractive and viable commuter settlement, a new railway network linking Guangming Smartcity to Hong Kong, Lo Wu and Guangzhou is proposed. Noise from the high-speed trains will be attenuated by the groves of lychee trees.

Three mass transit rail stations at strategically located positions serve the Intercity fast line (linking Shenzhen, Shiyan, Guangming, Huangjiang, Songshanhu and Guancheng), Incity fast line (serving Shenzhen, Longhua, Shiyan, Guangming and Gongming), and Local Area line (linking Shanjing and Guangming). These are integrated into the residential and commercial growth centres located at the Central Train Station Gateway, City Hall Plaza at the opposite end of the east–west town axis, and the Old Town settlement.

Guangming Central Bus Station is located to the east of the old town and caters for long-distance inter-city coach journeys including a shuttle service to Shenzhen International Airport 18km away. Three new local bus routes covering the perimeter of the site are proposed, linking into the existing network of Bao'an district by making full use of the proposed new underpasses.

At inter-local scale, Guangming Centre is provided with three new light rail (MTR) stations spaced approximately two kilometres apart that connect to the wider regional system. The Skybus, a cable car service, provides a quick hop service between principal tower-crater communities linking the Centres of Excellence. Intended for low cost everyday use as well as for recreational purposes, the elevated network allows the Smartcity's hybrid landscape and rooftop occupation to be fully appreciated.

The boardwalk that lines the urban beach and canal, with its interconnected network of pathways, is traversed by buggies and cycles that operate as a shared resource similar to the bike rental Product Service Systems (PSS) that have proved so successful in Lyon, Copenhagen and Barcelona. Buggy and cycle storage towers on a lowerator system are positioned at transport nodes and at the base of each tower community. The boardwalk will also be used for emergency vehicle access.

The communities are planned such that a commercial hub and inter-local scale transport hub are never more than 400m from a residence, the largest crater being 800m in diameter. Walking and cycling can therefore easily cover local scale transport. A funicular railway serves each tower and crater community, travelling directly from the base station to the public reservoir and suburb square at the summit. Radially distributed around the rings, escalators provide a secondary means of vertical transport within the towers and craters.

facing page: Hydrology – rainwater catchment area.

Construction Phasing

The city is to be built over a 13-year period. Four major construction phases are envisaged, each of which when complete will be a stable stage of development.

Phase 1: Three growth centres are established around Guangming Old Town, City Hall Plaza and the Central Train Station. Over the first four years, the major infrastructure elements are completed, and the first tower and crater communities built. Existing residential settlements are left undisturbed with minimal loss of agricultural land on the previously unconstructed zone. The Central Train and Bus stations, City Hall and their respective commercial centres are developed together with the new Fire, Police and Postal services. The existing schools and hospital will temporarily accommodate the increased demand.

After a year of detailed planning, work commences with the elevated Longda Expressway and main central station along the riverside. The urban light rail link (MTR) is constructed below ground level using 'cut and cover' technology. Construction roads are established from the existing ring road towards the city hall area of town as well as the three main underpasses, permitting the implementation of the new bus routes that will connect to Gongming and the Hi-tech Industrial Park.

The depressions and hills for the first two craters and three towers around the City Hall, and new Central Bus Station, are excavated and built up using a fixed system of mining excavation. Earth movement will be limited as far as possible with the 'fill' of the towers balanced by the 'cut' of the adjacent craters.

Photovoltaic structures, designed to be relocated as necessary, are erected at inception to provide energy for the construction work and to contribute to the national grid. The surface watercourses are redirected to feed into the central section of the canal and the balance tanks and water storage along the bank-sides installed. Main drainage and water treatment areas are established. Fresh water mains and hilltop reservoirs are built over the new topography. Buggy and cycle routes and their parking structures will be operational.

Communication within each area initially comprises local distributor roads constructed as circumferentially flexible pavements with diagonal ramps rising between each layer. Towards the completion of each ring segment, escalators will be installed. The first two Skybus links between the city hall growth centre and Guangming Central Train Station, and the southernmost tower and crater communities will be quickly placed once the hilltop public plazas are finished. The filled areas of the new towns are consolidated with ground improvement techniques and segmental development sites are then let to individual developers, introducing variation, character and local identity. A diversity of building types and public areas are set on the new ground level with shallow foundations or raised above simple reinforced concrete under-crofts.

Phase 2: Inhabitants of Gongming village and Loucun village are decanted into the newly completed Phase 1 housing that have a combined population capacity of 30 000. The communities between the city hall area and riverside are completed with additional Skybus lines serving them. The beach area of the town is built up, and lychee orchards established around the site boundary. The Phase 1 extension of the old hospital is underway.

The canal works are completed throughout the site. With full provisions for the Smartcity's water treatment and storage complete, the beach district is made using selected fill brought by river and deposited by water jet. Further excavation of towers and craters will take place.

The radial development of the first towers and craters is finished and the local routes pedestrianised allowing decanting of the village settlements to occur. The smaller tower and crater communities contain simple buildings bearing on existing ground, set alongside rings of elevated housing. As densities reach their intended level, prefabricated steel towers supporting vertical farms will be erected. These installations are coordinated with further photovoltaic arrays.

Phase 3: The area south of the central canal is developed. The internal vehicular circulation network (boardwalk) and recreational areas are completed. Residents of Guangming Old Town are decanted to completed housing. Phase 2 of the Hospital Extension is complete.

The longer runs of supply and drainage are now installed to open the entire site to urban development. Further blocks are developed together with extensions to the cable car network. Urban zones are created linking the city hall area to the riverside. Housing, agriculture, vertical farms and photovoltaic arrays are fully developed in this area of the site.

Phase 4: A final phase of community development in and around Guangming Old Town is implemented. There is a densification of all communities and renewable resource systems are fully deployed. Existing health and emergency service buildings are renovated and enlarged to service the increased population. Ground forming is complete and new development to plots in and around the old town is built. The segmental development of all urban areas is completed. Vertical farms distribution and energy collection reach their optimum level.

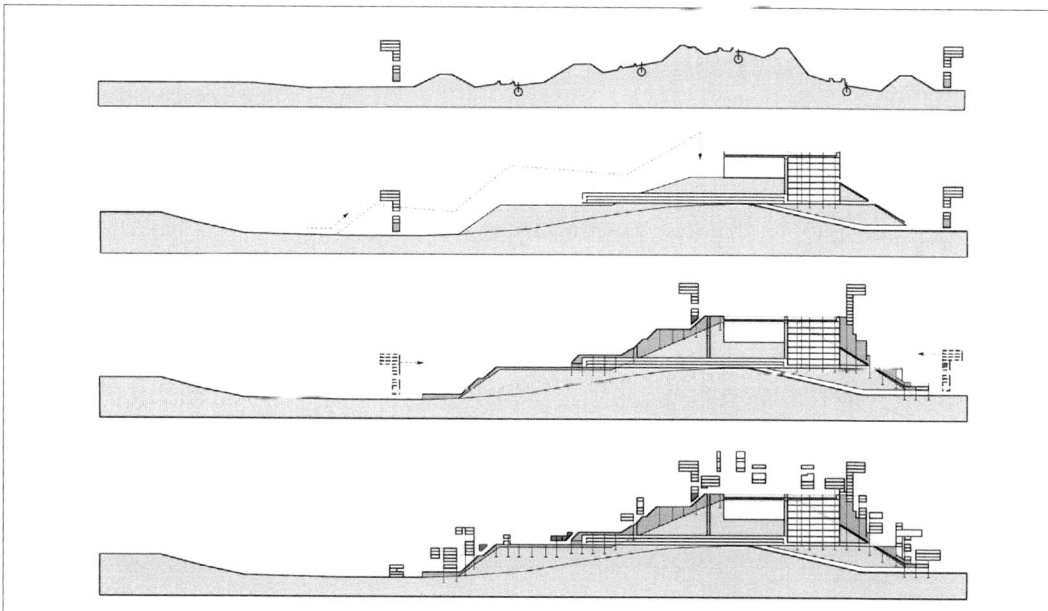

left: Construction phasing of a tower community.

following pages: Model of Guangming Smartcity.

89

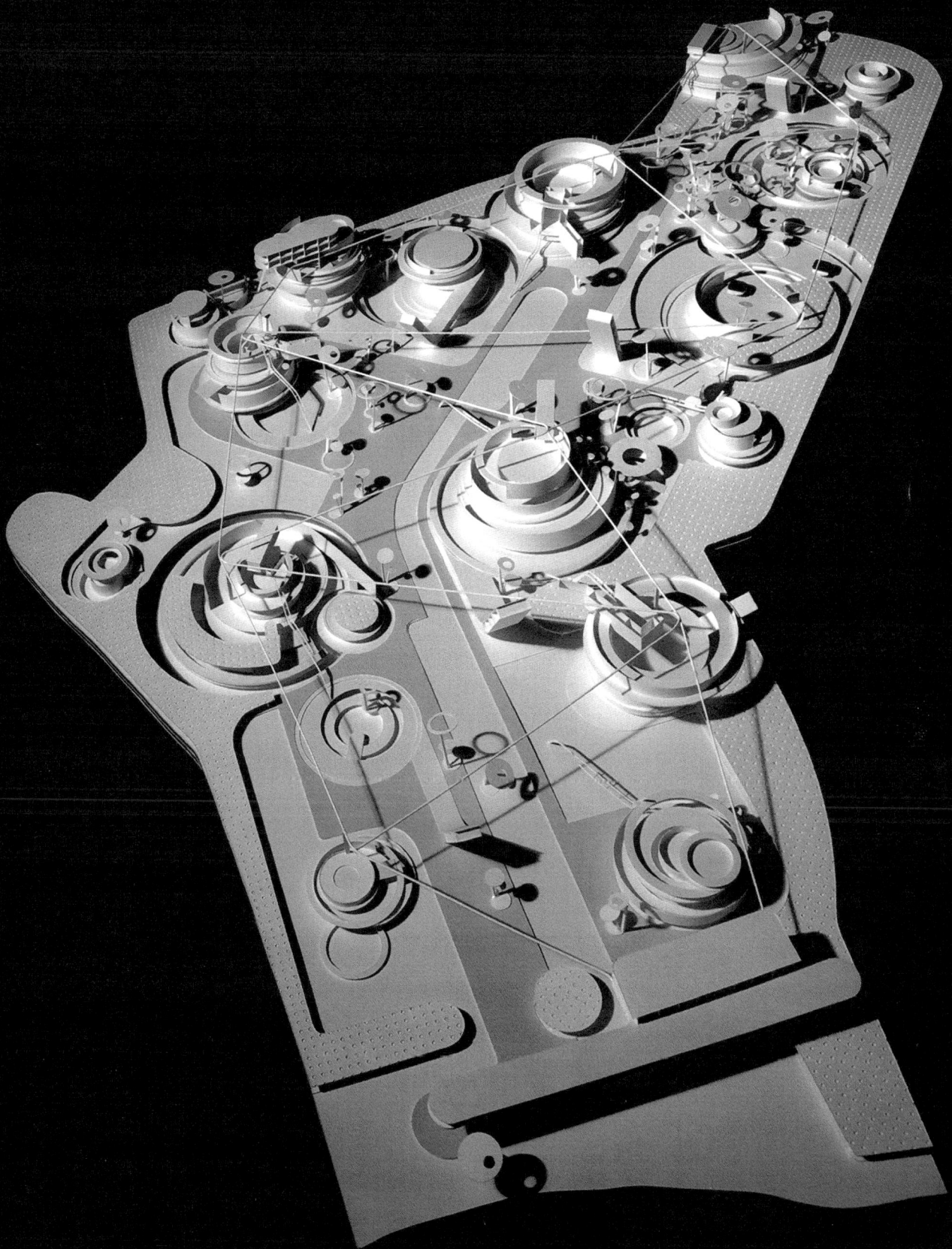

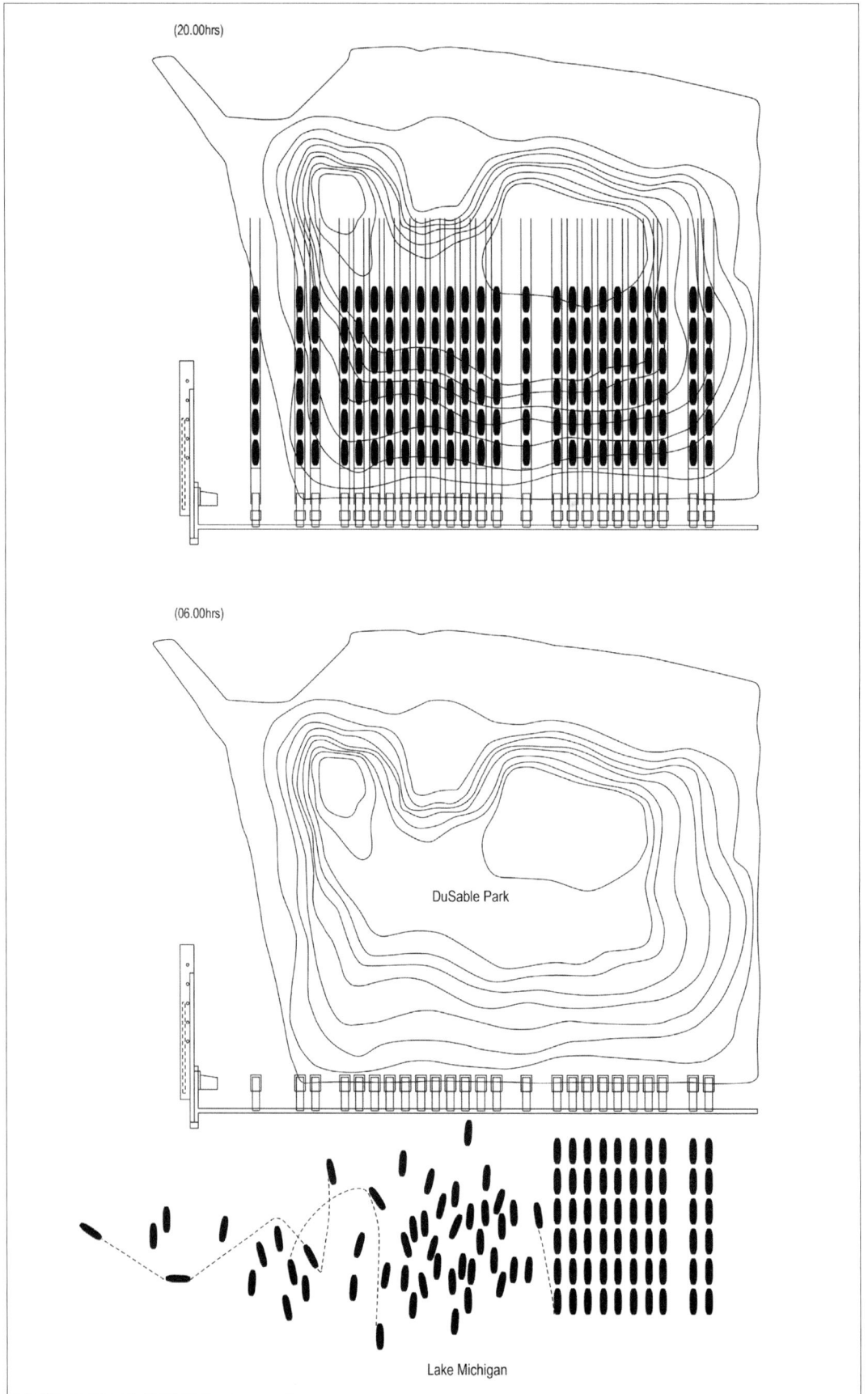

(20.00hrs)

(06.00hrs)

DuSable Park

Lake Michigan

92

DuSable Park USA

'Allotments are profoundly anti-capitalist spaces. The rents charged by your council for a plot often bear little relation to the actual land value. That's a wonderfully wilful rejection of today's dominant ideology of allocating resources via money and globalised markets.'

– George McKay, 'Radical Gardening', 2013

DuSable Park is a collection of allotments in waiting. Located on a peninsula accessible only by trespassing on private land, three acres on the shore of Lake Michigan in downtown Chicago were designated as parkland in 1987 under the administration of Mayor Harold Washington. These three acres represent a chronological, geographic and social anomaly. Abandoned for over two decades, the site is an overgrown meadow, surrounded by exclusive privatised urban space and colonised by opportunistic weed flowers, butterflies, songbirds and intrepid humans. The site attracted the attention of artist Laurie Palmer in 2001 who decided to invite artists and architects to conjure up multiple and co-existing manifestations of DuSable Park to provoke debate and lobby for action on the part of the Chicago Park District.

DuSable Park derives its name from Jean Baptiste Pointe DuSable, a Haitian-French pioneer who became the first permanent non-indigenous settler of Chicago in 1772, and is popularly known as the 'Father of Chicago'. Representing a key black historical figure, a park commemorating DuSable would go some distance to redressing long-standing inequities of privilege between the African American and Latino communities west of the lake, and the predominantly Caucasian population who have monopolised the waterfront following government policy that encouraged high-income development over public space.

In July 2000, a proposal to lease the land to a development agency as a 'temporary' parking lot was blocked by vociferous local opposition and later that year, radioactive thorium, with a half-life of 14 billion years, was purportedly found to contaminate the site, put forward as the reason for the Park Authority's failure to deliver on its promise of developing the land as a public recreational space. Two years later, three cubic metres of soil and the contamination problem were putatively removed, a claim which has since been repudiated.

Palmer notes that 'public space is ostensibly available to everyone, but someone is always excluded: the person who wants to sleep on the bench a little too long, set up camp for a few weeks, have sex in the tall grass, make loud noises, plant vegetables, roast a pig, roar her dirt bike in circles around the toxic hill, or have the whole place to himself for an ecological experiment'.[1] Any single proposition for the development cannot be wholly inclusive, but the proposal for the DuSable Park preserves the overgrown

facing page: The cyclical reconfiguration of DuSable Park.

1. L Palmer, P Philips & CR Reed, '3 Acres on the Lake: DuSable Park proposal project', WhiteWalls Inc., Chicago, 2003, p.6

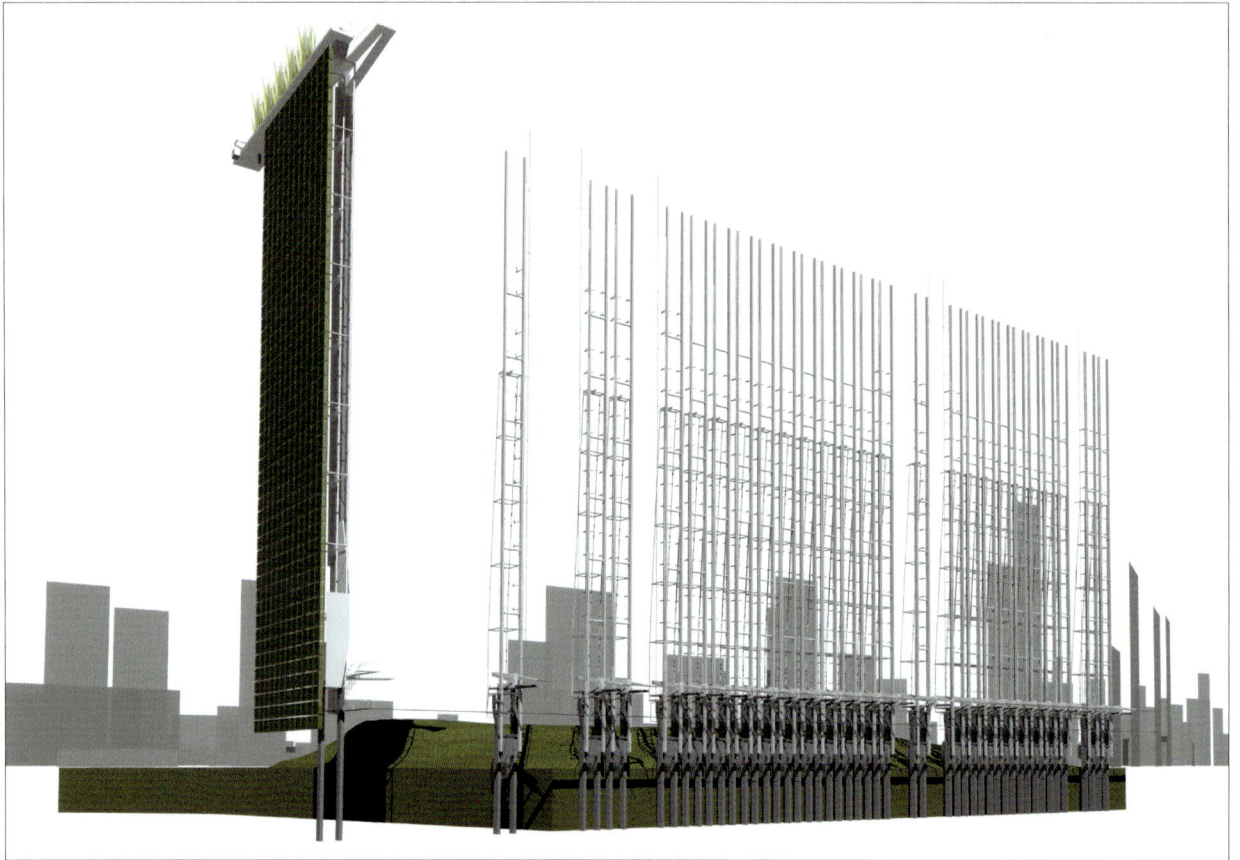

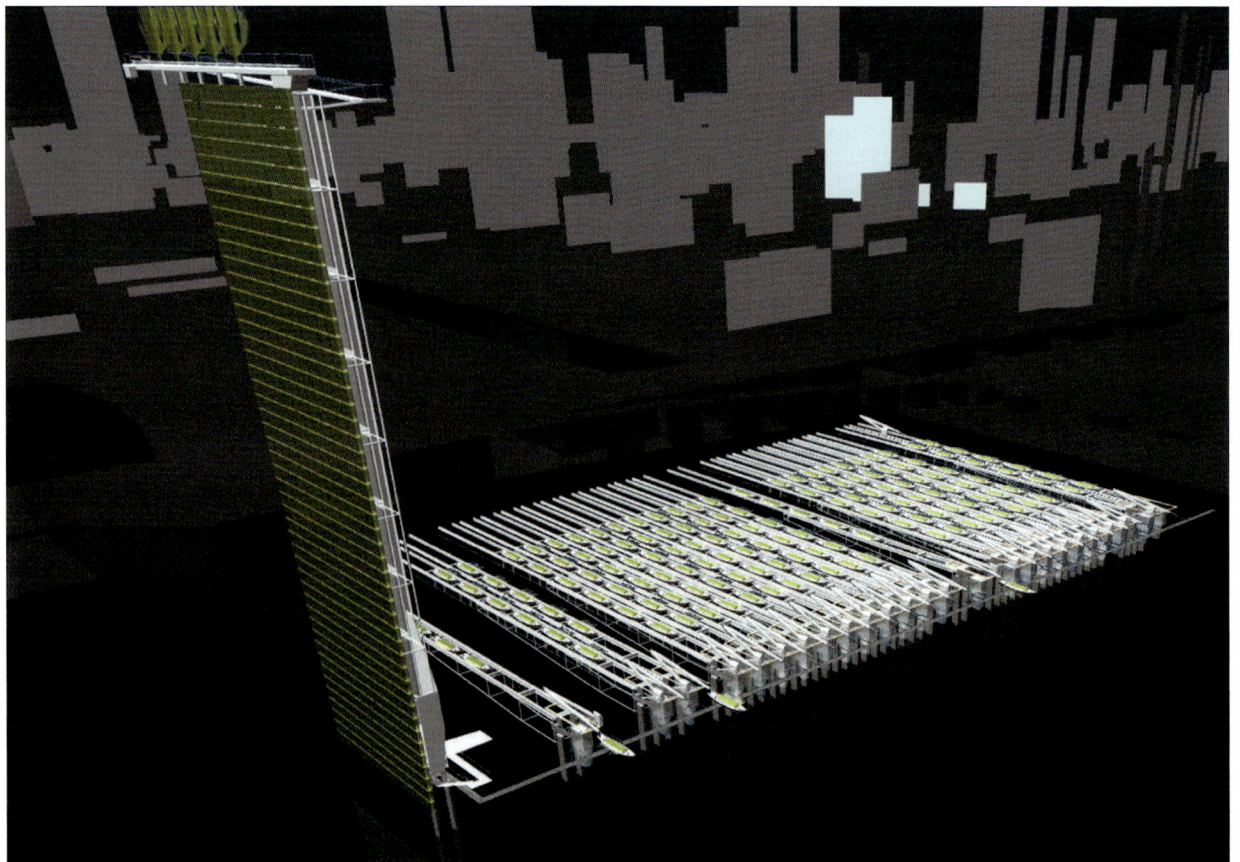

meadow and its rich human and non-human biodiversity by elevating a piece of community landscape over the existing ground plane, concurrently serving as a stark reminder of the environmental damage caused by human industrial activity.

This community resilient landscape is composed of an armada of hovering boats decked out in flora and edible produce, a skyscraper plant-nursery and a drawbridge linking the meadow with Grant Park. The skiffs that make up the floating garden symbolically celebrate the arrival of DuSable and other subsequent immigrants who have contributed to the city's cosmopolitan but segregated make-up.

Each floating boat-allotment can be leased to individual members of the local community, opening up the waterfront to other ethnic and deprived groups. Equipped with planting trays, clear frost-protection covers and lighting, the roving skiffs result in an endlessly expanding and contracting park, displaying a tapestry of non-indigenous vegetation and a multitude of colour change. The park gradually develops an ecological cycle of migrated plants, fostering the growth of new wildlife habitats.

The individual boat-allotments are secured on an array of lightweight pier structures pinned to the water's edge. By day, the floating gardens are deployed onto the lake by remote controlled cranes, releasing the pier structures into a vertical configuration to expose the meadow, mirroring the performance of drawbridges around the city. Operating on a diurnal cycle, the structures return to their horizontal positions at dusk, collecting the skiffs and rolling them back into place for the night. The boats are either navigated remotely or sailed by local residents into the lake. The choreography and arrangement of the park on the lake is infinitely variable.

The skyscraper nursery is an inhabitable south-facing glass structure, borrowing from the windy city's idiom of glass facades. The nursery cultivates non-indigenous flowers, vegetables and rare seedlings and supplies plants for the boat-allotments and outlying neighbourhoods. Individual glass seedling boxes are accessed via a vertical farming device similar to that of window-cleaning cradles on surrounding skyscrapers. The structure is capped by a sky garden of hydroponically grown trees and offers dramatic views over Lake Michigan and the city. Open to the entire community, the tower presents a spatial experience usually accessible only to the privileged few.

The skyscraper nursery not only serves as the entrance to the floating gardens of DuSable Park but also defines the boundary of Grant Park. The base of the vertical structure accommodates public washing facilities, gardening tool and material stores, a retractable open-deck market and a small kitchen. On Sundays at the end of each month, fresh produce from the floating gardens is sold at the market and the kitchen can prepare picnic hampers to be enjoyed in the park. With sufficient public awareness and pressure, the tableau of dining amongst the floating gardens of Lake Michigan on a clear midsummer's evening with Chicago city as the backdrop could well become reality.

facing page top: The skyscraper nursery defines the boundary of the meadow.

facing page bottom: At night the boat-allotments are secured on an array of lightweight pier structures.

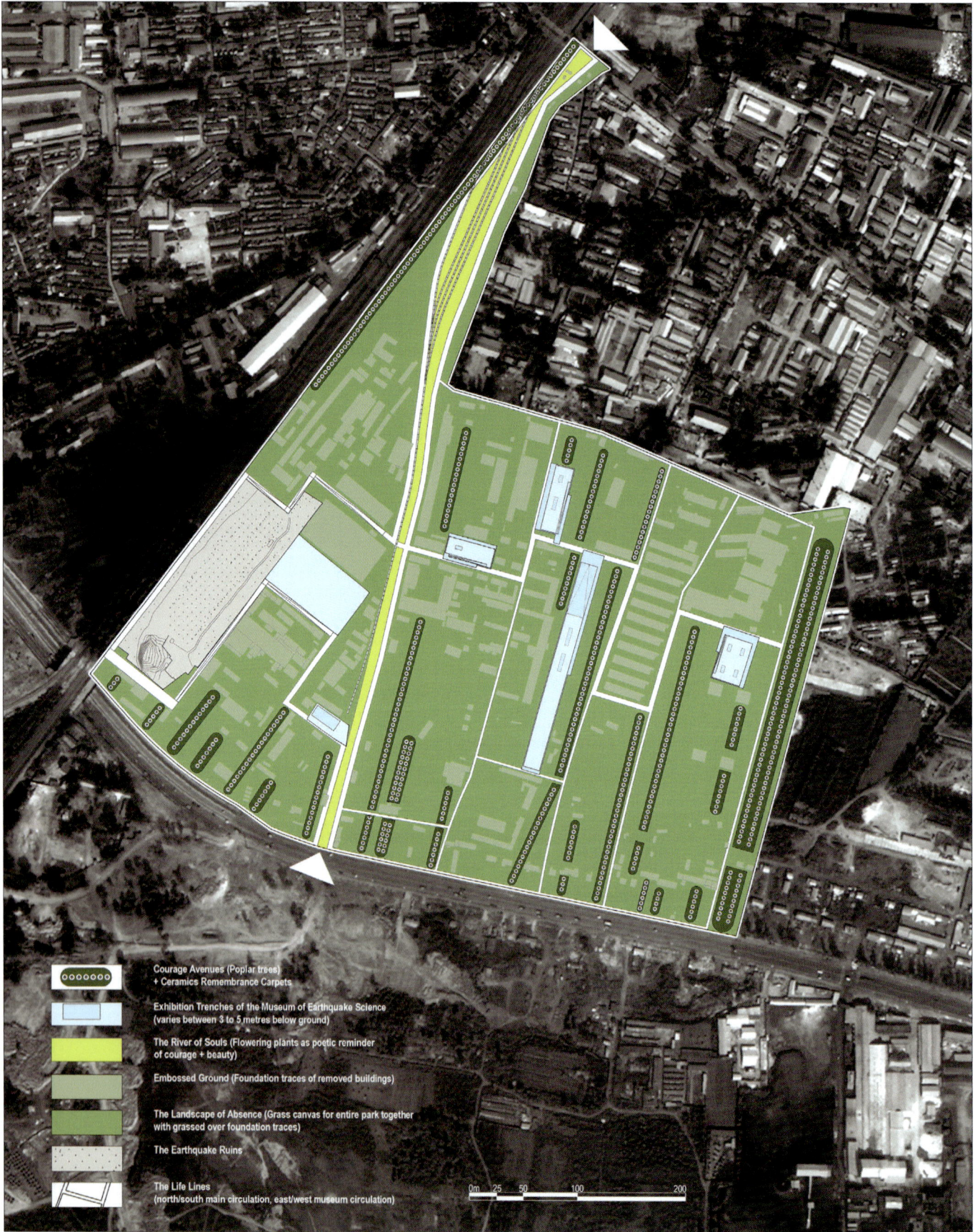

Courage Avenues (Poplar trees)
+ Ceramics Remembrance Carpets

Exhibition Trenches of the Museum of Earthquake Science
(varies between 3 to 5 metres below ground)

The River of Souls (Flowering plants as poetic reminder
of courage + beauty)

Embossed Ground (Foundation traces of removed buildings)

The Landscape of Absence (Grass canvas for entire park together
with grassed over foundation traces)

The Earthquake Ruins

The Life Lines
(north/south main circulation, east/west museum circulation)

0m 25 50   100   200

Tangshan Earthquake Memorial Park China

'We planted the trees to save the children in the school. We are still planting the trees because we are still worried about our children. We are planting the trees because there is nothing else we can do. See? We are not crying here, we are planting trees.'[4]

– Haitian villagers, post Hurricane Emily 1987, 'Greening in Red Zone', 2015

A year had already passed since the hurricane, and yet these women and older men were still planting trees on the uphill side of a school that clung precariously to a precipice, closed ever since the disaster hit. Seen as 'a waste of time' by the rest of the community, the reason behind why they planted became poignantly clear at the very moment when the building collapsed and slid down, announcing the failure of their efforts. And yet, amid great cries and wailing, the tree planters waited only a few moments before returning, persistent and dogged, to their task. The mood spread; an impromptu collective formed by the hillside, planting among the rubble of the fallen school.[1]

The situation in Haiti is a motivating study about 'greening responses' post calamity – work recounting numerous cases from the greening of the Berlin Wall, to the planting of community gardens in war zones, from reforestation efforts in refugee camps to landscaping projects in correction facilities. The study points to the phenomena where 'brave people combine their own fate with that of the animal, tree, flower, forest or garden that lives or dies'.[2] This act of mingling our fate with the environment's is eloquent in art and literature. In multiple traditions from Chinese mourning poetry composed at sites of loss to the green language of 18th century romantic poets grappling with the onslaught of industrialisation, nature is the generous interlocutor that both speaks and listens to humans afflicted by catastrophe.[3]

The artistic record speaks to the ways disasters, experienced in sudden environmental cataclysms, or in the collective psyches of cities ravaged by disease or wars, or in the imperceptible yet inexorable impact of climate change, intertwine the agencies and destinies of human beings with each other and with the ever-changing world. Our countless conversations with nature are imprinted and painted in the archive of human thought and actions. Walt Whitman famously turned to a 'spear of summer grass' to forge an iconic American identity out of the tumult dividing the US nation in the period leading up to the Civil War.[4] Both literally and figuratively, the natural world aids and reflects the human capacity for creation and interconnectedness.

The literal and poetic acts infuse our environments with symbols that code the values of space referred to by French philosopher Gaston Bachelard as a practice of topophilia – the recognition and reading of

facing page: Tangshan Earthquake Memorial Park masterplan.

following page top + bottom: Axonometric views of the park.

1+2. KG Tidball & ME Krasny (eds.), 'Greening in the Red Zone: Disaster, resilience and community greening', Springer, 2015, p.xiii + p.xi

3. R Williams, 'The Country and the City', Vintage, 2016, p.127

4. W Whitman, 'Song of Myself', The Walt Whitman Archive, 1855 [https://whitmanarchive.org/published/LG/1891/poems/27], retrieved 9 July 2018

5+6. G Bachelard, 'The Poetics of Space', M Jolas (trans.), Penguin Books, 2014, Bachelard, p.48

spaces that must be 'defended against adverse forces, the space we love'.[5] Through 'poetic shadings' a place becomes 'eulogised space... seized upon by the imagination, [it can no longer] remain indifferent'.[6] From immigrants sewing seeds into the hems of their garments to children in Hiroshima preserving survivor trees in their playground, greening tactics in spaces sited with ruin activate these poetic shadings imbricated in the people's attachment to the space itself.[7]

Tangshan Earthquake Memorial Park employs symbolism to cast shades of meaning. Resilience scholarship stresses distinct human abilities that conceptual work of making and interpreting symbols enact: the ability to self-organise, to understand and cope with change, to appreciate diversity, to combine and weave various types of knowledge.[8] Symbolic interpretation of the natural world enables conceptual awareness that is essential to resilience because the mental faculties are urged towards open-ended, creative and autonomous acts of thought and from here, action.

A Chinese garden for the contemporary imagination, Tangshan Earthquake Memorial Park both draws from and elaborates an intricate tradition of representation. A traditional governing metaphor of the Chinese garden as microcosm of the natural world provides a malleable structure of meaning for the typical elements of rock, water, plantings and architectural elements to newly correspond to singular counterparts of the site and its history. Therefore, in Tangshan Earthquake Memorial Park, there are no statues, no monuments and no inscribed names to commemorate the 240 000 lives lost in the 1976 earthquake. However, each visitor is conscious of every turn of the past, from the earthquake ruins to the subtle ground imprints reminding the user of the geologic catastrophe. The river of flowers, the battalion of poplar trees and the ceramic memorial carpets, these collective vignettes metaphorically propagate the park's strategic ideas of absence and traces to uncover an archaeology of memories. Like the unrolling of a horizontal scroll painting, the visual experience of the space captures the incremental unfolding of meaning and is characteristic of Chinese gardens: 'rarely if ever visible at once, [the Chinese garden] is discovered scene by scene, one scene leading to the next which as it is revealed replaces the earlier and now invisible vistas which we first perceived.'[9] The park is read by way of 'nature revealing itself slowly, part by part, moment by moment'[10] which is traditionally valued over any one singular dominant picturesque scene.

Visitors are led by the 'lifelines', the two main circulation routes for the park. The north to south lifeline provides a direct route through the park connecting the two main entrances, running along the existing railway tracks and the river of flowers. The east to west exhibition lifeline gently navigates between the earthquake ruins and the exhibition trenches. The choreographed routes form a spatial dialogue with the earth, and navigate visitors amongst the traces of history and memories represented in the new landscape. At night, the two main paths are illuminated by recessed ground lighting. Viewed from above, the lit lifelines mark the location of the park within Tangshan city. Meanwhile, a series of secondary paths enable wandering and serendipitous encounters to catch glimpses of the temporality of light, seasons and a complex of sensorium.

The Museum of Earthquake Science is correspondingly structured as a dispersal collection of vignettes across the park. The exhibition is experienced not in one single building but through a series of open exhibition trenches carved into the earth's surface. Every surface of the exhibition trench except one is clad with local ceramic tiles – the remaining wall, through thick clear acrylic sheets, posits the mysterious conditions of earth's strata through a magnificient sectional cut of the ground. The new landscape further accomodates curations of cultural, historical and scientific interest in environmentally controlled glass cabinets, assigning an architecturally modest glass enclosure to house the ticket and

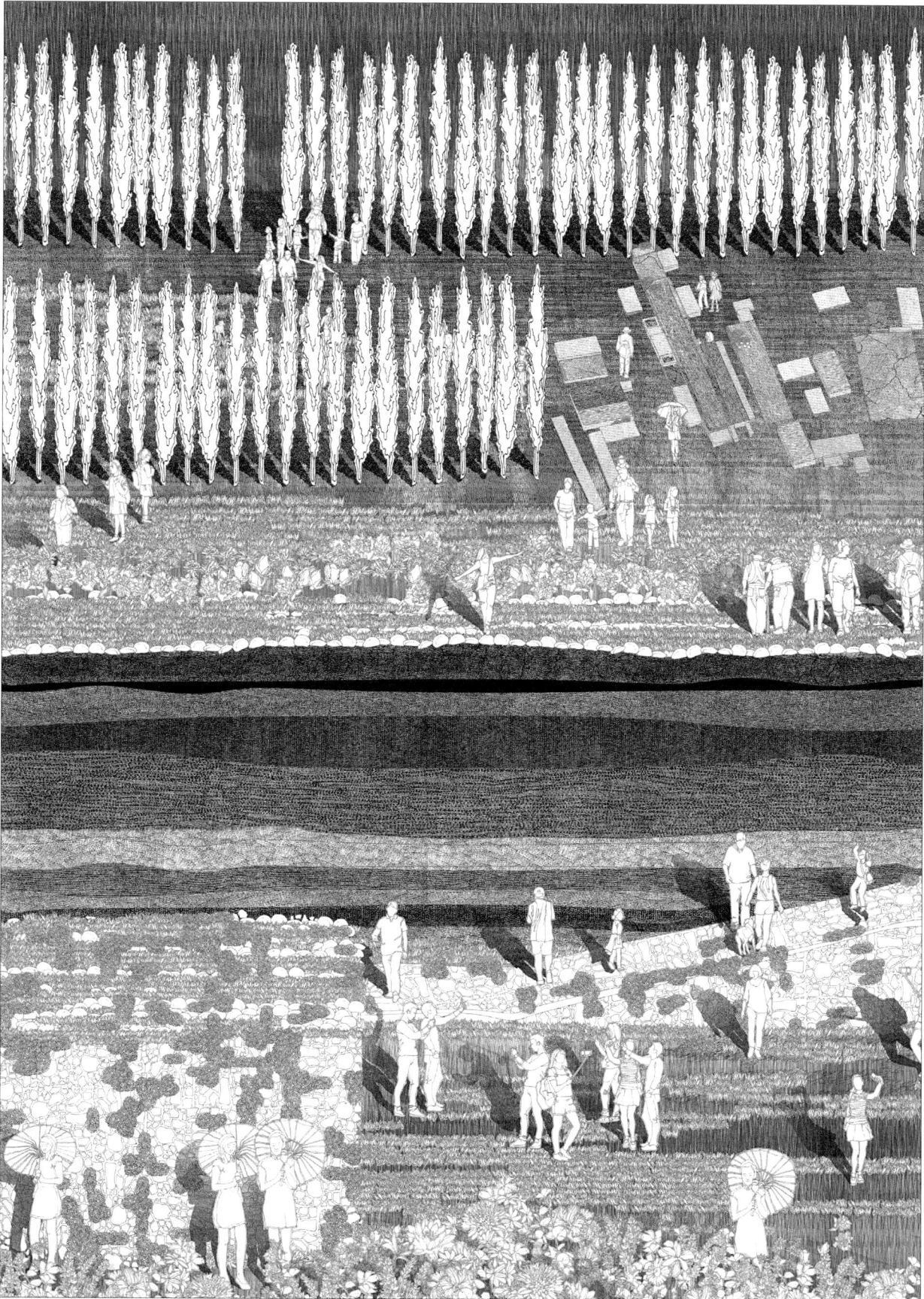

information office, gift shop and teahouse. Hundreds and thousands of pieces of broken ceramic from local factories are ceremoniously recycled to create a series of remembrance carpet pavings. The ceramic carpets provide areas for rest, shade and contemplation in the park – a metaphoric representation of the reconstruction of shattered lives and devastated city after the earthquake. In a gesture of reserve, the earthquake ruins, the main highlight, are left untouched.

Inspiring a self-aware practice in the use and occupation of space, the park greets visitors with the passage of time as a bricolage of human history and memory along with the geological. History, as a communally forged representation of the past, relies on consensual sharing of meaning by survivors and the generations thereafter. Memory, on the other hand, is a 'phenomenon tying [the survivor] to the eternal present'.[11] The park showcases a shared history, where coherence and recognition between human and forces of nature become possible. It also codes a shared memory, where imagination and expression blend the hurt and recovery of the survivors with the earth itself.

Apart from the preserved earthquake ruins, all remaining workshop structures and buildings on the site are removed; the structural foundations are the only traces remaining, and over time will be grassed over. This embossed jade-green canvas of the entire park connotes an embrace of the ghost of past activities, while presenting a resilient landscape to the city and its future generations. The bravery of lives lost in the earthquake is commemorated in the 950-metre long river-like flowerbed, within the existing railway tracks whose width varies from 20 to 60 metres. Additionally, the floral river carries heat exchange capillary tubes to elevate soil temperatures, creating a microclimate that allows an extended growing season. Thus, even in severe winter conditions, spring and summer flowering plants provide blossoms and scented fresh air: a poetic reminder of courage and beauty in adverse conditions. On the south, battalions of 'trembling' poplar trees stand upright, boldly protecting the park; the unwavering edge filters the noise and air pollution from surrounding highway traffic. The name evokes the way the poplar's leaves tremble in even slight breezes, generating a soft quaking sound – a poetic reminder of the wrath of nature.

The naming of natural features extends the traditional completion of the landscape through poetry. In Cao Xueqin's 'Dream of the Red Chamber', one the great Chinese classical novels of 18th century, Jia Zheng solicits his son Bao Yu to compose names and verses for his garden, maintaining, 'All those prospects and pavilions, even the rocks and trees and flowers will seem incomplete without the touch of poetry.'[12] The exuberant naming process did not conclude even after 20 pages, as each suggestion gives rise to counter-suggestions, emendations and embellishments. Such practices constitute a capacity to symbolically 'redefine the ruin as source of life'[13] – the withered tree becomes a motif of extraordinary energy and spirit in Chinese paintings precisely because 'while displaying signs of death and winter, they also offer hope for rebirth'.[14]

The history of the Chinese garden shows how its makers coded the natural world with values that allowed them to cope with the mutable and disenfranchised reality. The gardens had served as the site of refuge for the political elite to retreat from dynastic collapse, and a sanctuary for artistic expression for the contemplative philosopher – a spatial analogue juxtaposed against the failure of human order. In the modern garden, the gathering place is made more inclusive, where all can come to remember and collectively heal. In keeping with its genealogy in Chinese gardens, Tangshan Earthquake Memorial Park is a place that responds directly to the lived realities of upheaval and cataclysm as well as imagining an elsewhere only apprehensible as an ideal. The resilience of Tangshan's community is emblematised by the park, and what is captured in the landscape of absence is a new didactic space between past and future.

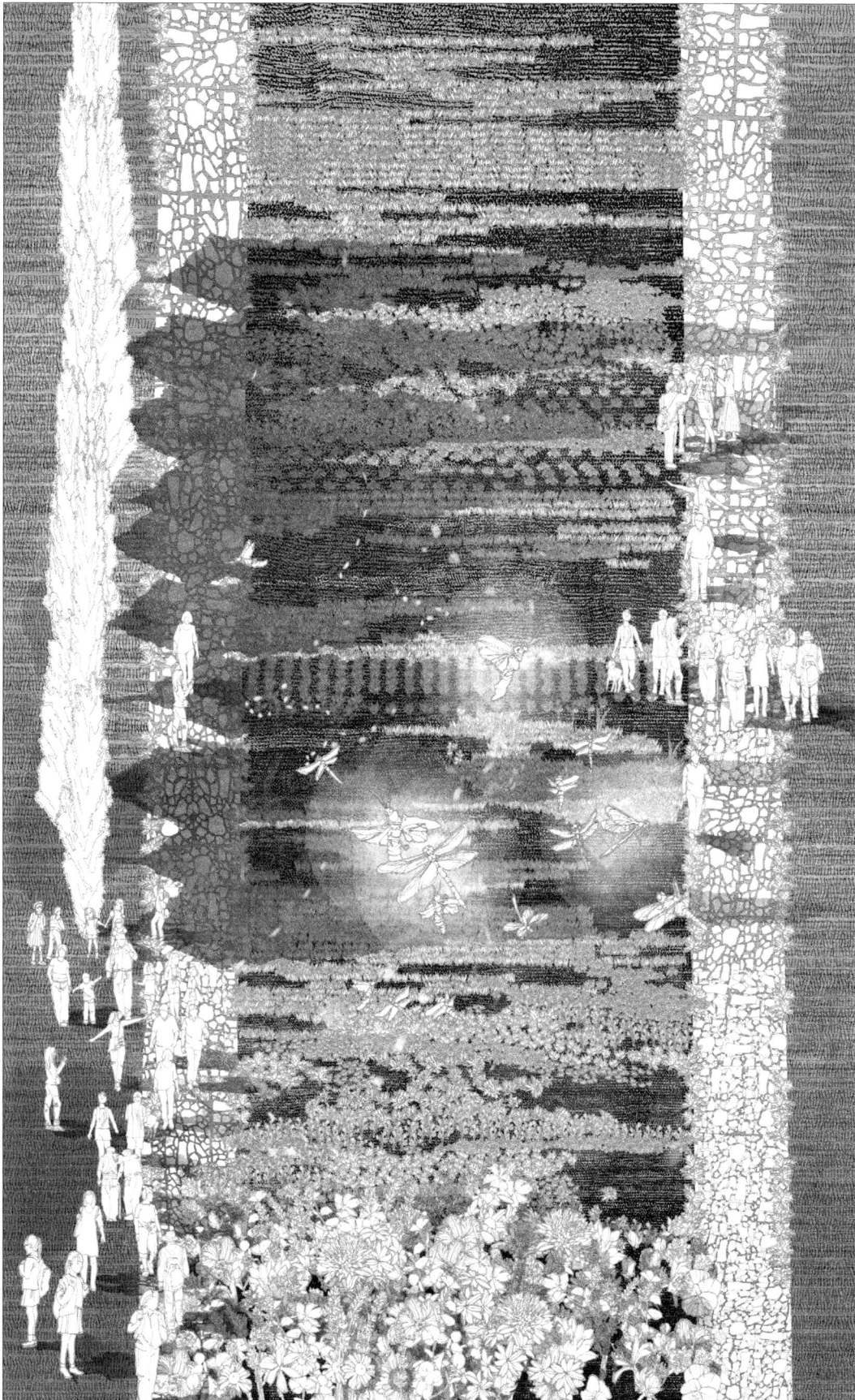

previous page left: Over time, the structural foundations will be grassed over, creating an embossed jade-green canvas.

previous page right: The earth's strata is presented through a magnificient sectional cut of the ground.

left: The floral river is a poetic reminder of courage and beauty in adverse conditions.

7+8. KG Tidball & ME Krasny (eds.), 'Greening in the Red Zone: Disaster, resilience and community greening', Springer, 2015, p.9 + p.27

9+10. C Thacker, 'The History of Gardens', University of California Press, 1997, p.43

11. H Wu, 'A Story of Ruins: Presence and absence in Chinese art and visual culture', Reaktion Books, 2012, p.38

12. C Thacker, 'The History of Gardens', University of California Press, 1997, p.50

13+14. H Wu, 'A Story of Ruins: Presence and absence in Chinese art and visual culture', Reaktion Books, 2012, p.172

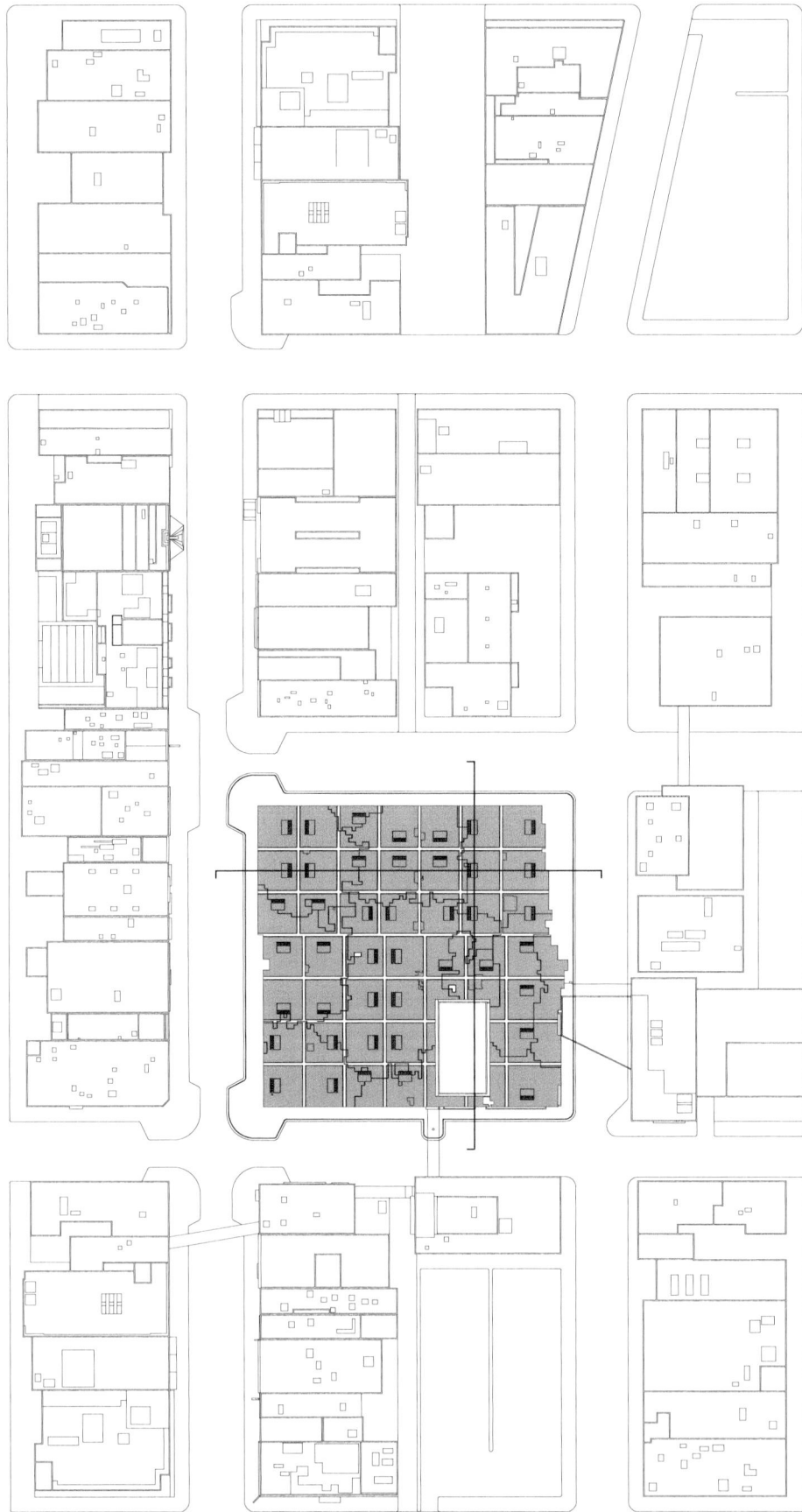

Remembering the Great American Plains USA

'At the time of the persecution when the Mormons struck out into the wilderness to find a place where they could worship God in their own way, the members of the first exploring party scattered sunflower seeds as they went. The next summer, when the long trains of wagons came through with all the women and children, they had a sunflower trail to follow. The legend has stuck in my mind, and the sunflower-bordered roads always seem to me the roads to freedom.'

– Willa Cather, 'My Antonia', 1918

In the novel 'My Antonia', the narrator's memories of the wild grasslands of the Great Plains pay homage to the foreign born whose grit transformed the wilderness of the prairie into settlements – the social pariahs and the escapees of persecution who in turn became the makers of the American Dream. One could imply that the recent phenomenon of urban farming in downtown Fargo possesses an echo of this agrarian history, as the motto of Little Free Garden project attests 'a lot can happen with a little land'.[1] These 21st century farmers, with modest plantings throughout the city, evoke the history of the Homestead Act of 1862 that settled the US wilderness with the belief that a quarter of land (160 acres) is what it takes to start a new life. Throughout the course of homesteading, over seven million people settled on 235 million acres of their own land, with the tangle of 'multiple, overlapping, interconnected, and mutually beneficial relationships' rooting the settlers into the land and into networks of shared expertise and care.[2] The contemporary longing for green space is tied to this remembrance of the great effort to turn land into homes. Collective memory becomes a perennial spring of awareness that leads to a community's resilience.

In downtown Fargo, the inherited relationship to the rural landscape is manifested in an urban-infill proposal for a square for spring, a square for winter. The new public squares focus an array of activities that negotiate modern day living, working environments and public place making at different scales, presenting urban configurations in a contemporary vision of the pioneering communities of 19th century America. The strategy includes changeable open spaces for the community at varying times. The adjustable physical states afforded by this innovative inhabited vertical landscape attracts people from outside the area, stimulating wider social capital and encouraging urban renewal. Drawing pedestrians in and up, bringing residents out and down, the two squares are symbiotic expressions of the longing for the wilderness of the great plains and the democratising possibility of the public square.

The rectilinear shaped site situated in the heart of Fargo shares graphical similarity with that of the actual county map in which the city resides. The spring and winter squares literally mimic the

facing page: Plan of the urban-infill in downtown Fargo, inspired by the Great Plains.

1. J Knight et al., 'Grow. Take. Share.', Little Free Garden [www.littlefreegarden.com]

2. R Edwards, 'Invited Essay: The New Learning about Homesteading', Great Plains Quarterly, vol.38, no.1, 2018, p.14

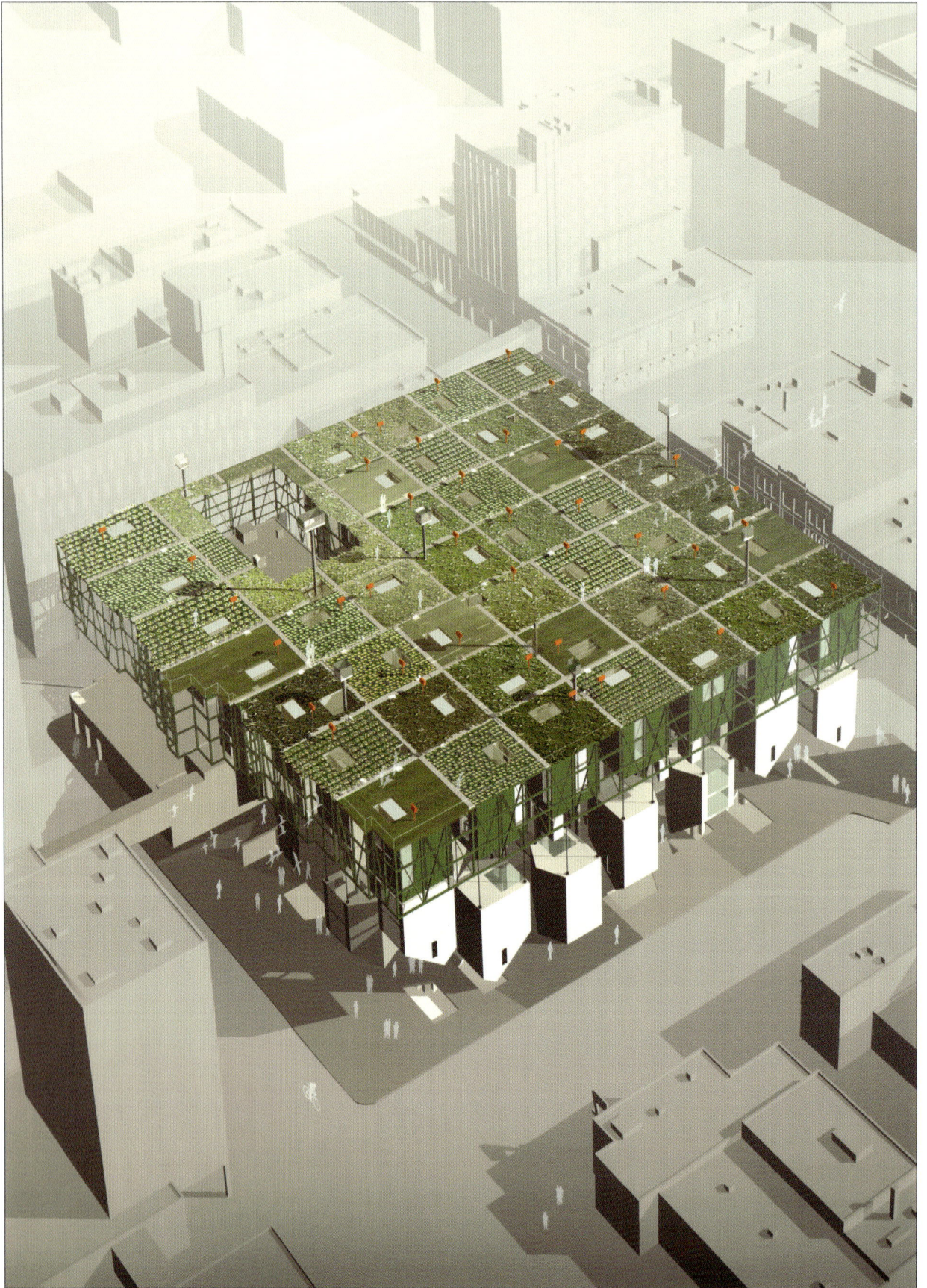

geographical hierarchy of Cass County in their formation, taking the local features of the county aerial map and suitably displaying them through their built form. Cass County townships, city locations, school district boundaries, railway lines, and agricultural field configurations have all been used as the basis for designing the proposal. The inimitable patchwork of agricultural fields is abstracted into a spatial logic that subdivides the square house plots, creating the grid that defines the plan of each house. Walls, doors, staircases, windows and bathtubs are all multiplications based upon the 'agricultural' grid, a move that applies the local landscape to create a clear well-proportioned urban habitat. It not only forces a new reassessment of conventional residential and retail typologies to produce a sustainable environment but also provides an opportunity to physically re-affirm the city of Fargo as the geographical crossroads and economic centre of southeastern North Dakota.

The ground-level 'winter' square provides sheltered commercial retail and office space, and multifunctional flexible space for a retreat from severe winter conditions. The public landscape encourages local markets and cafés to convene on the weekends in the centre of the city. The open nature of the square facilitates pedestrian activities between 5th Street and Broadway while integrating street connections in Downtown Fargo. Use begets use, as American urbanist William H Whyte found – the chatting on café chairs or chance encounters at the market stands – in other words, the diverse uses of space turn into proprietary interests that help preserve the publicness of the square.[3]

The 'spring' square on the roof has a different character; it is a green habitat offering tranquil temporary withdrawal from the frenetic noise and urban activity. The plan encourages biodiversity and invites nature into a sustainable urban environment that reflects the vision for a new Fargo. The resilient landscape is a site of storytelling in the city – urban gardens, beyond food cultivation, are cultural and social expressions of loss, of aspiration for new beginnings or of transformation. As Pierrette Hondagneu-Sotelo's study of gardens reveals: culture and gardening practices have in fact defined landscapes as graffiti might announce a neighbourhood's personality or tattoos imprint the body.[4] By reconnecting with nature, Fargo residents have an opportunity to mark the diversity of their culture and heritage; the urban landscape broadens community awareness not only of their own pasts, but also of each other's.

The roof landscape draped with allotments has a continual symbolic relationship with the county's surrounding plains and is the crowning emblem of the built articulation of nature. A theatre for the seasons, the roof becomes a year-round staging ground for ecological assemblages, as varied plantings bring in the ecological diversity inherent in the county. Fargo's short growing season, in fact, attunes urban gardeners to the constraints and also possibilities of their micro-prairie. If seeded with juniper, highbush cranberry, native roses or the Maximilian sunflower that retain fruit and seed well into the winter, for example, the gardens will come alive with the warbler's birdsong or the swallow-tail's rustling flight in the bush. Evanescent colour from purple clover or blue blanket flower can blend with the greens of the buffalo grass or dropseed – cueing the motion of the prairie biota immanent in the environment.[5]

Furthermore, the green roof also responds to the environmental needs of the other spaces, by providing shade from the hot summer temperatures and shelter from the winter snow for the public square

facing page: Aerial view of a resilient landscape – the 'spring' square.

following pages: Diverse lives in the inhabited vertical landscape.

3. SJ Kayden, 'Privately Owned Public Space: The New York City experience', John Wiley & Sons, 2000, p.302

4. P Hondagneu-Sotelo, 'Paradise Transplanted - Migration and the making of California gardens', University of California Press, 2014, p.4

5. USDA Natural Resources Conservation Service, 'Living Landscapes in North Dakota: A guide to native plantscaping', USDA NRCS, 2006, pp.9–16

1. James approaches the city...

2. ...through traffic to a downtown urban block

3. Finally arriving back at his home, James drives into the underground car park

4. He locates a parking space and gets out of his car, gazing across the light filled car park

5. A hydraulic lift will take him to...

... the 'Winter Square'

7. In the vibrant 'Winter Square', a Saturday market is taking place

8. James walks towards the suspended shop

9. He enters a florist...

10. He finds a Poppy plant that he can plant with his housemates in their wild meadow

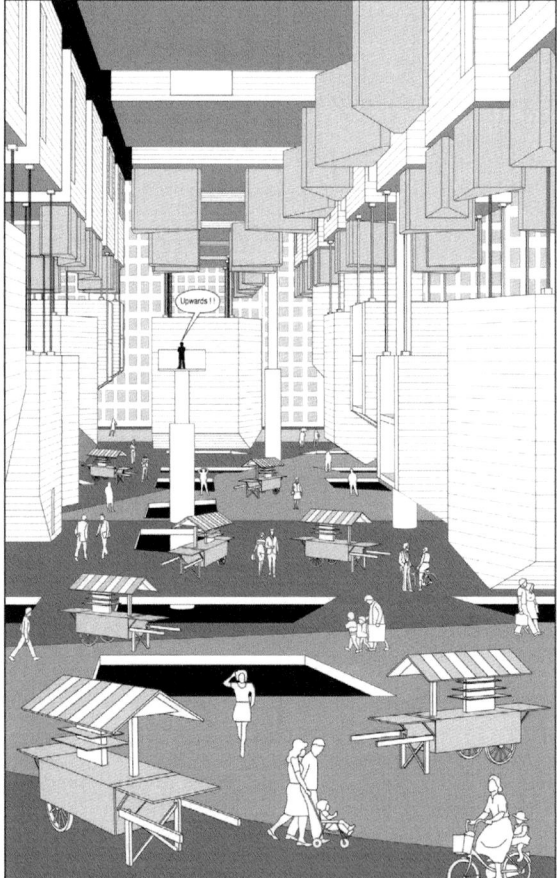

11. He ascends through the urban atrium looking back across the bustling market

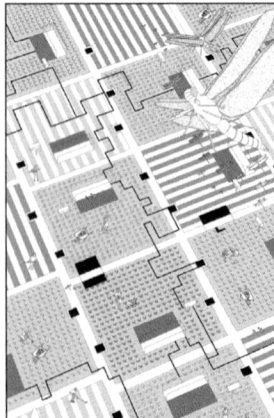

12. Excitement builds as he is lifted upwards!

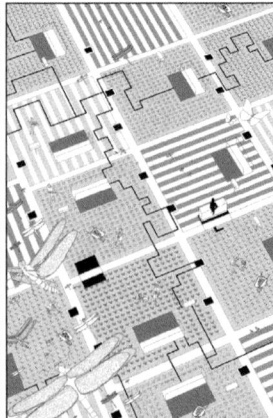

13. ...and amongst the colourful meadows

108

14. James gets out of the lift and walks amongst the agricultural fields to find his apartment

109

17. His housemates wave from the lounge as he arrives through the top of the house

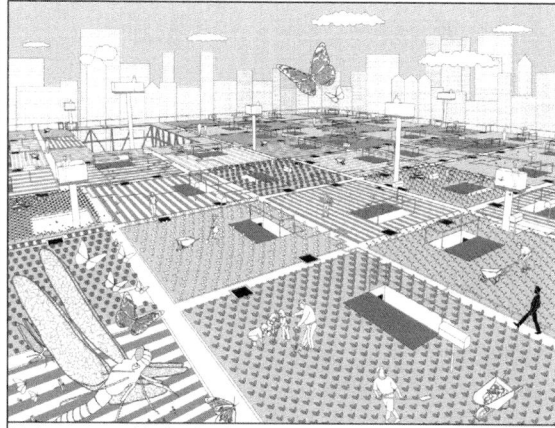

16. Found dragonflies downstairs

15. The entrance is amongst the meadows

18. With his housemates, James planted the gift in the meadow of the 'Summer Square'

19. College begins in September!

20. Through the long hard days of October...

21. During November...

22. Slept through the chilly autumn nights

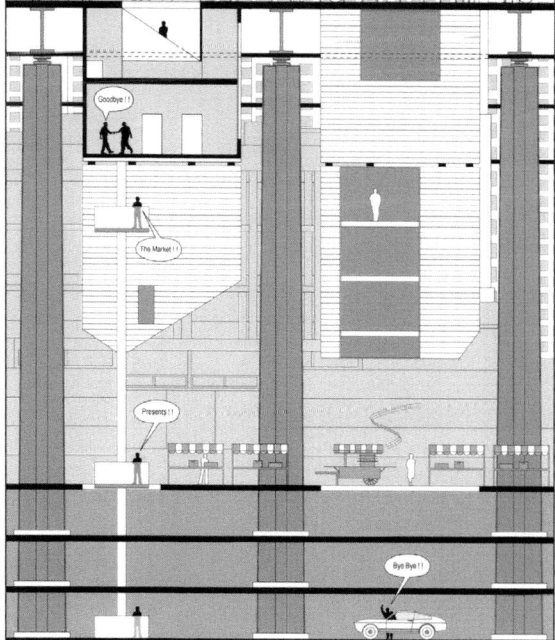

23. James closes his books and leaves his desk for the Christmas holidays!

24. James bought Christmas gifts in the 'Winter Square', before leaving for home!

beneath. The roof reduces heat penetrating to the living spaces while collecting and recycling melted snow water for the residential units and allotments. A series of gaps between each residential unit allows a rich amount of sunlight to permeate between the individual growing plots, projecting a calming atmospheric light to the winter square below. Prevailing wind filters through the roof to facilitate cross-ventilation for the entire development and extend maximum visual permeability across downtown Fargo. The vibrancy of public spaces relies on movement and environmental comfort, and has the potential to acquire diverse social and spatial possibilities, if reimagined as a form of symbiosis between human and nature. The winter and spring squares encourage a dynamic three-dimensional matrix of everyday movement and connectivity: from people to cultivated food, even to the flight of pollinators and migrant birds, all express a blended urban choreography set to the seasonal music of the North Dakota landscape.

111

While the Hanging Gardens of Babylon is only locatable in myth, the tale of its existence nonetheless compels listeners with its story of intertwined love and longing. It hardly matters whether Nebuchadnezzar II really did build the 'pensile paradise' in the air for Queen Amytis of Media. What persists is the notion that homesickness for nature brought the wondrous urban landscape to being. Nebuchadnezzar's ingenuity, it is said, lifted the very trees skyward to assuage Amytis's desire to see the mountains of her home, seeding the high walks on stone pillars with plantings of fragile wildflowers, making the terraced wonderland fragrant with the roses of her youth.[6]

As political activist Simone Weil pointed out, 'to be rooted is perhaps the most important and least recognised need of the human soul'.[7] Longing for home – nostalgia, from 'nostos' for home and 'algia' for longing – is what one reads in the queen's desire for nature in the city. Imagine the future of the Rust Belt regions if 'Make America Great Again'[8] could re-establish a longing for the nation's lost landscape?

facing page top: The 'winter' square from spring to summer.

facing page bottom: The 'winter' square during winter months, adapted for community activities.

bottom: An urban development invoking the nation's lost landscape.

following page left: The 'spring' square from spring to summer.

following page right: The 'spring' square during winter months.

6. M Soderstrom, 'Green City: People, nature and urban places', Véhicule Press, 2006, p.2

7. S Weil, 'The Need for Roots: Prelude to a declaration of duties towards mankind', Routledge, 2010, p.40

8. 'Make America Great Again' is the 2016 presidential campaign slogan popularised by Donald Trump.

Floating Mobile Hubs - Organic Farmers Market,
Town Hall + Library, Arts + Education Centre

Suburb Squares - Main Street with Shops and
Municipal Facilities located around Water Reservoirs

Commercial + Office Towers - Towers ranging from
30 to 100m with 3 Floors of Commercial and
Office Space on top

Housing Blocks + Villas - Each Block houses an
average of 50 Flats (ranging from 40 to 120m2)
and 5 Villas (160m2 each) for a total of 170 residents

Nursery

Nursery Playground

Flower Garden

Swimming Pool

Playground on top of Lawn Ribbon-Platform

Flower Beds on top of Lawn Ribbon-Platform

Skybus Stop

Sports Fields

Arable Kitchen Garden

Lawn Ribbon-Platform with Suspended Monorail Tram
System (Biogas Skybus) connecting the Suburbs

Fruit Orchard

Beach

Board Walk - extending along the embankment and
culminating into the Centre; it is also the Emergency
Access. Buggies provide Services and Transport.

Road - leading to Underground Car park

Urban Plaza - stretching across the Centre, it provides
a flexible, pedestrian Platform for Towers and
Housing Blocks

0m   125   250   500

Nordhavnen Smartcity Denmark

'I have only one dream. It is the oldest of humanity, of man, in time. It is paradise. I would like to give paradise to everyone.'

       – Frei Otto, Pritzker Prize, 2015

Nordhavnen, a harbour area on the coast of the Øresund in Copenhagen, represents Scandinavia's largest metropolitan development project. Covering 200 hectares of land reclaimed in the late 19th century, Nordhavnen is surrounded by water on three sides and is the potential site for Northern Europe's first Smartcity.

Nordhavnen Smartcity takes the philosophies and infrastructural elements of Guangming but situates them within a Western context on a site defined by water. In contrast with Shenzhen, Copenhagen is already a world leader in sustainable living, and hosted the 2009 UN climate change summit. The National Technical and Environment Administration has recently formulated a strategy aimed at making Copenhagen the world's leading environmental capital, and Nordhavnen is seen as a key development in this ambition.

The population of the city is expected to increase by 45 000 by 2025 and will require new housing, workplaces and community facilities to counter the undesirable trend of increased commuting in the region and to reduce car traffic congestion. Nordhavnen Smartcity will provide housing and workplaces for 40 000 people and herald a new breed of individual, the citizen farmer.

Denmark is a net exporter of food and has a tradition of allotment or colony gardens stretching back to the 18th century. Legislation in 2001 secured the future of communal gardens by granting them 'permanent' status, which protects them from urban development. The implementation of urban agriculture at a truly urban scale, however, would introduce a new dimension and further environmental synergies to programmes in low-energy transport, waste recycling and wind-based facilities that are already world-leading. The Smartcity will not be suburban in character; it will overlay a rural blanket onto the form of a city, benefiting both from the vitality and social bonds of dense diverse communities and from the recreational wellbeing that comes from large tracts of open space.

The increased local sufficiency engendered will lead to a modification of attitude towards nature and communities, and make use of immigrants' expertise in cultivating new non-indigenous cuisine. Differences in age, race, wealth and class can be celebrated instead of being the cause of friction.

facing page: Nordhavnen Smartcity masterplan.

following page: Aerial views of the Smartcity on the Øresund coast.

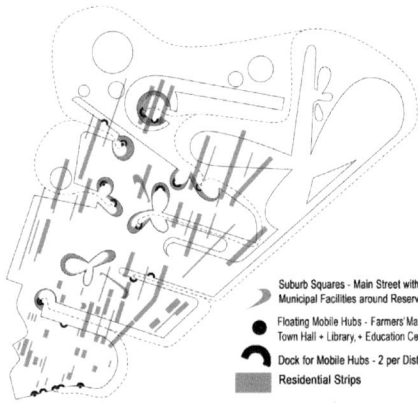

Suburb Squares - Main Street with Shops,
Municipal Facilities around Reservoirs

Floating Mobile Hubs - Farmers' Market,
Town Hall + Library, + Education Centre

Dock for Mobile Hubs - 2 per District

Residential Strips

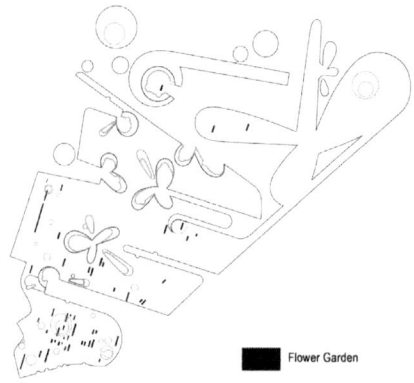

Flower Garden

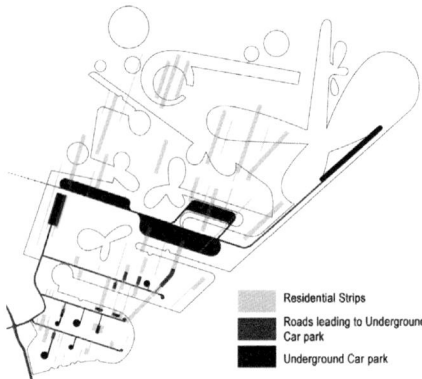

Residential Strips

Roads leading to Underground
Car park

Underground Car park

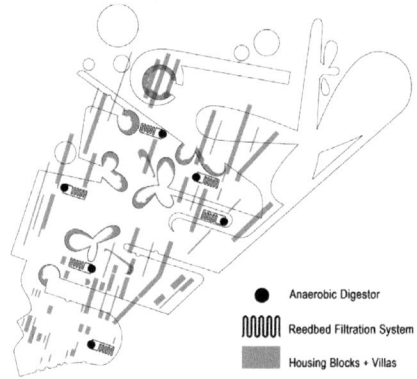

Anaerobic Digestor

Reedbed Filtration System

Housing Blocks + Villas

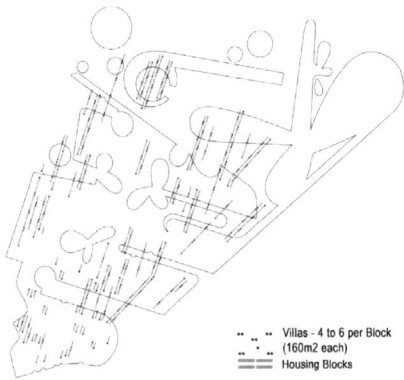

Villas - 4 to 6 per Block
(160m2 each)

Housing Blocks

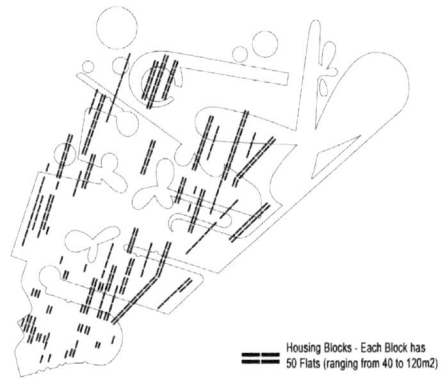

Housing Blocks - Each Block has
50 Flats (ranging from 40 to 120m2)

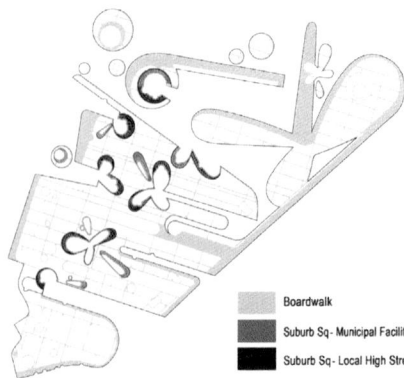

Boardwalk

Suburb Sq - Municipal Facilities

Suburb Sq - Local High Street

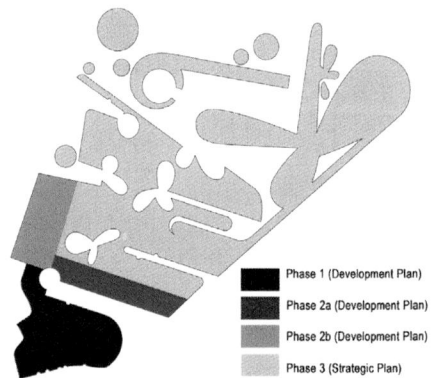

Phase 1 (Development Plan)

Phase 2a (Development Plan)

Phase 2b (Development Plan)

Phase 3 (Strategic Plan)

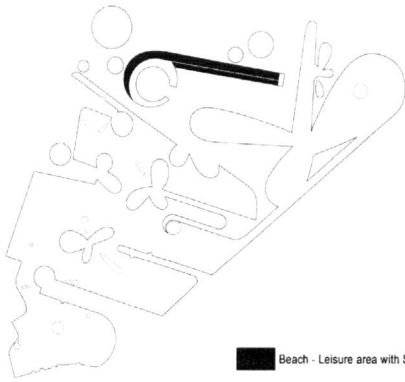

Beach - Leisure area with Sports

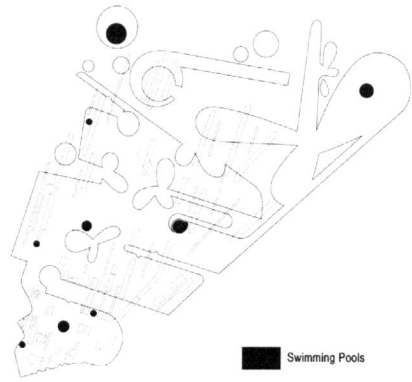

Swimming Pools

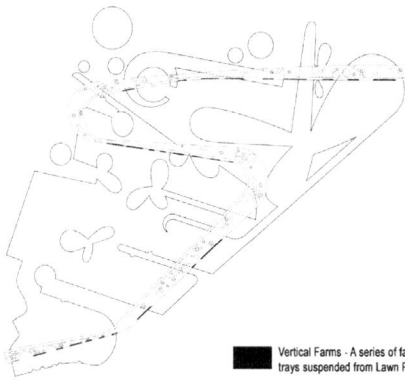

Vertical Farms - A series of farming
trays suspended from Lawn Platform

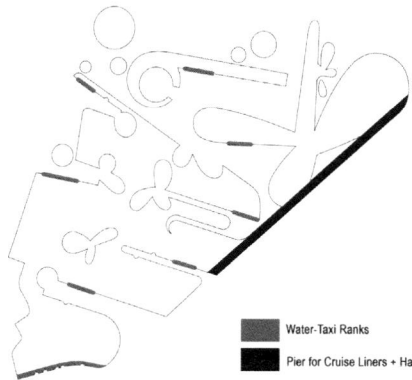

Water-Taxi Ranks

Pier for Cruise Liners + Harbour

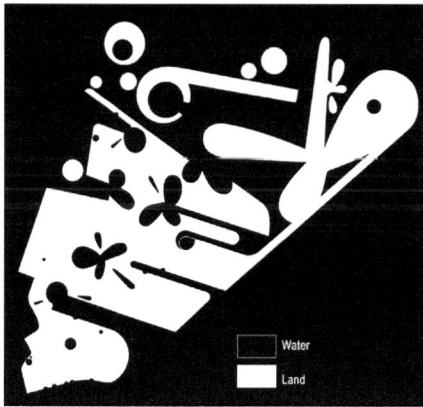

Water

Land

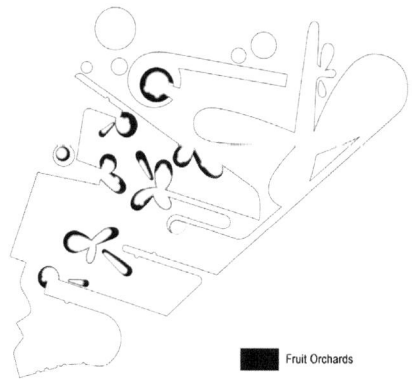

Fruit Orchards

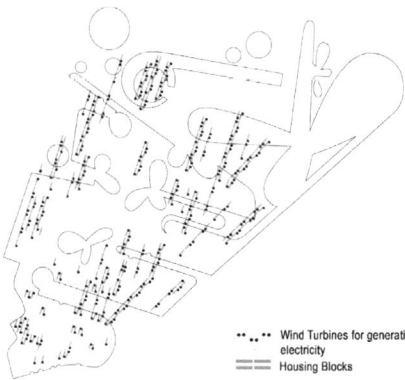

Wind Turbines for generating
electricity

Housing Blocks

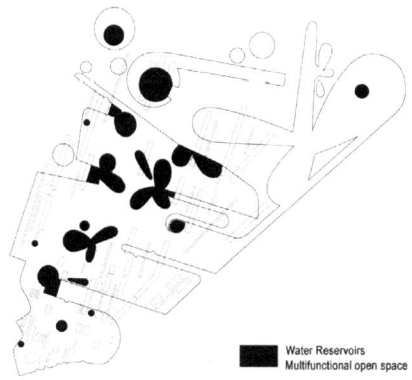

Water Reservoirs
Multifunctional open space

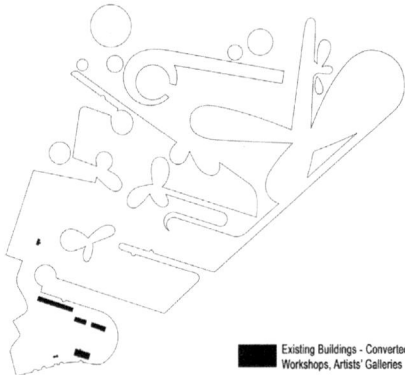

Existing Buildings - Converted into
Workshops, Artists' Galleries

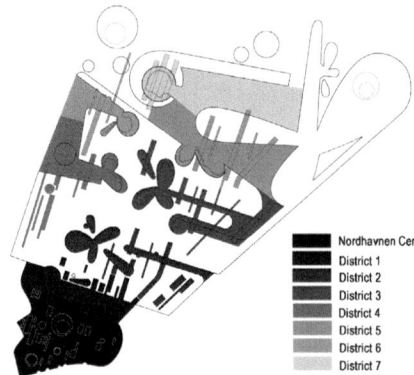

Nordhavnen Centre
District 1
District 2
District 3
District 4
District 5
District 6
District 7

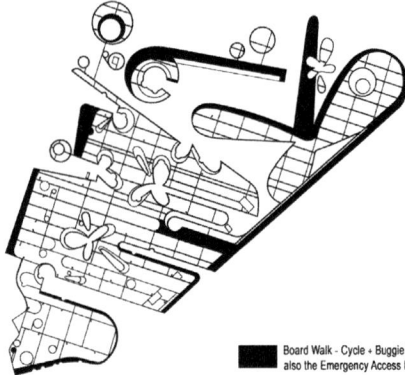

Board Walk - Cycle + Buggie Path
also the Emergency Access Route

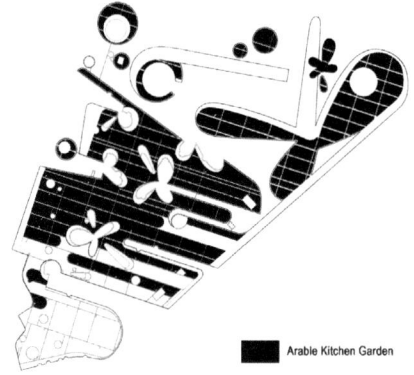

Arable Kitchen Garden

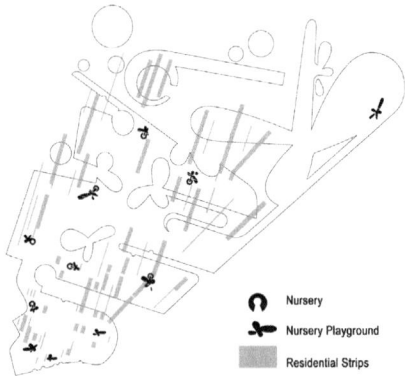

Nursery
Nursery Playground
Residential Strips

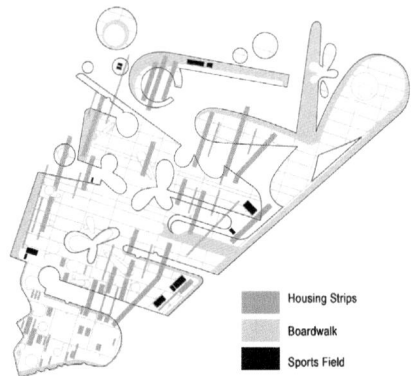

Housing Strips
Boardwalk
Sports Field

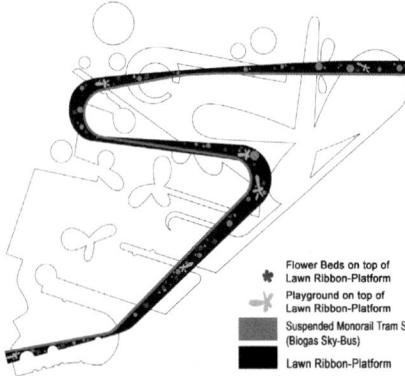

Flower Beds on top of
Lawn Ribbon-Platform
Playground on top of
Lawn Ribbon-Platform
Suspended Monorail Tram System
(Biogas Sky-Bus)
Lawn Ribbon-Platform

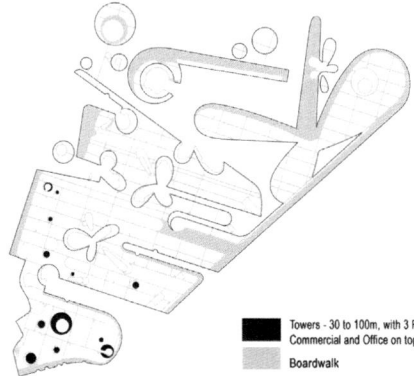

Towers - 30 to 100m, with 3 Floors
Commercial and Office on top
Boardwalk

City Framework

The water landscape of Nordhavnen has striking qualities that have the potential for conversion into unique recreational facilities. In outer Nordhavnen, Copenhagen meets the Øresund, offering views of the sea, Sweden, the historical fortifications of Copenhagen and a beautiful northern coastline characterised by beaches and forests. Currently, the site is used for harbour-based activities that will be retained and complemented by specialised infrastructure:

Arable Kitchen Garden Park: The new landscaped carpet of arable fields is carefully ordered both to provide a variety of environments and to take advantage of symbiotic sustainable polycultures. Over 70 per cent of the ground is dedicated to vegetable farming with occasional zones assigned to livestock grazing. The presence of generous scattered water bodies allows the establishment of a cyclical farming system similar to the mulberry dyke fishpond model used in China's Pearl River Delta.

The biomass of fast-growing shrubs will provide nutrients for arable crops with manure retrieved from fishponds where carp, perch and pike are farmed. Mild winters in Copenhagen together with nutrient rich water will combine to produce healthy yields of fish that receive their nutrients from the leaves of the cultivated plants. The integration of urban waste streams from the city's buildings will take this continuous culture a step further by permitting a real world application of the Integrated Food and Waste Management System (IFWMS) theorised by George Chan.

Irrigation of the arable land will require large quantities of water. Despite its omnipresence, water from the Øresund cannot be used to grow conventional crops and desalination is exceedingly energy inefficient. The kitchen garden park will therefore specialise in halophytic crops such as samphire and edible seaweed species including sea lettuce, carrageen moss, dulse and kelp. Such marine algae are a natural food, rich in gastronomic and nutritional qualities that are a key component of Asian cuisine and the European diet in the past. Samphire is rich in unsaturated oils and protein, making it additionally suitable for animal feed and biofuel. Effluent from the carp and perch will accelerate crop yields.

Live-Work Clusters: Residential and office developments are arranged into clusters overlaid above the newly cultivated landscape. The housing forms, inspired by the existing docks, are arranged as three-dimensional streets that are mixed tenure and encourage communal interaction. Each cluster is equipped with an energy station that controls a bank of PV arrays, wind turbines, and a combined heat and power plant that services and is fuelled by the adjoining farming zone. The tower components of the housing units contain villas for short- and long-term visitor accommodation.

Municipal Facilities: The communities of each cluster are served by a suburb square located along the canal that provides basic high street facilities, a reservoir and an event space that hosts a farmers' market. This urban arrangement encourages interaction within and between community clusters.

Ribbon-Platform: The live-work clusters are physically stitched together by a ribbon-platform elevated 12m above ground which is accessed by vertical circulation cores. Uninterrupted stretches of

previous + facing pages: Infrastructure plans of Nordhavnen Smartcity.

recreational green space on the platform provide surfaces for picnics, sunbathing and sport. An ideal vantage point to watch the shifting canvas of sea and sky, the raised manicured grass plane presents spectacular panoramic views over the productive landscape and the Øresund. Gymnasia, cultural facilities, agritourist accommodation and the biogas-driven Skybus are suspended from the ribbon. Hydroponic curtains drape the sides of the platform, yielding a further farming site.

Lifelines: The Nordhavnen development is virtually car-free. The main intra-regional transport system is the Skybus monorail that connects live-work clusters while defining new civic territories in the sky. The network of surrounding roads extends into the Smartcity via the commercial gateway district and rapidly softens into decked pathways that weave across the site. These are the primary lifelines providing circulation for both vehicles and pedestrians. A boardwalk next to the natural beach stretches along the coastline, creating a threshold between the land and water that culminates at the heart of the Smartcity.

The Gateway: Phase I of the development, containing the creative and business district, is located at the southwest edge of the site towards the centre of Copenhagen. Refurbished warehouses accommodate craft workshops and artists' studios that retain an industrial character and present a transitional threshold to the cultivated city.

Water Interfaces: The harbour and adjoining sand beaches along the northern edge of Nordhavnen will be a regional tourist attraction offering marine-based recreation including sailing and deep-sea fishing. The Øresund flows into the site via a series of water avenues traversed by water taxi. These are complemented by trefoil-shaped lagoons, whose convoluted form increases the length of the land–water interface and divides up the water bodies into discrete activities. Residents will be in contact with water immediately outside their doors; recreation in the Smartcity will be where people live rather than a destination. Visitors will have a different relationship with the water, jogging, walking, cycling or sunbathing alongside it, or swimming, rowing or paddling in it.

Mobile Hubs: Three mobile floating hubs containing an organic farmers' market, town hall and library, and an arts and education centre navigate around the canals providing services to each town cluster on a shared basis. These floating hubs double as mobile public plazas for events and periodically perform outreach visits to other parts of Denmark, spreading the ideals of the Smartcity.

Organic Food + Ecogastronomy: Nordhavnen would benefit from a twin town relationship with Guangming Smartcity, making it a North European capital for Food and Agriculture. Its location in Copenhagen has a number of inherent advantages where the ratio of organic food consumed is the highest in the world. Where every other European city has submitted to the monopoly of supermarket chains, Nordhavnen will offer an alternative by re-establishing traditional markets as a part of Danish culture and social communication. The city is also increasingly being recognised internationally as a gourmet destination and hosts Copenhagen Cooking, a food festival that takes place every August in various urban locations. Copenhagen's proximity to global cities and an international airport would make Nordhavnen Smartcity an ideal international conference location.

Transport: The ring road at the Smartcity's southwest boundary provides the main regional road traffic connection with Central Copenhagen. Secondary roads branch off the expressway into the site below ground level, connecting into public car parks. The Smartcity is also connected at this point with the main railway line between Copenhagen and Elsinore. Harbour buses and a new pedestrian bridge connect areas on both sides of the main harbour basin. The CityCirkel electric bus fleet will also extend its route into the commercial area.

Inter-local scale travel will be provided in the form of the Skybus monorail and a shared cycle network. Whenever a resident leaves their home or an employee leaves their workplace, he or she will pass a bicycle station before encountering a bus stop, followed by a metro station and lastly a car park. Copenhagen is already one of the most bicycle-friendly cities in the world and municipal policies are aiming for 50 per cent of all commutes to work, school or university being by bicycle by 2025. A fine mesh network of pathways surrounded by canals, basins and the coastline will be conducive to both cycling and walking, while water taxis ferry groups along the aquatic avenues.

Environmental Sustainability

The Danish Environmental Protection Agency (EPA) has reported that approximately 99 percent of waste from the food industry is recycled into animal feed or fertiliser. Treatment of solid organic waste from domestic households, however, has focused more on hygiene and protection of aquatic ecosystems than energy efficiency. Organic human waste constitutes just one percent of household waste by volume, but contains approximately 85 per cent of the nutrients. In usual urban environments, conversion of this waste into energy and fertiliser is unviable due to the distance between urban and agricultural locales. At Nordhavnen, human waste can be conveniently combined with organic farm waste and be broken down in anaerobic digesters to produce digestate to fertilise the adjoining farmland as well as biogas and biomass. The latter will be added to municipal solid waste to fuel combined heat and power plants local to each cluster community. Copenhagen already possesses one of the world's most efficient waste handling systems, recycling 90 per cent of construction waste and incinerating 75 per cent of domestic waste to generate electricity and district heating. The Smartcity will sustain and improve these efficiencies using local expertise.

Precipitation averages between 40mm and 70mm a month and will be captured and stored in freshwater zones of the trefoil lagoons. Greywater and blackwater will also be recycled for crop irrigation and should be sufficient in quantity, given that marine plants are also farmed and do not require fresh water.

Further renewable energy is generated through wind and solar collection. Smartcity residents will be eligible for tax exemptions by investing in local energy production. The prevailing winds at Nordhavnen are westerly or southwesterly, although from February to May and from October to November, there may be easterly and southeasterly wind. High energy potentials can be expected as the site is bordered by water on three sides. The construction of the cluster buildings is sufficiently robust to mount significantly sized turbines without causing adverse vibration effects.

The cluster buildings will also be equipped with photovoltaic arrays. The annual solar energy available per square metre of open surface areas facing south is a little more than 900kWh, ranging from 30 to 100kWh in certain months.

following page: View of Scandinavia's largest metropolitan development.

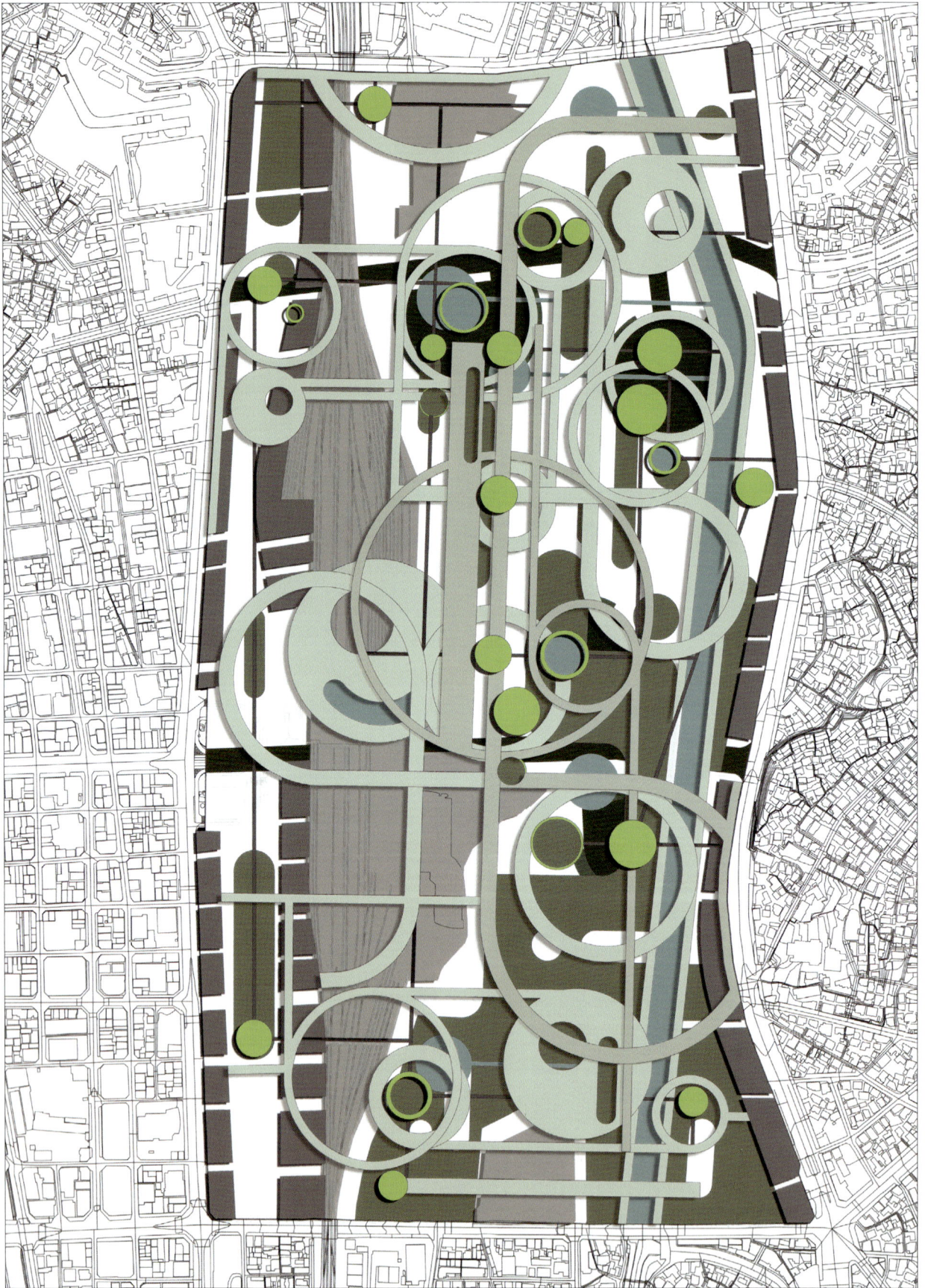

Daejeon Urban Renaissance South Korea

'The concept of vehicular/pedestrian segregation is now an accepted part of planning theory. But once one accepts this and the idea of multi-level single buildings, it is only logical to conceive of multi-level cities. The organisation of, say, New York, which tolerates multi-level components connected by only two horizontal levels (street and subway), and both of these at the base, is archaic.'

– Editorial, 'The Metropolis Issue', Archigram 5, 1964

The Daejeon Urban Renaissance (DJURe) zone in South Korea, covering an area of 0.89km2, suffers from classic doughnut or inverted concentric zone complex. Daejeon Old City District, once the thriving heart of the city, has since been overtaken by surrounding towns assimilated into Daejeon by the Government in 1990. The DJURe area is currently cluttered with single-storey detached commercial, industrial and residential buildings that have fallen into disrepair and are fenced in by four- and six-lane highways. Businesses and homes have migrated to newer suburbs outside the Old Town with superior infrastructure and flourishing communities.

At the same time, DJURe is the beneficiary of excellent transport links, containing a high-speed rail (KTX) and regional rail transit station linking Seoul with Busan, a subway line running across the city, and major road connections. A new scale of development is proposed for the regeneration of Daejeon's obsolete core. Occupying a far smaller area than Guangming Smartcity, DJURe will explore diverse vertical densities in order to free up open space for recreation and agriculture. Introducing new leisure programmes to conventional commercial, office and residential mixed-use towers will attract residents from outside the zone, preventing an island culture and stimulating wider social capital.

The height of the buildings in the DJURe Micro Smartcity will exceed that of its immediate surroundings, matching that of other major international cities, and will accommodate an increase in the district's population from 7500 to 13 000. The scale brings with it a new dynamic and diverse intensity, and presents a bold iconic image for Daejeon. This second wave of life and development will eventually infiltrate and be adopted by the surrounding towns as a model for sustainable urban growth and resilience.

The boundary of Daejeon Smartcity Centre is arranged into mixed-use human-scale boulevards that are conceived as the inhabitable walls of an urban courtyard garden. The walls contain residential and commercial developments overlooking a verdant landscape punctuated by a community of towers. At the centre of the urban courtyard is a city plaza made up of hard landscaping and multi-level green open spaces. Deadong Creek, running along a north–south axis across the DJURe site, feeds a chain of ponds

facing page: Daejeon Urban Renaissance masterplan.

following page right: Plans of the layered city.

following page left: Aerial views of an urban courtyard garden city.

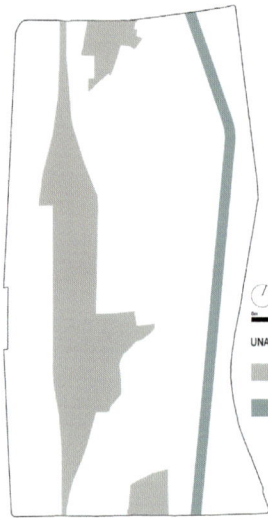

UNADJUSTABLE PLOTS

Existing buildings unadjustable plots

Existing canal on site

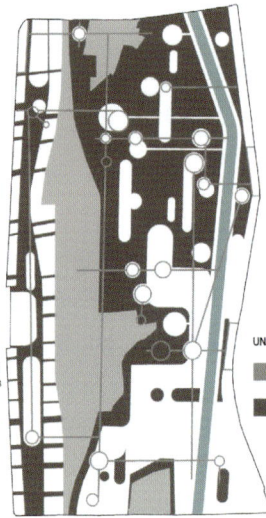

UNADJUSTABLE PLOTS

Structural Network including services

Public Plaza

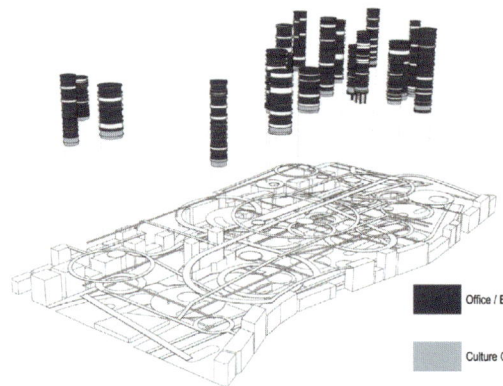

Office / Business Use

Culture Centres

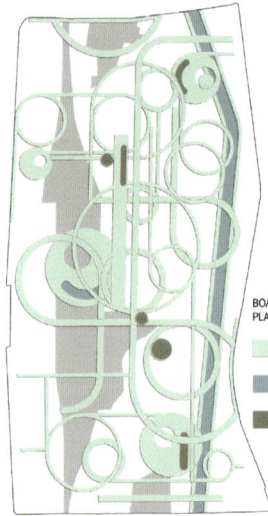

BOARDWALK/PUBLIC PLATFORMS

Agriculture Boardwalks

Ponds

Gardens

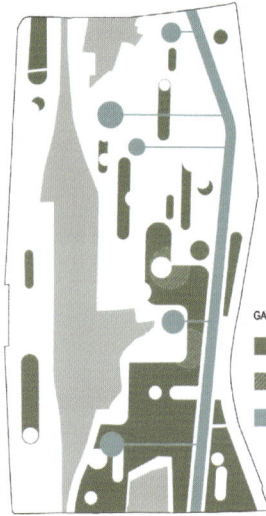

GARDENS / PONDS

Public Gardens at ground level

Green Voids

Water at ground level

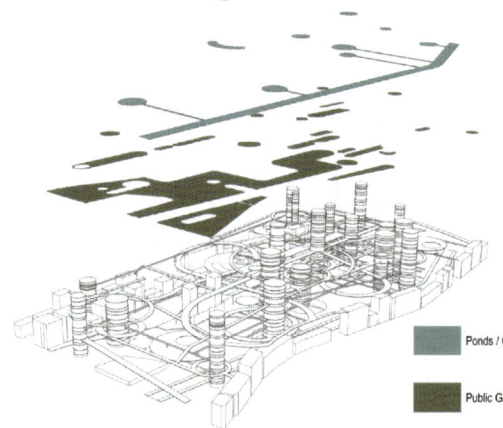

Ponds / Creek

Public Gardens

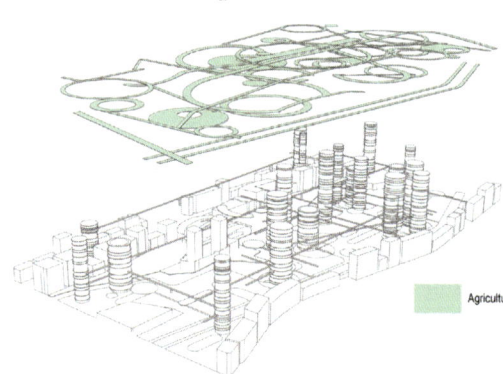

Agriculture Broadwalk

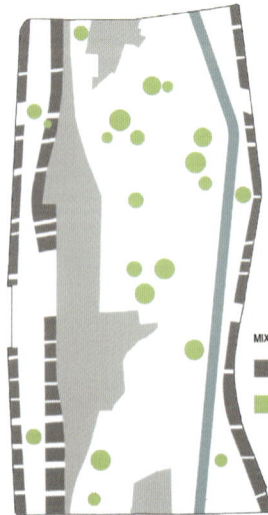

MIXED USE & OFFICES

Mixed-use Buildings

Multi-storey Office Buildings

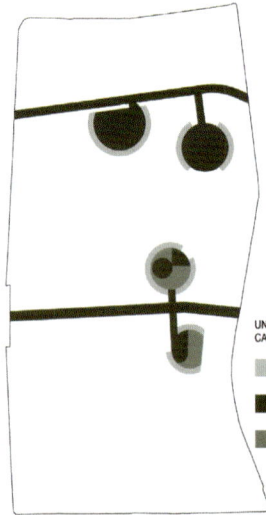

UNDERGROUND ROADS / CAR PARKS

Underground car park

Underground road network

Green areas at subterranean level

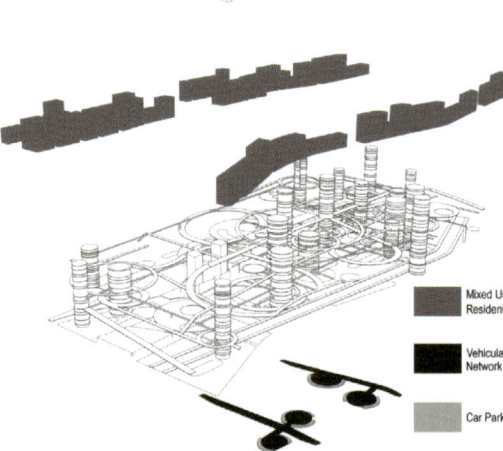

Mixed Use: Residential & Service

Vehicular Transport Network

Car Parks

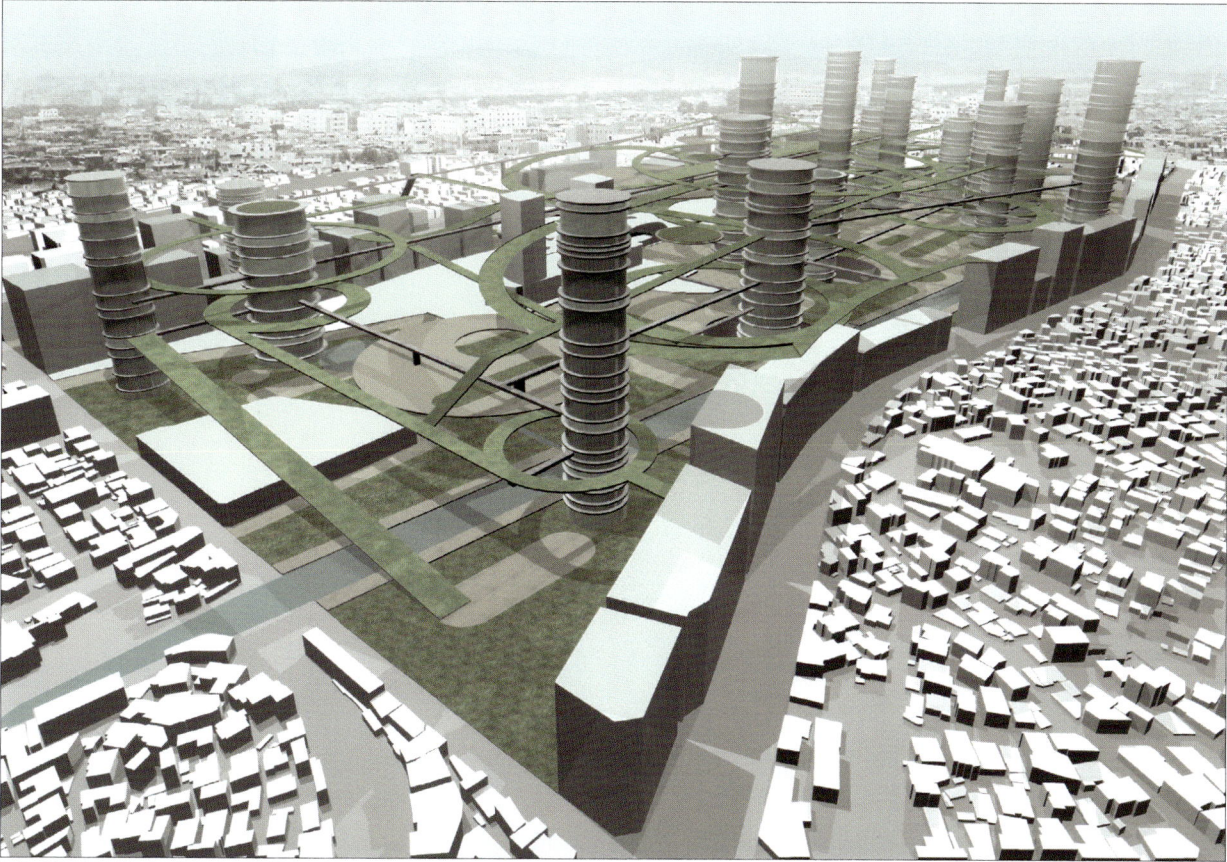

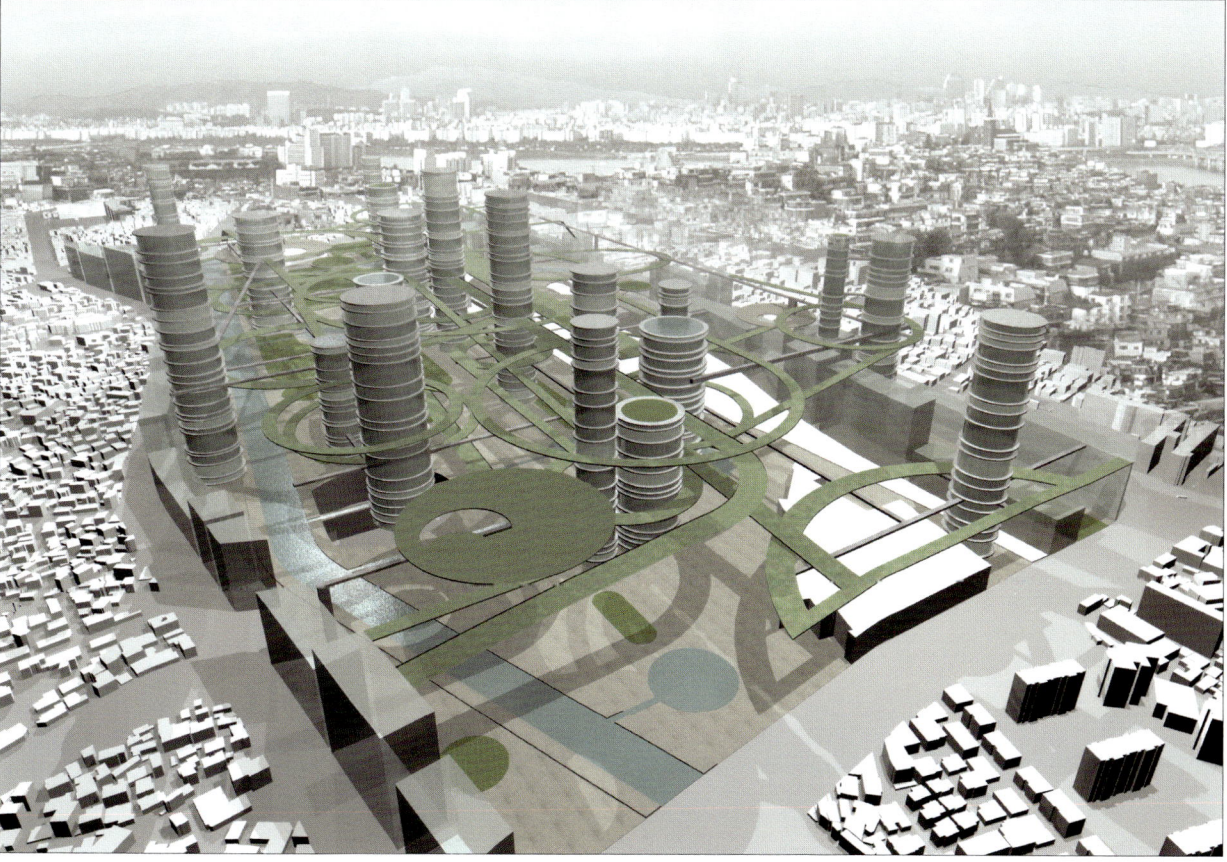

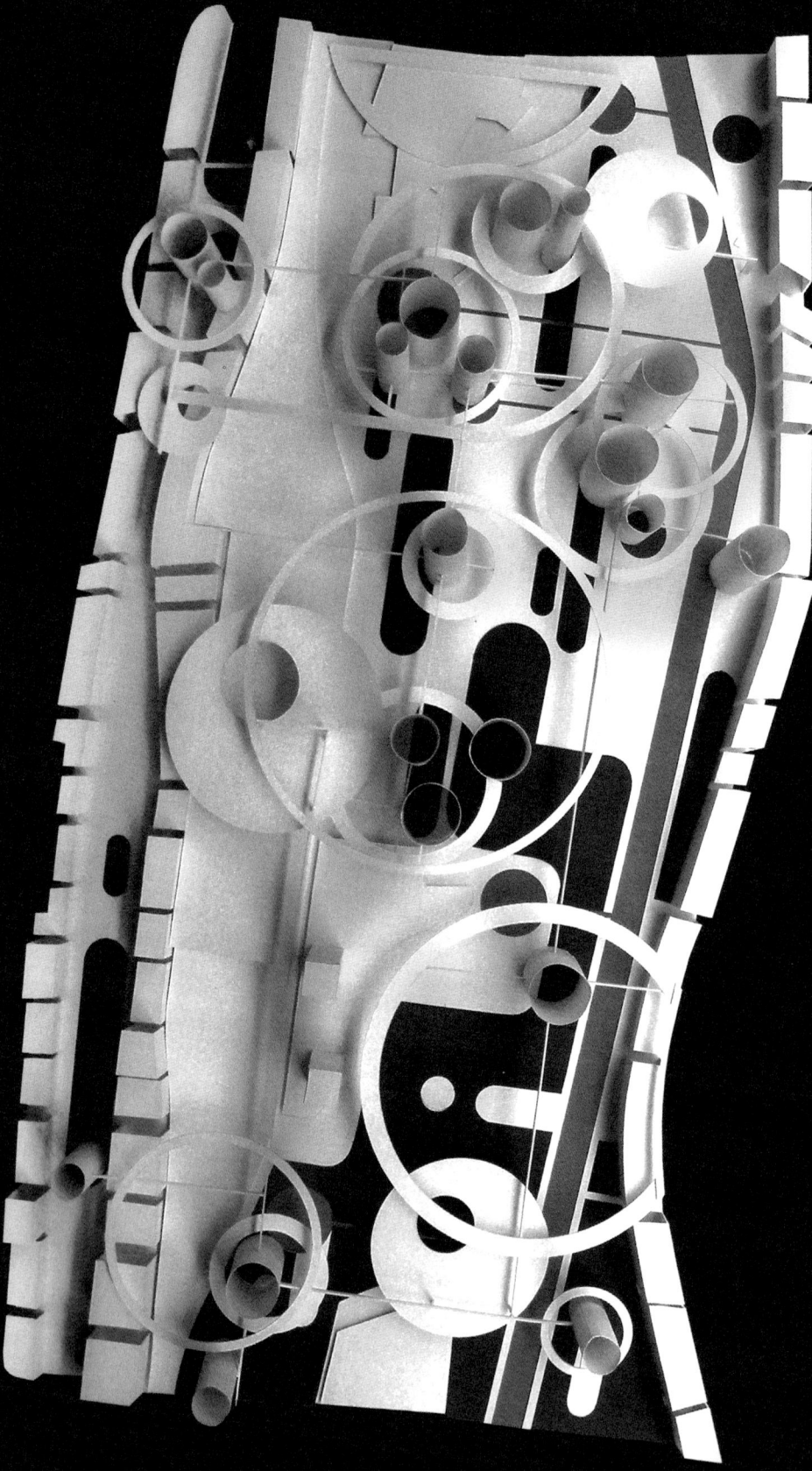

and reservoirs that store water for irrigation and maintain what precious biodiversity remains in the city. Accessed by foot through the inhabitable wall or via the underpasses, the city plaza is experienced as a secret garden, appearing as an unexpected oasis of vegetation and natural harmony within a concrete sea, while promoting vital cultural and commercial activities.

Twenty-one skyscrapers are distributed across the city centre and collectively they constitute the city's business hubs. Each tower contains a cultural facility and a sky plaza, linked at high level as well as on the ground. Associative function spaces such as the concert hall, music academy and recording studios are located next to each other to make use of beneficial adjacencies whilst ascribing individual identity to the vertical communities.

133

Le Corbusier imagined skyscrapers as 'streets in the sky', a compelling metaphor that has so far achieved only a one-dimensional reality. Streets are not islands; they connect to other streets, forming a lattice of complex relationships that engage in rich and unpredictable synergy. Even mixed-use high-rise buildings are connected to the rest of the city only at ground level; the disconnection from urban life increases as the storeys in a tower rise, culminating in a FTSE100 company boardroom or exclusive penthouse apartment.

facing page: Exploring vertical densities for agriculture and mixed-use programmes.

The network of elevated streets in the Central District of Hong Kong that connects flagship commercial office space, hotels, shopping malls and the general post office has proved remarkably successful in creating a second horizontal tier of public activity. This arrangement will soon be replicated in Shanghai's Pudong district where a ring-shaped walkway will link most of the skyscrapers in the main financial and business district. The primary objective in both these installations, however, is the separation of pedestrian and vehicular traffic. Consequently, the horizontal coupling exists only a storey or two above the existing ground plane and offers little in the way of establishing new public territory in a third dimension.

At DJURe, the towers will be linked at several levels by 'green' streets, incorporating SPIN farming and recreational activities to multiply social connections, as well as improving safety in terms of structural stability and fire evacuation. Sky gardens are located at major nodes at which more than two streets in the sky connect. The gardens are intensively planted and the elevated streets are draped with hydroponically grown vegetation to shape a green network that provides shading for activities in the urban courtyard below. The 'streets in the sky' that Le Corbusier envisioned will be horizontal as well as vertical, giving a civic shape to the urban development, reversing the monopolisation of the skyline by global capitalism. Jane Jacobs was resolute that 'the more successfully a city mingles everyday diversity of uses and users in its everyday streets, the more successfully, casually (and economically) its people thereby enliven and support well-located parks that can thus give back grace and delight to their neighbourhoods instead of vacuity'.[1]

1. J Jacobs, 'The Death and Life of Great American Cities', Vintage, 1992, p.111

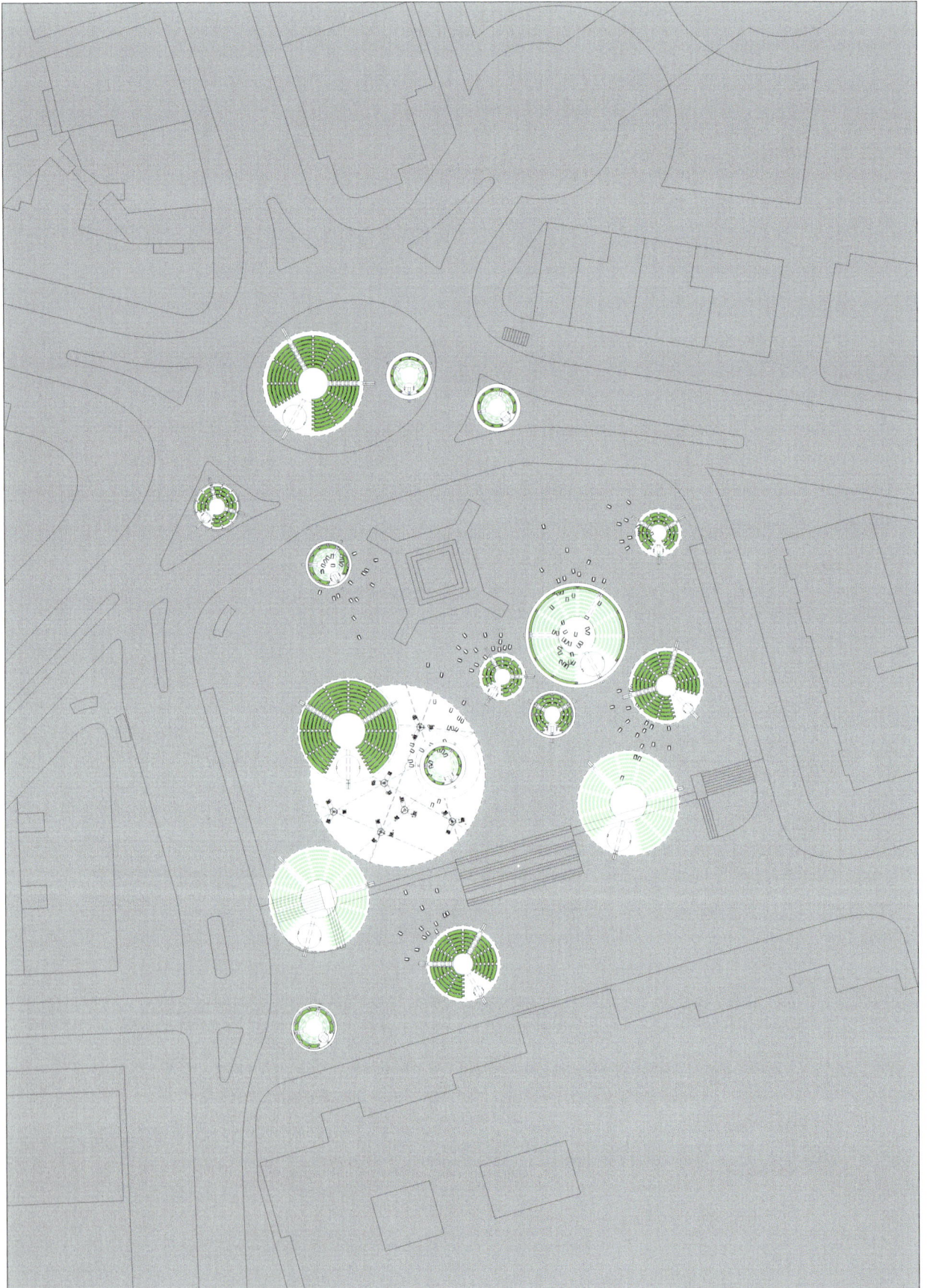

The Tomato Exchange UK

'A technological revolution, centred around information technologies, began to reshape, at accelerated pace, the material basis of society. Economies throughout the world have become globally interdependent, introducing a new form of relationship between economy, state, and society, in a system of variable geometry.'

– Manuel Castells, 'The Rise of the Network Society', 1996

135

Manuel Castells, widely regarded as the authority on communications research and information societies, has argued that the global city is not a single city – not London, New York, Tokyo or Johannesburg – but the fragment of each that is connected to its analogous counterparts in other world cities. Where suburban and rural areas have well-developed community infrastructures, the community of a metropolis tends towards a network of telematic relationships.

Community centres in the modern metropolis are a rarity. Conceived in the early 20th century in America, community centres were established to provide facilities for gatherings, group activities, social support and public information, premised on the idea that local communities are the permanent homes of most of their residents. The mobility and peripatetic nature of modern society has led to dispersion and diversity, challenging the relevance of the traditional community centre. Cities such as London celebrate their cosmopolitan spirit through events and processions throughout the calendar year, but these festivities tend towards the monocultural, and there is no permanent forum for intercultural exchange celebrating the ethnic diversity of world cities.

The Tomato Exchange is a 21st century community centre for the Smartcity that redresses this oversight in the form of 16 gleaming glass bell-like structures hovering over London's Trafalgar Square. The undercroft of these concave tubular towers provides shelter for sharing tomato-based delicacies and recipes specific to a variety of non-indigenous cuisines. Surrounded by circular tiers of sprawling solanum vines peppered with colourful fruit, scent and visual lushness pervade each campana structure.

Ranging in size from 5mm diameter tomberries to 5-inch 'big boys' and encompassing a colour spectrum of yellows, oranges, pinks, purples, greens, blacks and whites, tomatoes become the raw materials for intercultural exchange. Urban agriculture has a rich history in London, providing precious food in the world war victory gardens that took over public spaces such as Hyde Park and the moat around the Tower of London. In addition to providing food towards the war effort, the gardens demonstrated the worth of growing and sharing meals as a catalyst for social cohesion.

facing page: Plan of a 21st century global community centre – the Tomato Exchange.

following page: London's Tomato Exchange in Trafalgar Square.

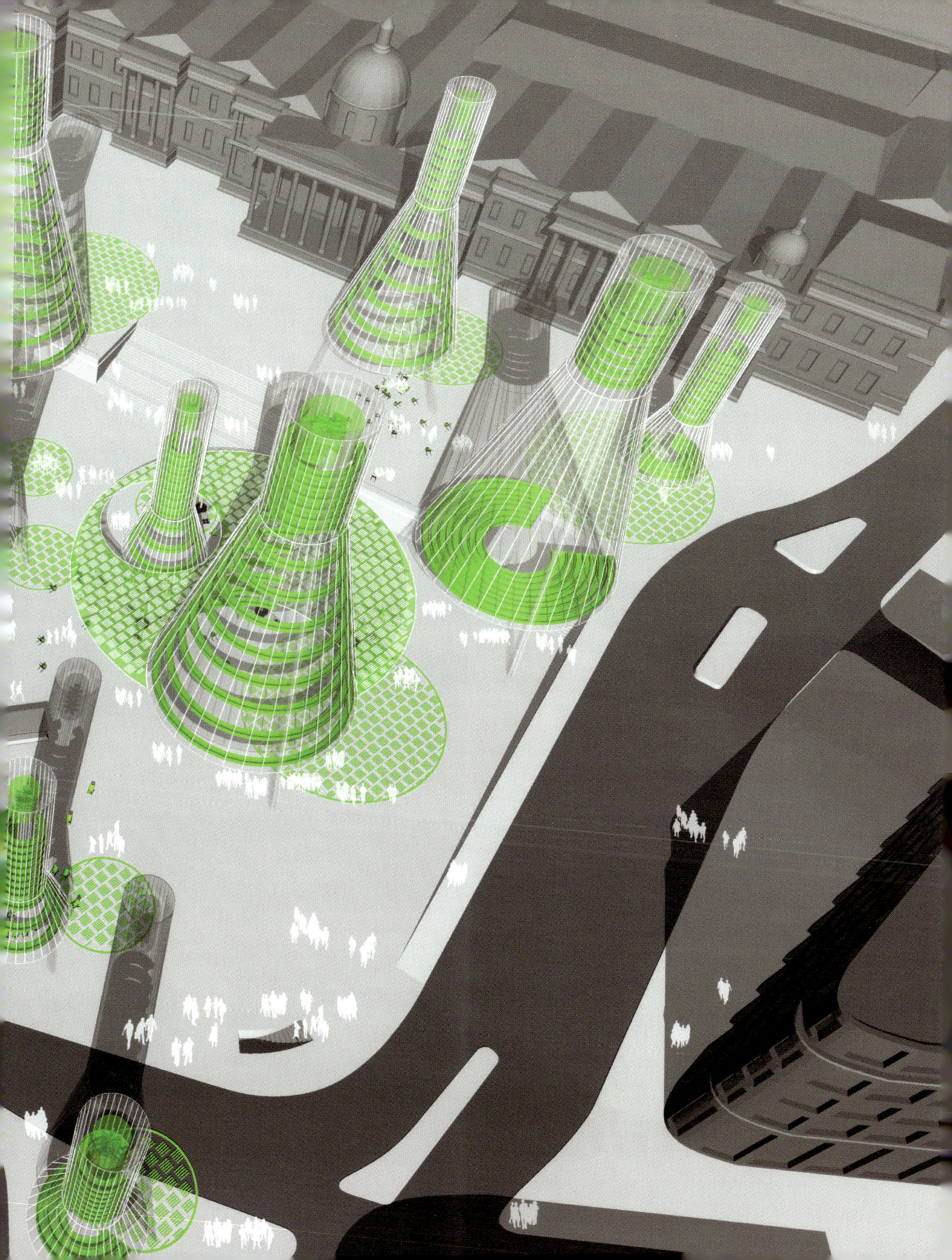

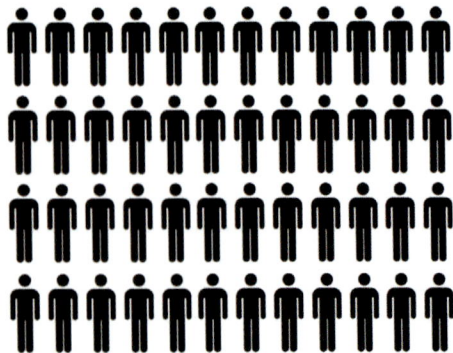

TIANANMEN SQUARE BEIJING	SAFRA SQUARE JERUSALEM	RED SQUARE MOSCOW	OLD TOWN SQUARE PRAGUE
DONEGALL SQUARE BELFAST	MARKET SQUARE KRAKÓW	RADHUSPLASSEN OSLO	AZADI SQUARE TEHRAN
PLAZA DE MAYO BUENOS AIRES	TRAFALGAR SQUARE LONDON	PLACE de la CONCORDE PARIS	NATHAN P. SQUARE TORONTO
TAKSIM SQUARE ISTANBUL	CITY SQUARE MELBOURNE	CHURCH SQUARE PRETORIA	MCPHERSON SQUARE WASHINGTON DC

A tinted glass lift car, doubling as the quarters of the bell's custodian, is a beacon at the top of each structure and doubles as a seed bank. Raised high above the ground, the custodians are brought toe-to-toe with Vice Admiral Nelson on his column – eco-warriors for the 21st century.

The semi-enclosed glass skin of the towers traps the sun's energies to create an ideal microclimate and provides protection from winds and pests. High plant density is achieved through a hydroponic nutrient system, facilitating the growth of over 7000 heirloom tomato species and ensuring the survival of rare and threatened strains. Improved crop yields are achieved by inverting the plants, which has the added benefit of enhancing the visual spectacle from below, and facilitating the harvesting process.

139

At periodic intervals, the concentric planting trays, suspended in a concertina arrangement, are lowered to ground level in a piece of dramatic spatial theatre. The fruits are then harvested by the capital's community of able- and disable-bodied, young and old, local and migrant, slowly transforming into salsa, gazpacho, ketchup, ragu, borscht, chutney, relish, marmalade, bloody mary, chow chow and jambalaya under the supervision of the preparation's cultural originators adopted by the city. Responsible for tending the square's edible produce, the custodians of the bells also disperse seeds from the nursery to visitors and the suburban populace who in turn propagate and share the plants, fruit, recipes, knowledge and stories to their own communities thereby creating a secondary network for the exchange.

Dissemination extends at a global scale through replication in other metropolitan squares round the world including Tiananmen in Beijing, Paris' Place de la Concorde, and Moscow's Red Square. Collectively, the exchanges become nodes of an international resilience network that simultaneously encourage social integration and celebrate ethnic diversity.

Castells believes that the physical infrastructure which we 'collectively consume', such as public transport, social housing and city squares, are symbiotic with rather than in competition against global virtual networks.[1] The Tomato Exchange trades in both social commodities, employing transient and permanent populations as an exchange medium and transforming physical monuments into meaningful diverse social spaces.

facing top: The semi-enclosed glass skin traps the sun's energies to create an ideal microclimate.

facing middle: The Tomato Exchange in Tiananmen Square, Beijing; and in Red Square, Moscow.

facing bottom (left to right): Global citizens and local communities; Global networks for social integration and ethnic diversity; Inverted tomato plant culture.

left: Sixteen glass bell-like structures hover over London's principal square.

following page: Night views of the Tomato Exchange.

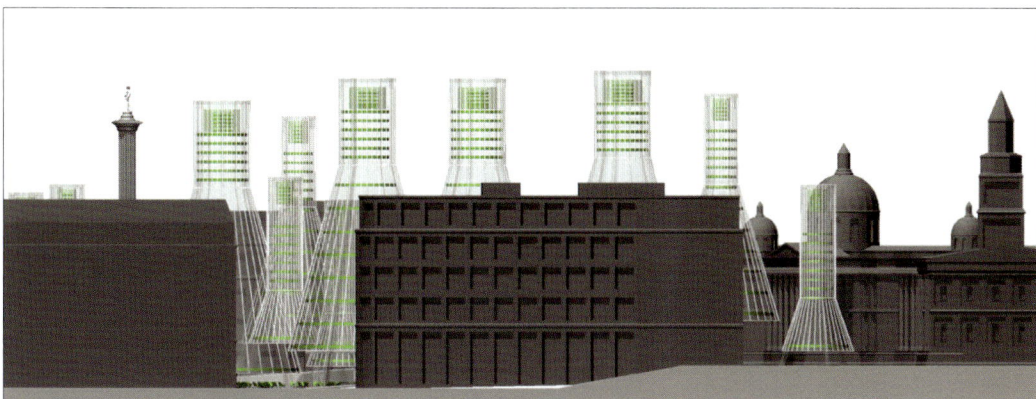

1. M Castells, 'The Urban Question' (1977), developed the term 'collective consumption' as a critical concept for explaining urban changes in the post-war era.

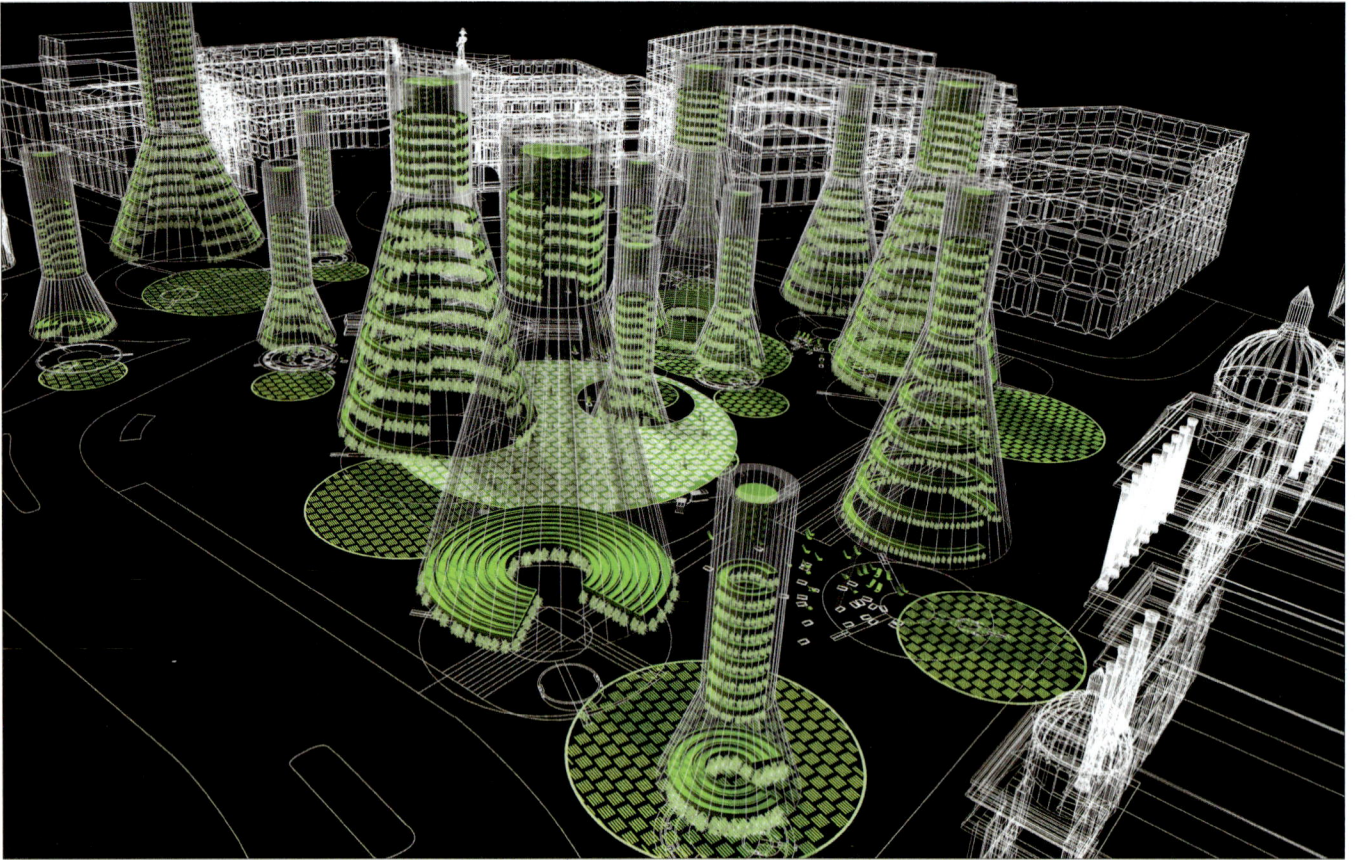

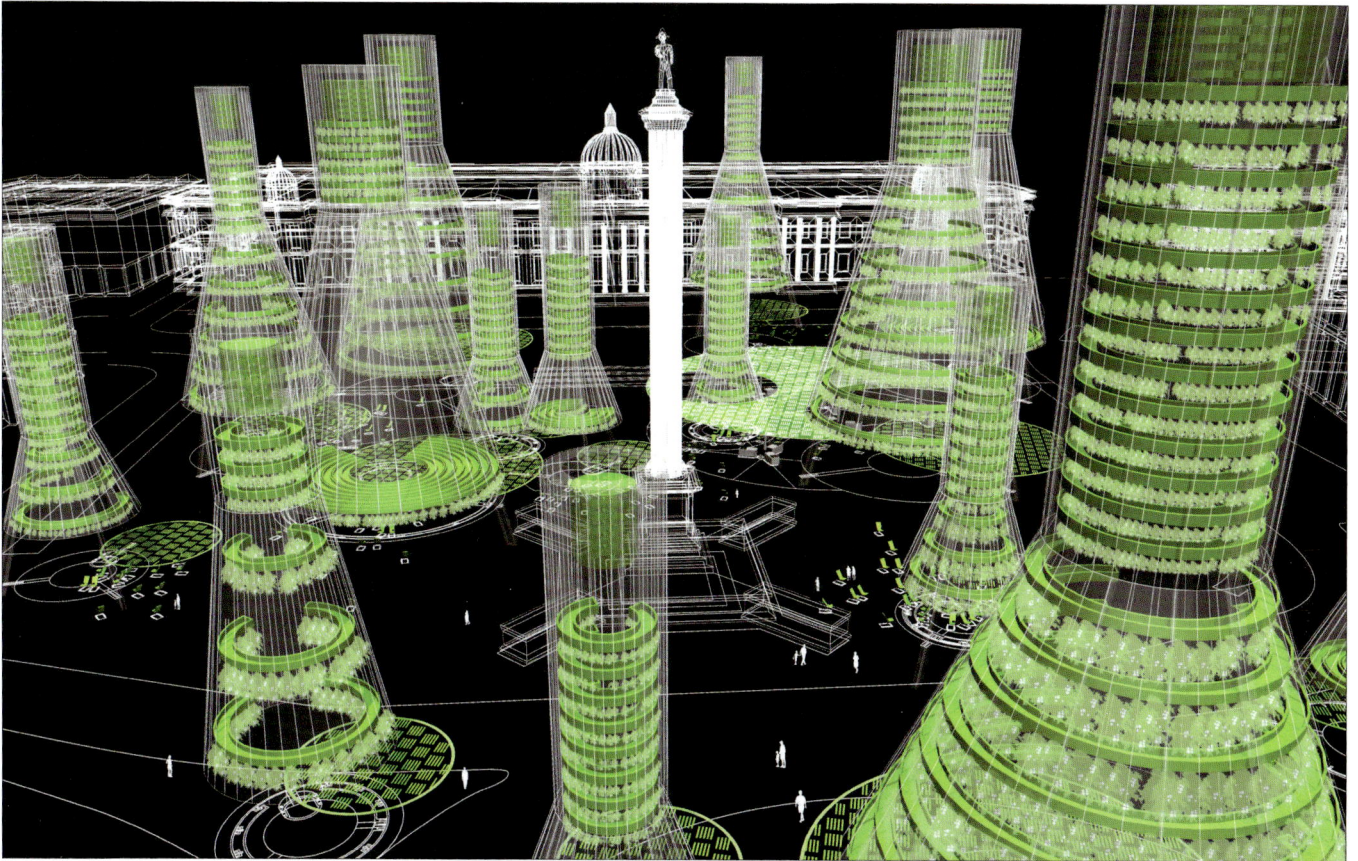

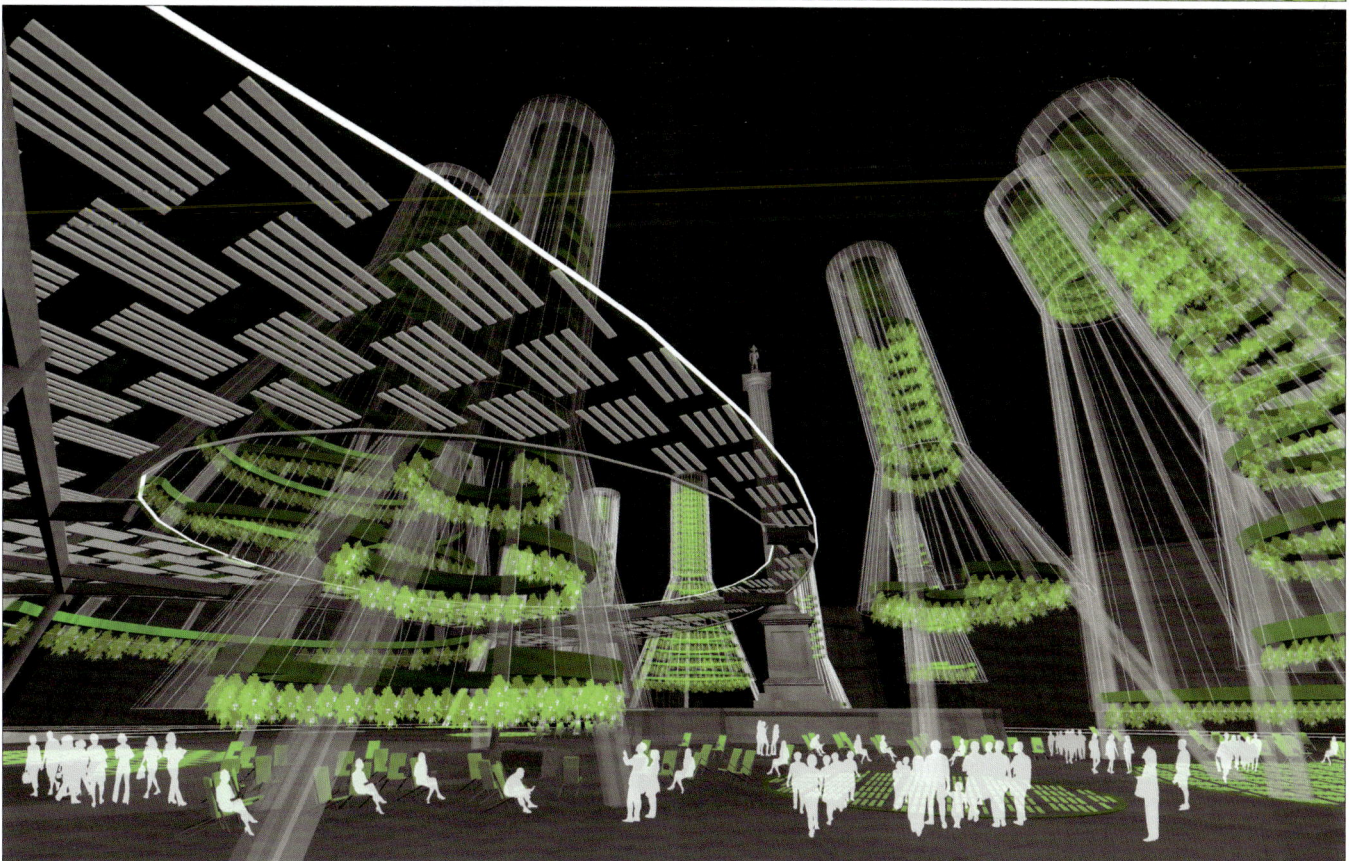

Central Open Space: MAC South Korea

'**Most of us live in ignorance of the effort it takes to feed us.**'

– Carolyn Steel, 'Hungry City', 2008

The Republic of Korea is in the process of building a new Multi-functional Administrative City (MAC) to alleviate excessive concentration of the Seoul Metropolitan area by relocating the government ministries and to promote balanced national development. Located in Chungcheongnam-do Province, 150km from Seoul, MAC will cover an area of 72.19km2 and takes the form of a ring that symbolises the government's principles of non-hierarchy and decentralisation.

The MAC is intended to be a model city for sustainable growth, enhancing the quality of Korean urban environments by acting as an exemplar development. Commissioned by the Korea Land Corporation and the Multi-functional Administrative City Construction Agency, the Central Open Space (COS) is to be the green hub at the heart of the city that connects various cultural facilities and represents the government's philosophy and vision. Occupying an area of 6.982km2, the COS is a great plain that aligns with Mount Jeonweol-san and Mount Wonsu-bong, and has the Geum-gang River running through it. As befits a model city, the COS transcends traditional notions of the park as a verdant isolated island within the metropolis to become a dynamic environment that engages in dialogue with the city through regeneration, nature and culture.

The adoption of an urban agriculture programme is eminently suited to the COS, offering a true model of sustainability for a global 21st century city and re-establishing a meaningful and fluid relationship between fresh food production and the city's population. The proximity of government ministries to cultivated land sets out a clear position regarding Korea's commitment to food security. COS incorporates cultural institutions such as a performing arts complex, a history and folk museum and a design museum, presenting opportunities to exploit the poetic juxtaposition of civic activities within a picturesque but functional backdrop. In contrast to Guangming Smartcity where urban fabric is integrated into farmland, fragments of the Multi-functional Administrative City are inserted into the farmscape of the COS. Artists' studios, libraries and villas for ecotourists float amongst the canopies of peach and pear trees, bringing city dwellers into unexpected contact with the verdant and scented bosom of mother nature.

The procurement of a new park of this scale is an immense undertaking and the strategy for the development of the COS is designed to minimise land movement and to keep the land in use throughout the various phases of construction. The development area is currently arable land and the existing

facing page: Central Open Space masterplan.

following page: Infrastructure plans of the agricultural and cultural city.

New Artificial Lake
(PAT)

Geum-gang River

EXISTING WATERCOURSES

EXISTING ARABLE FIELDS
seasonal organic vegetables

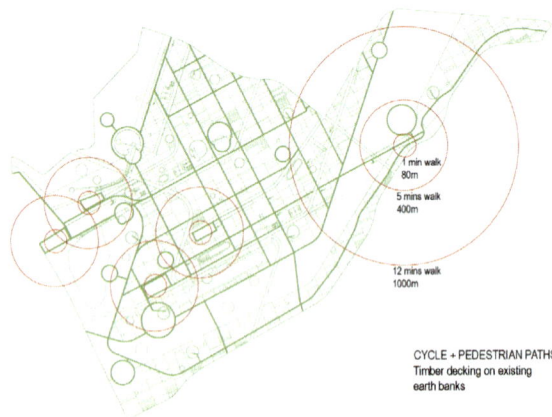

1 min walk
80m

5 mins walk
400m

12 mins walk
1000m

CYCLE + PEDESTRIAN PATHS
Timber decking on existing
earth banks

CIRCULATION
Vertical (lifts + stairs) and
Inclined (ramps)

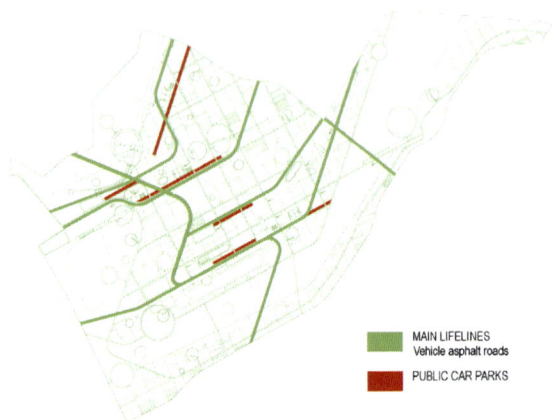

MAIN LIFELINES
Vehicle asphalt roads

PUBLIC CAR PARKS

SPORTS FACILITIES

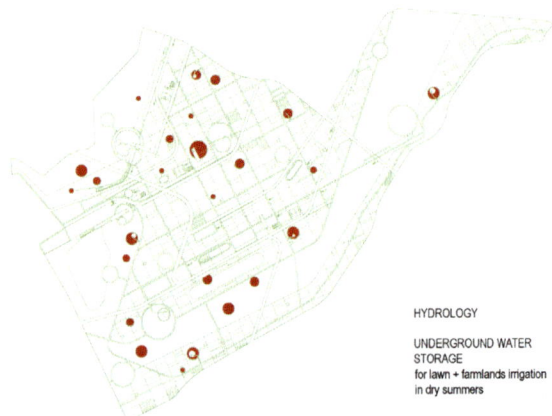

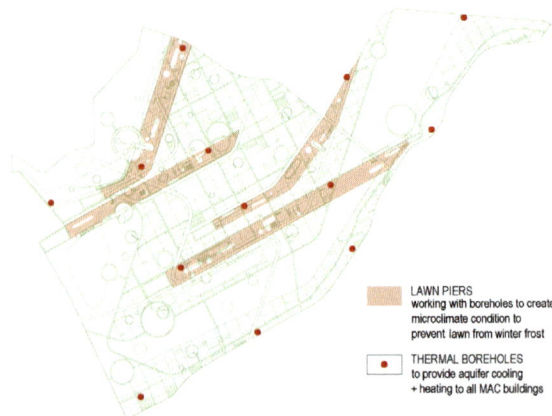

HYDROLOGY

UNDERGROUND WATER
STORAGE
for lawn + farmlands irrigation
in dry summers

LAWN PIERS
working with boreholes to create
microclimate condition to
prevent lawn from winter frost

THERMAL BOREHOLES
to provide aquifer cooling
+ heating to all MAC buildings

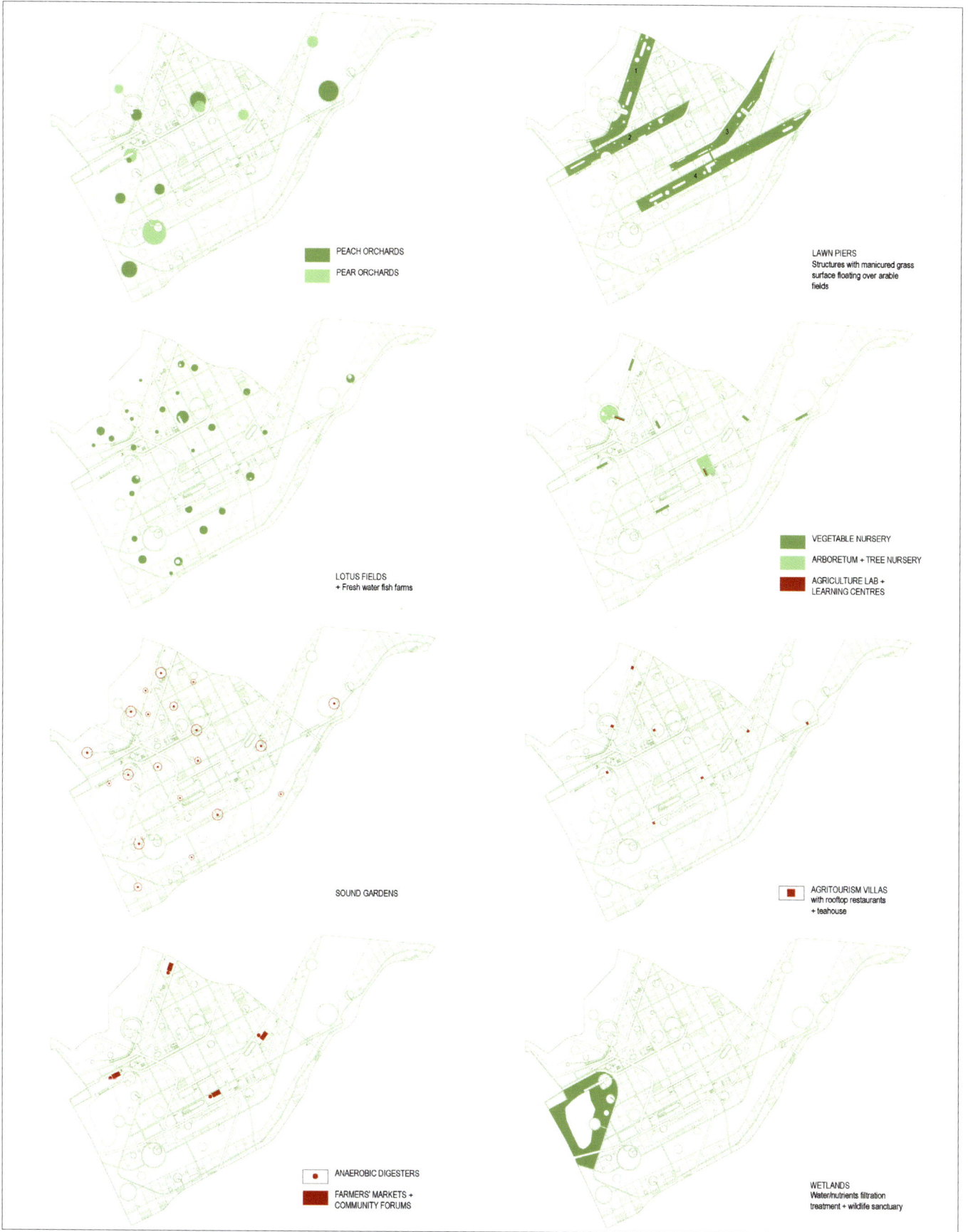

PEACH ORCHARDS
PEAR ORCHARDS

LAWN PIERS
Structures with manicured grass
surface floating over arable
fields

LOTUS FIELDS
+ Fresh water fish farms

VEGETABLE NURSERY
ARBORETUM + TREE NURSERY
AGRICULTURE LAB +
LEARNING CENTRES

SOUND GARDENS

AGRITOURISM VILLAS
with rooftop restaurants
+ teahouse

ANAEROBIC DIGESTERS
FARMERS' MARKETS +
COMMUNITY FORUMS

WETLANDS
Water/nutrients filtration
treatment + wildlife sanctuary

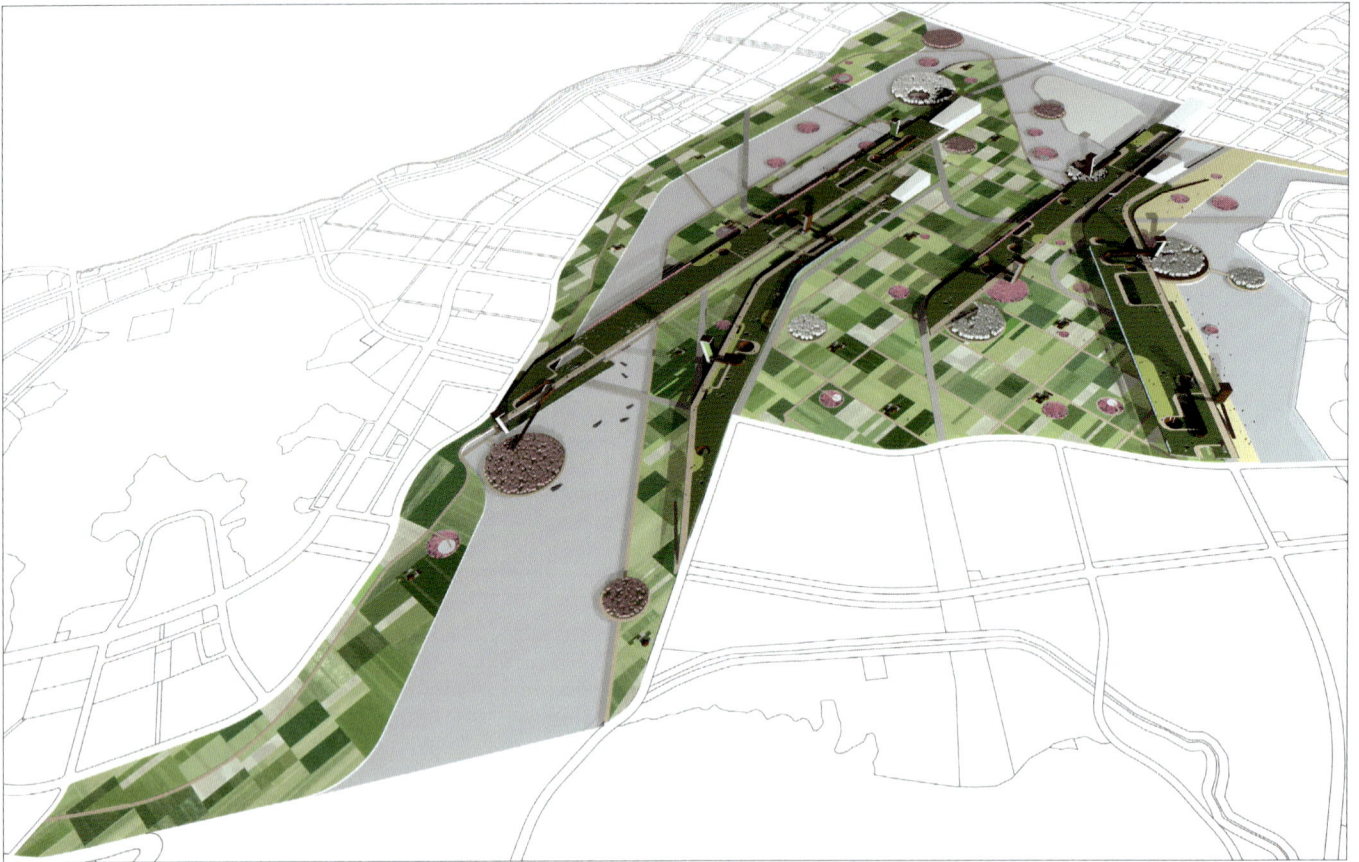
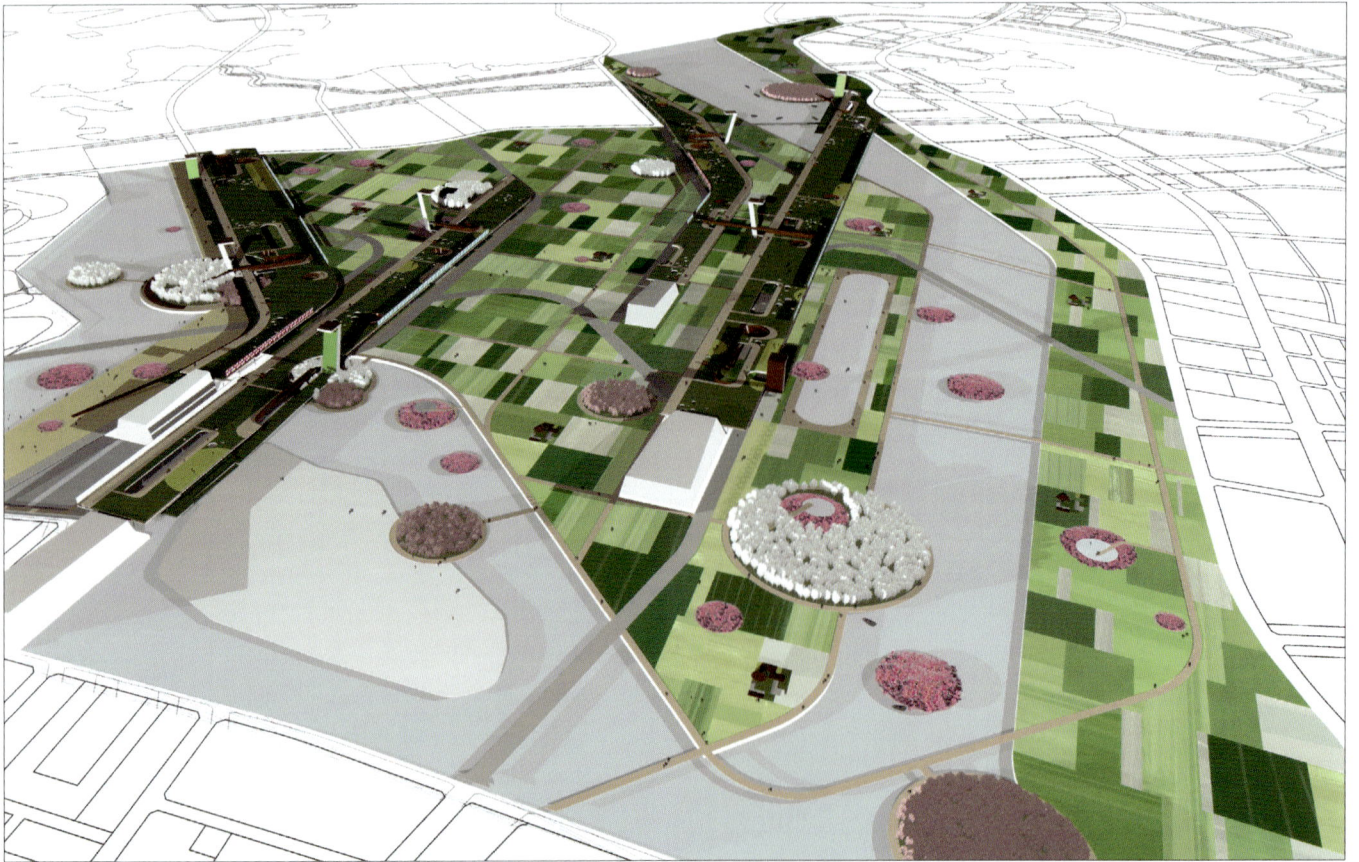

local practice of vegetable farming will be retained along with all the local traditions of food production, although the plain will be reconfigured into striated bands of seasonal colour, bringing into being a chromatic spectacle viewed from four elevated inhabitable piers. The pier structures will house new programmatic functions and extend out over the fields.

Urban Agriculture

During the second half of the 20th century, the urban and peri-urban agriculture movement in Korea grew rapidly due to an influx of farmers to the cities who could command high prices within a large consumer market, cultivating land that was still relatively inexpensive. Rapid industrialisation and spiralling land prices have resulted in a shift in urban agriculture, with a new focus on green tourism and nutritional benefits. In comparison to the West, Koreans have evinced greater disquiet over the sedentary lifestyle of their children and the disjunct between the city and nature, resulting in the growth of farm-stay schemes in which city dwellers holiday on farms and reconnect with their rural heritage. Due to the high population density of the country, the only viable form that urban agriculture can take is intensive horticulture, and city cooperatives concentrating on improving quality in order to compete with low-cost imports have found new markets by setting up framework agreements to supply schoolchildren with high quality produce. The urban agricultural strategy at this exemplar city follows this model, prioritising quality and education over quantity and a disposable culture. Less can indeed be more.

The existing farming community will play a crucial role in the upkeep of the open space and be called upon to instruct and supervise other residents who take up a support role in tending the land, disseminating cultural practices and a breath of fresh air to a generation of desk-bound urbanites. The farmers, too, will need training to integrate nutrient waste recycling and energy generation systems made possible by the farms' access to the city. Local food production and the exchange of skills between the farming community and native city-dwellers will create new social bridging capital. Encounters between these two disparate groups will no longer be a rarity and the former, so long taken for granted, will be empowered by the arrangement. Fresh produce will be sold directly to the public and commercial interests will not be permitted any control on the produce farmed that would otherwise negatively influence the colour and texture of the COS.

Agritourism

Farm-stay programmes have demonstrated a yearning for a calmer and more rewarding lifestyle. For people who are more interested in how their food is produced, the COS is the place to go, without the inconvenience of long congested journeys into the countryside. Vegetable fields, fruit orchards and watercourses offer scenic beauty and green space. Visitors will be able to stay in villas on the farm where they can assist with farming tasks and enjoy the fresh country air. Hiking and trekking paths will be marked out on the Jeonweol-san and Wonsu-bong mountains. At the peak of each mountain, viewing posts in the forest clearings will present stunning vistas back over the COS. The new urban beach sandwiched between the Public Administrative Town (PAT) and arable farm-scape, is a more convenient leisure destination than the northern coastline, making the COS an ideal weekend resort.

147

facing page: Aerial views of a sustainable food production city.

following page: Views showing the relationship between lawn piers and existing agricultural carpet.

Culture

The new landscape of COS aims to be a public recreational green space, simultaneously mixing culture with ecology. A 50 000m2 Performing Arts Complex including music halls and theatres for the performance of operas and traditional Korean music, a 25 000m2 History and Folk Museum celebrating the nation's cultural heritage, a 50 000m2 Design Museum and a fourth cultural facility are planned for construction within the COS.

Interactive technologies will be introduced to increase the interface between nature and culture. Soundtracks will be composed to stimulate the senses and choreograph mood to complement the landscape. Sound gardens and gathering wells break up the plain, providing areas to sit, converse and relax. Changes in the topography, textural variation and ephemeral acoustic boundaries demarcate zones of occupation. Seated in these garden oases, rich in haptic sensation, visitors will experience the vanishing sound of wind chimes and rhythmic melodies whilst viewing some of the country's historic artefacts positioned at the focal point of the wells.

Organisational Framework

The restructuring of the landscape starts with preserving the area's historical and cultural identity; minimal modification is made to the existing fields. A new planting schedule displaying seasonal texture and colour change will be implemented to this hybridisation of kitchen garden and park. Strong horizontal spatial moves are introduced to accentuate the flatness of the site. A matrix of five interacting systems are laid onto the inhabitable vegetative canvas:

(1) A network of paths radiating from the five principal roads surrounding the site extends into and weaves across the open space, designated as the primary lifelines. These are used for vehicular circulation and parking, and connect the cultural buildings and leisure nodes.

(2) The existing network of mud banks and ridges are made good and covered with timber decking. These constitute the secondary lifelines, available for cycling and walking.

(3) Four lawn piers accommodate the non-farming activity infrastructure and are distinguished by the four designated cultural museums and arts venues. Built in lightweight steel and timber construction, the piers float over the arable plain making reference to the urban roof garden as well as the ringed form of the Governmental Complex of Administrative City. The expanses of green lawn provide a multi-functional surface for picnics, sunbathing and ballgames while the elevated plane presents spectacular panoramic views. Sports facilities, farmers' markets, tourist accommodation and other future urban space-forms will emerge beneath these four living earth structures.

(4) A scattering of fruit and lotus orchard hubs provide new natural habitats and introduce a new dynamic to the arable plain. The region is celebrated for its peaches and pears, and the trees impart a dramatic seasonal colour change from emerald green to white and pink. The lotus, a versatile plant, is used in a number of cooking preparations and represents classical notions of beauty that contrast with the large-scale composition of arable striations.

(5) A nexus of watercourses reinforces the hydrology and ecological dynamics of the site, linking a new artificial lake with the Geum-gang River and offering opportunities for fresh water fish farming, recreational fishing and boating.

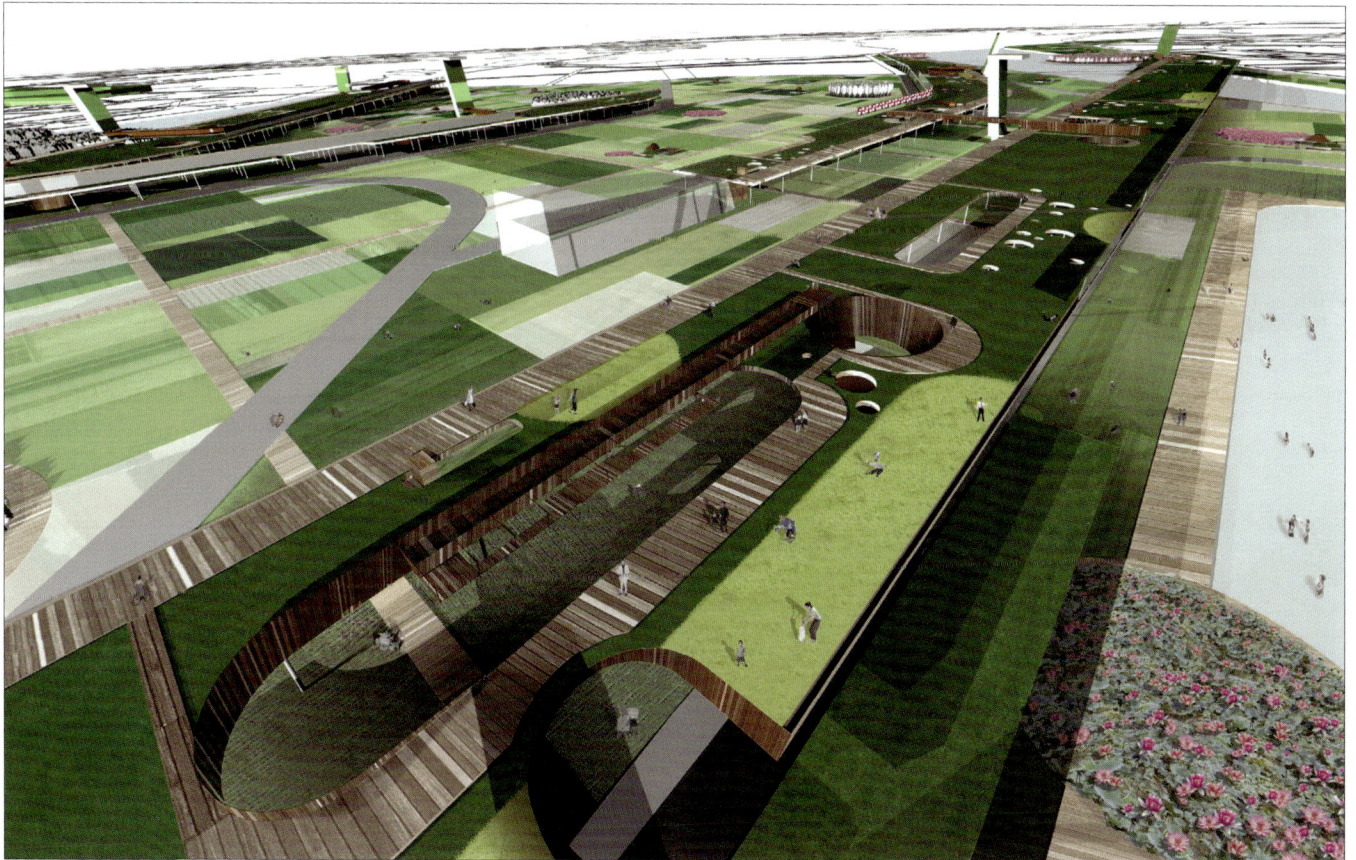

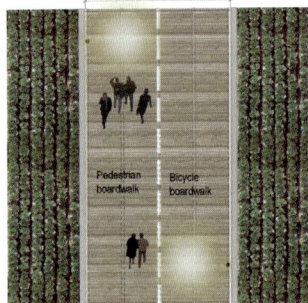

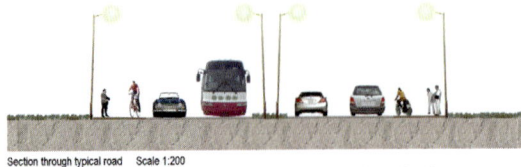

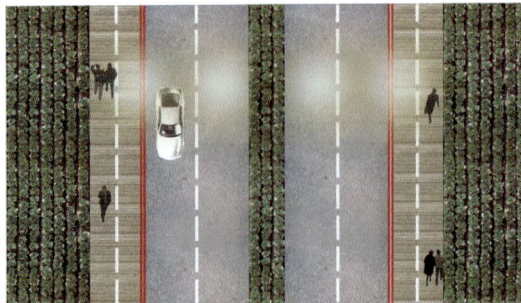

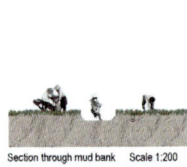

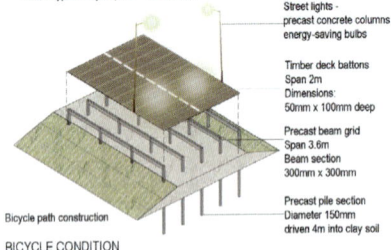

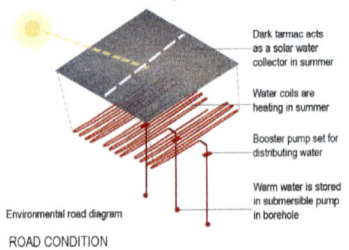

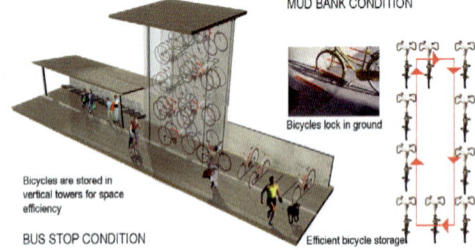

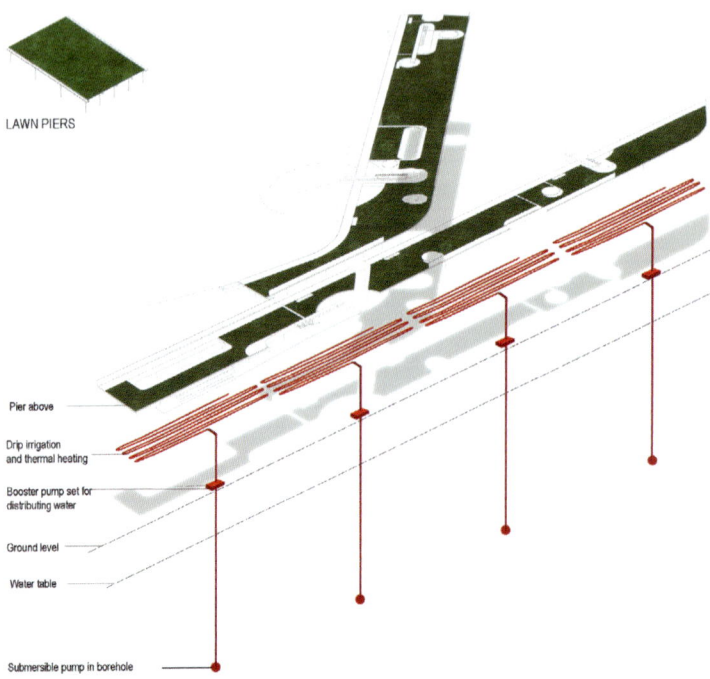

Section through typical bicycle path   Scale 1:200

Section through typical road   Scale 1:200

Section through mud bank   Scale 1:200

Plan of typical bicycle path   Scale 1:200

Pedestrian boardwalk

Bicycle boardwalk

Section of typical road   Scale 1:200

Section of typical mud bank   Scale 1:200

MUD BANK CONDITION

Street lights -
precast concrete columns
energy-saving bulbs

Timber deck battons
Span 2m
Dimensions:
50mm x 100mm deep

Precast beam grid
Span 3.6m
Beam section
300mm x 300mm

Bicycle path construction

Precast pile section
Diameter 150mm
driven 4m into clay soil

BICYCLE CONDITION

Dark tarmac acts
as a solar water
collector in summer

Water coils are
heating in summer

Booster pump set for
distributing water

Warm water is stored
in submersible pump
in borehole

Environmental road diagram

ROAD CONDITION

Bicycles are stored in
vertical towers for space
efficiency

BUS STOP CONDITION

Bicycles lock in ground

Efficient bicycle storage

LAWN PIERS

Pier above

Drip irrigation
and thermal heating

Booster pump set for
distributing water

Ground level

Water table

Submersible pump in borehole

HEATING AND COOLING

As part of the park's sustainable park management program, the MUSCO system proposes a thermal store capable of storing heat so that heat can be collected in the summer by providing cooling, allowing the piers to remain green all year around.

In winter, the heat is upgraded using heat pump technology to supply the temperatures required by the pier lawn with the heat pump configured to extract the heat until chilled water or chilled ground stores cooling used the following summer.

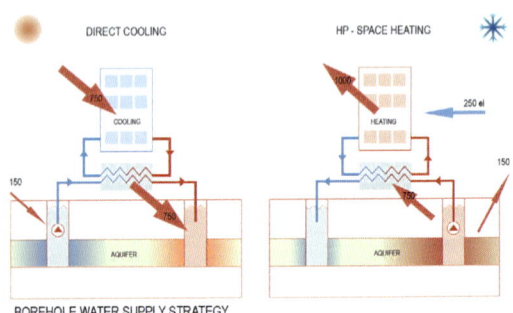

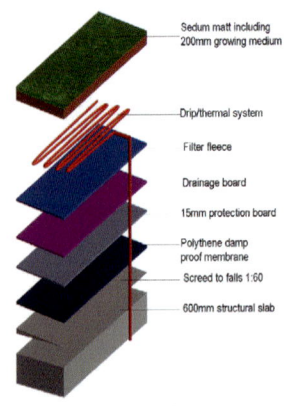

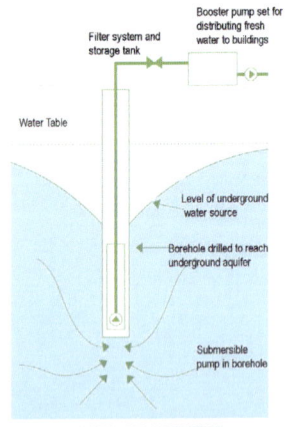

DIRECT COOLING

HP - SPACE HEATING

BOREHOLE WATER SUPPLY STRATEGY

Sedum matt including
200mm growing medium

Drip/thermal system

Filter fleece

Drainage board

15mm protection board

Polythene damp
proof membrane

Screed to falls 1:60

600mm structural slab

PIER CONSTRUCTION DIAGRAM

Booster pump set for
distributing fresh
water to buildings

Filter system and
storage tank

Water Table

Level of underground
water source

Borehole drilled to reach
underground aquifer

Submersible
pump in borehole

BOREHOLE WATER SUPPLY STRATEGY

Sustainable Energy + Resource Management

The opportunities to provide an exemplar open space in the middle of a new planned city include methods of providing not only sustainable energy and resource management for the park, but also commercial possibilities where the park itself can offer these facilities to the city as a whole. The Central Open Space will therefore adopt a commercial sustainable energy and resource management system in the form of a MUSCO (multi-utility service company) that holds the following advantages over more traditional single supply commercial models:

• A model of local governance can be applied to energy and water supplies.

• Local high-level employment and training possibilities are presented.

• Efficiencies in management costs and possible income from low carbon grants and tax advantages offered by the South Korean government should enable a well-designed and efficiently operated supply system to offer services which are lower carbon and cheaper tariff than those offered by single supply models, reducing fuel poverty and attracting commercial organisations to the area.

• Forms of supply can be tailored to meet the precise needs of the customers. The Public Administrative Town will require high quality office and commercial space. Where many National Grid systems have proved unable to provide a high quality electricity supply essential for new technology businesses, the MUSCO would be able to ensure low-fluctuation electricity.

• Electricity generation from the combustion of fossil fuels inherently produces far more waste heat than electrical energy. This heat, when produced and rejected in summer, can worsen urban heat island effects and when released in winter is wasteful when the MAC depends on winter heating for thermal comfort. The MUSCO arrangement will provide combined cooling, heat and power to enable electricity to be sold alongside waste heat, either as district heating or cooling via an absorption chiller installation. Costs and the carbon content of heat, cooling and electricity will be reduced.

• An average annual precipitation in excess of 1500mm provides more than enough water for urban expansion. However, the irrigation necessary for the arable kitchen garden park constitutes a good business case for a separate irrigation quality. Where groundwater is available from the alluvial deposits in the area, it can be treated to drinking water quality and water can be abstracted from river water in the treatment plant on site. Wetlands adjoining the river can be made beautiful and still serve in a final cleansing process. Studies have shown that the plant and distribution network servicing the Central Open Space could be efficiently scaled up to supply the immediately surrounding city. Excess urban nutrients that would otherwise be pumped into the river and lead to eutrophication can be dealt with by tertiary filtering using the wetlands together with sustainable surface drainage features such as gravel bed hydroponic networks.

previous pages: Model of the COS – knitted texture of the existing agricultural carpet.

facing page top: Solar collection paths.

facing page bottom: Borehole Water Supply strategy.

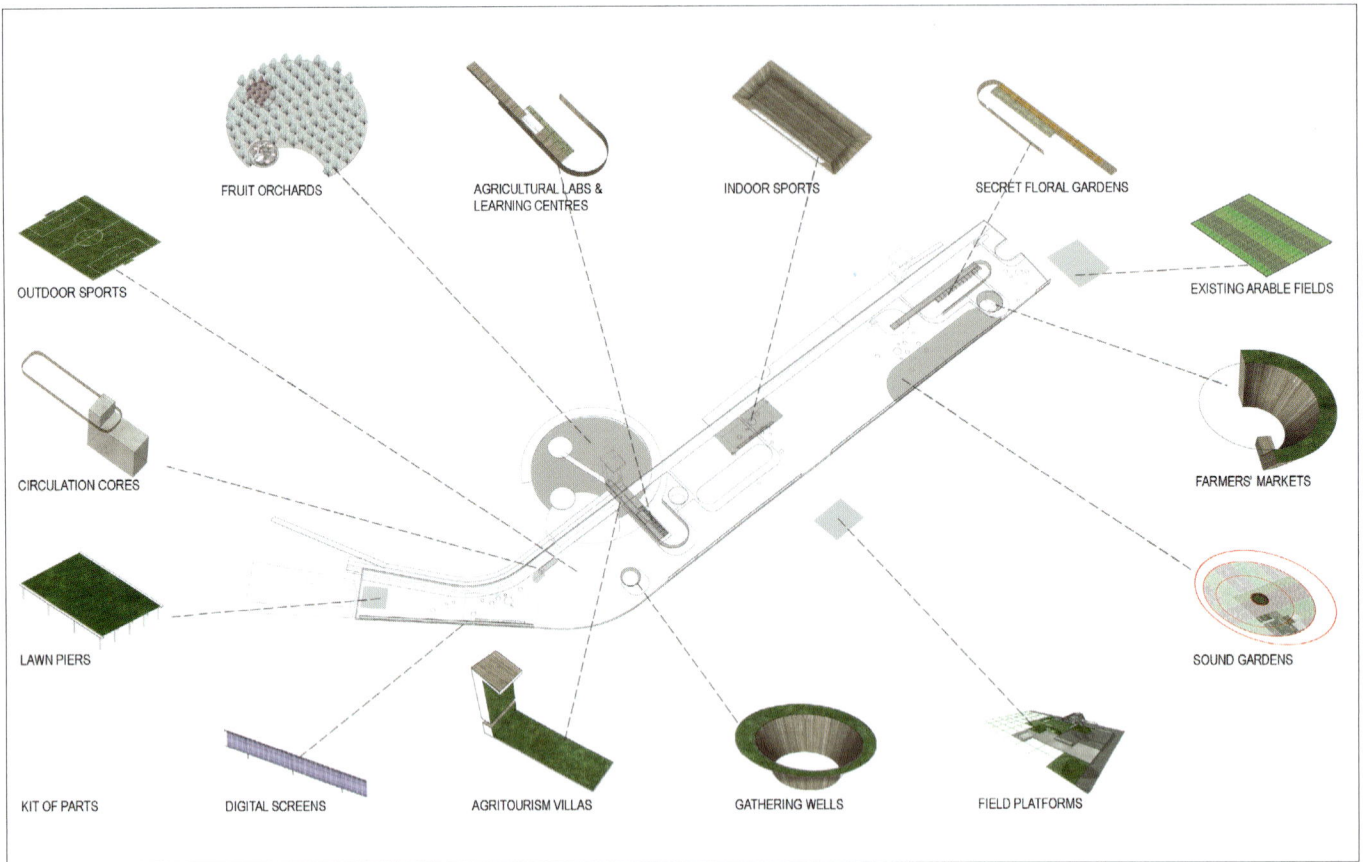

FRUIT ORCHARDS

AGRICULTURAL LABS & LEARNING CENTRES

INDOOR SPORTS

SECRET FLORAL GARDENS

OUTDOOR SPORTS

EXISTING ARABLE FIELDS

CIRCULATION CORES

FARMERS' MARKETS

LAWN PIERS

SOUND GARDENS

KIT OF PARTS

DIGITAL SCREENS

AGRITOURISM VILLAS

GATHERING WELLS

FIELD PLATFORMS

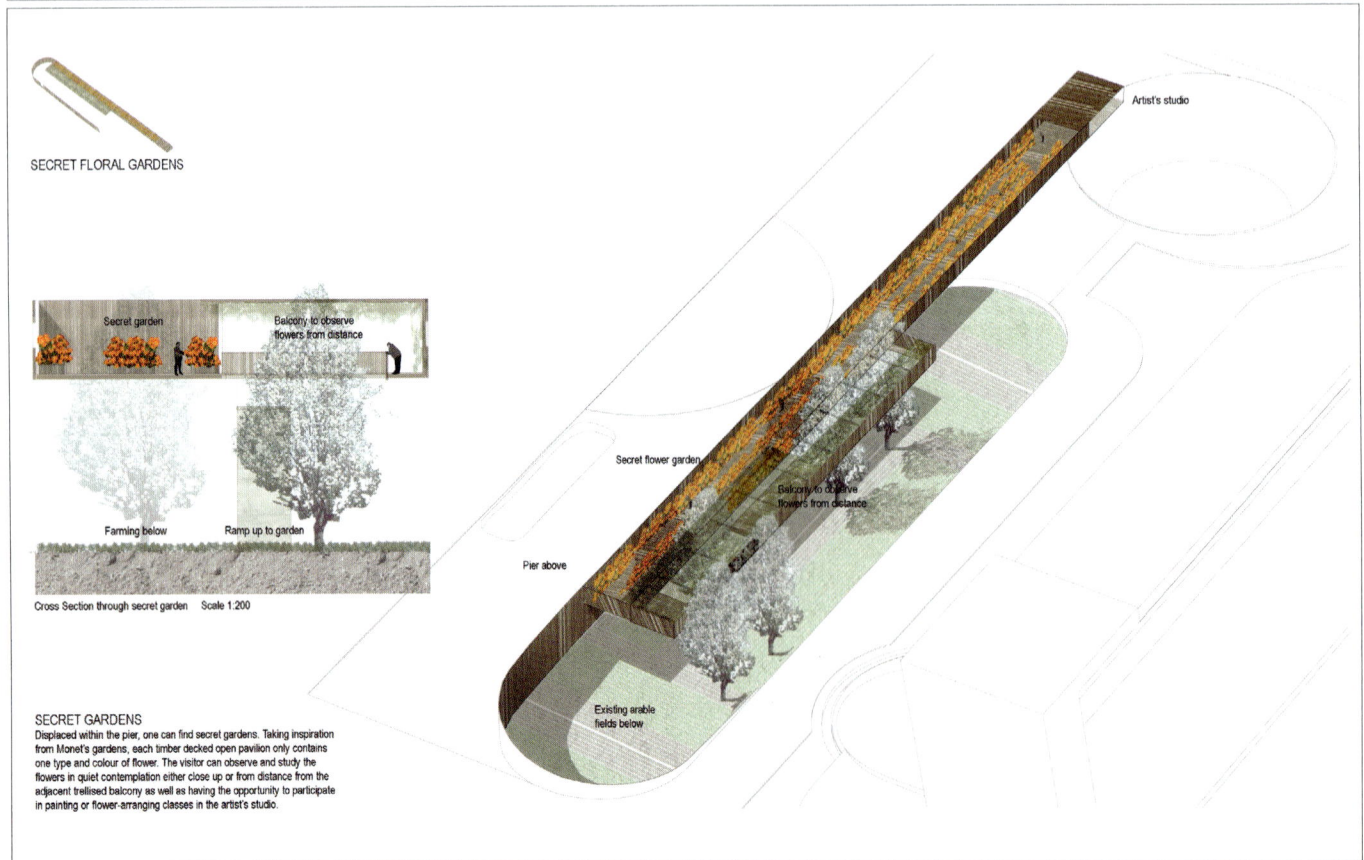

SECRET FLORAL GARDENS

Secret garden

Balcony to observe flowers from distance

Farming below

Ramp up to garden

Cross Section through secret garden   Scale 1:200

Artist's studio

Secret flower garden

Balcony to observe flowers from distance

Pier above

Existing arable fields below

SECRET GARDENS

Displaced within the pier, one can find secret gardens. Taking inspiration from Monet's gardens, each timber decked open pavilion only contains one type and colour of flower. The visitor can observe and study the flowers in quiet contemplation either close up or from distance from the adjacent trellised balcony as well as having the opportunity to participate in painting or flower-arranging classes in the artist's studio.

• Nutrients produced in the arable plain and orchards will leave the COS as food consumed by the local community. In order to establish a classic urban nutrient cycle, the COS will act as the local organic waste treatment authority, receiving an annual income for treating organic waste using anaerobic digesters to yield biogas that can be converted into saleable energy. Remaining nutrients can be returned to the soil as fertiliser.

• The management/public interface of the MUSCO will take the form of outlets on each pier set to maximise the farming income run by the city's utilities management company, the setup similar to that established in Guangming Smartcity.

Electricity + Biomass

Energy from photovoltaic panels will serve localised demand within the COS and illuminate streets, bus stops and bicycle stations in the neighbouring city. Integrated PV panels along the lifelines will provide shading and further increase electricity supply.

Biomass waste from landscaping and agriculture is carbon neutral and can be burnt to generate energy, processed into biogas or converted into high-quality compost. In addition to district heating and district cooling via the use of biomass CCHP installations, inter-seasonal thermal storage will be employed.

COS is underlain by quaternary alluvial deposits suggesting the possibility of aquifer thermal energy storage (ATES), the most advanced and lowest energy of inter-seasonal thermal storage forms. This system uses boreholes to access different sections of the aquifer to create separate warm and cool zones. A single thermal borehole into a sandstone aquifer could yield 2MW of cooling or heating. As aquifer water is merely pumped from one part to another, there is no risk of aquifer depletion.

A similar system can be utilised by turning the earth itself into separate large-scale heat and cooling stores. There will be an element of cut and fill earth movement in planning the open space, providing an opportunity to install circulating coils within the fill.

If the buildings served by these inter-seasonal thermal storage systems are not in equilibrium – that is if too many commercial buildings are served and too much heat begins to build up towards the end of summer or if buildings requiring heat outweigh those requiring cooling and, overall, additional heat is required – there are new opportunities at the scale of landscape for the redistribution of stored temperature. New pathways, roads and hard surfaced areas will incorporate circulating plastic pipework taking heat from close to the surface during the summer (asphalt is approximately 60 to 70 per cent efficient as a solar collector) to 'top up' heat in the warm thermal store. Similarly, the same surfaces can be used to reject heat in the winter ensuring that they remain ice and snow-free.

Operating on a similar principle, thermal coils fixed to the underside of the piers above multi-use games areas will emit radiant heat, extending the time available for sporting activities.

facing page top: The lawn pier accommodates non-farming infrastructure.

facing page bottom: Secret gardens hover over the fruit orchards.

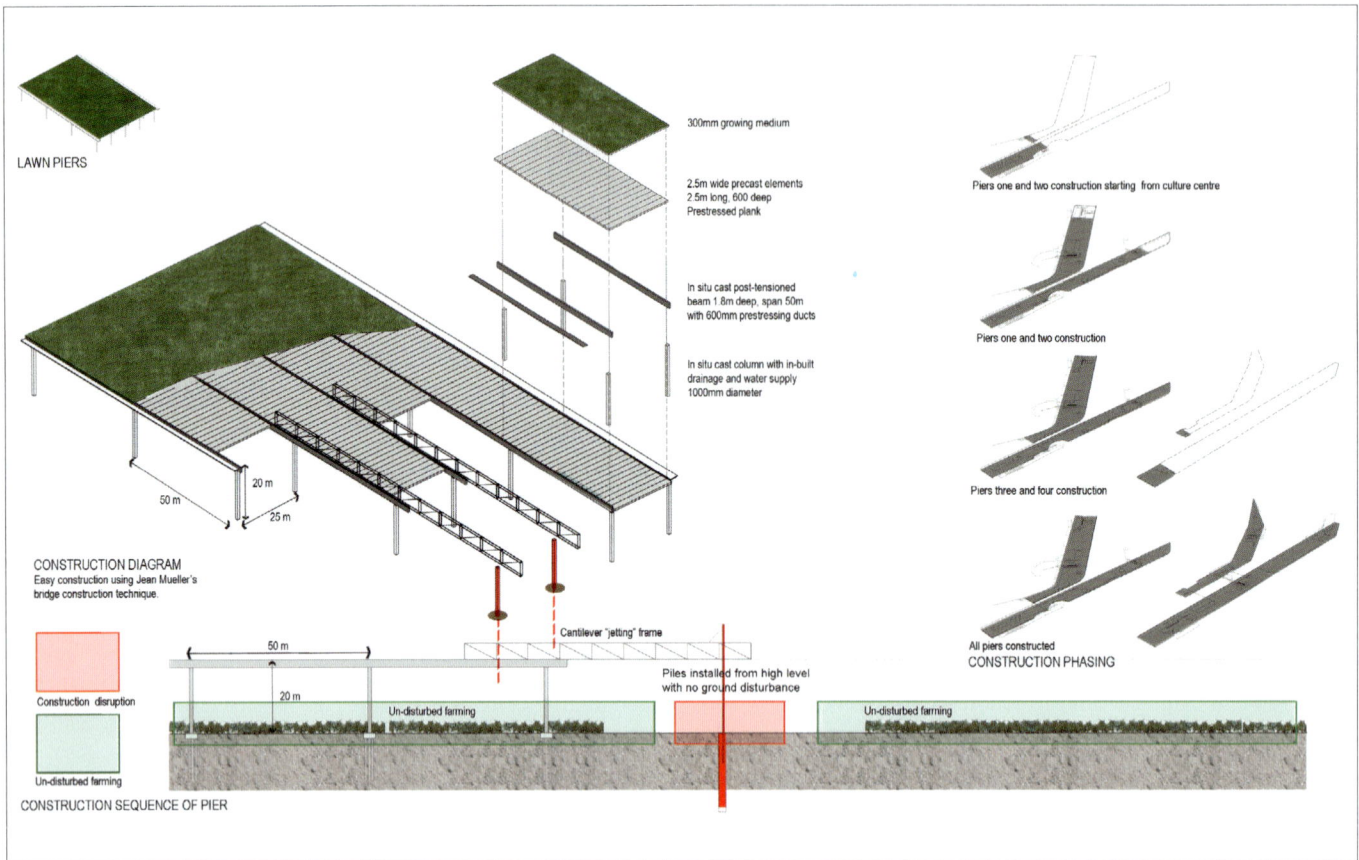

LAWN PIERS

300mm growing medium

2.5m wide precast elements
2.5m long, 600 deep
Prestressed plank

In situ cast post-tensioned
beam 1.8m deep, span 50m
with 600mm prestressing ducts

In situ cast column with in-built
drainage and water supply
1000mm diameter

Piers one and two construction starting from culture centre

Piers one and two construction

Piers three and four construction

All piers constructed
CONSTRUCTION PHASING

50 m
20 m
25 m

CONSTRUCTION DIAGRAM
Easy construction using Jean Mueller's
bridge construction technique.

Construction disruption

Un-disturbed farming

50 m

20 m

Cantilever "jetting" frame

Un-disturbed farming

Piles installed from high level
with no ground disturbance

Un-disturbed farming

CONSTRUCTION SEQUENCE OF PIER

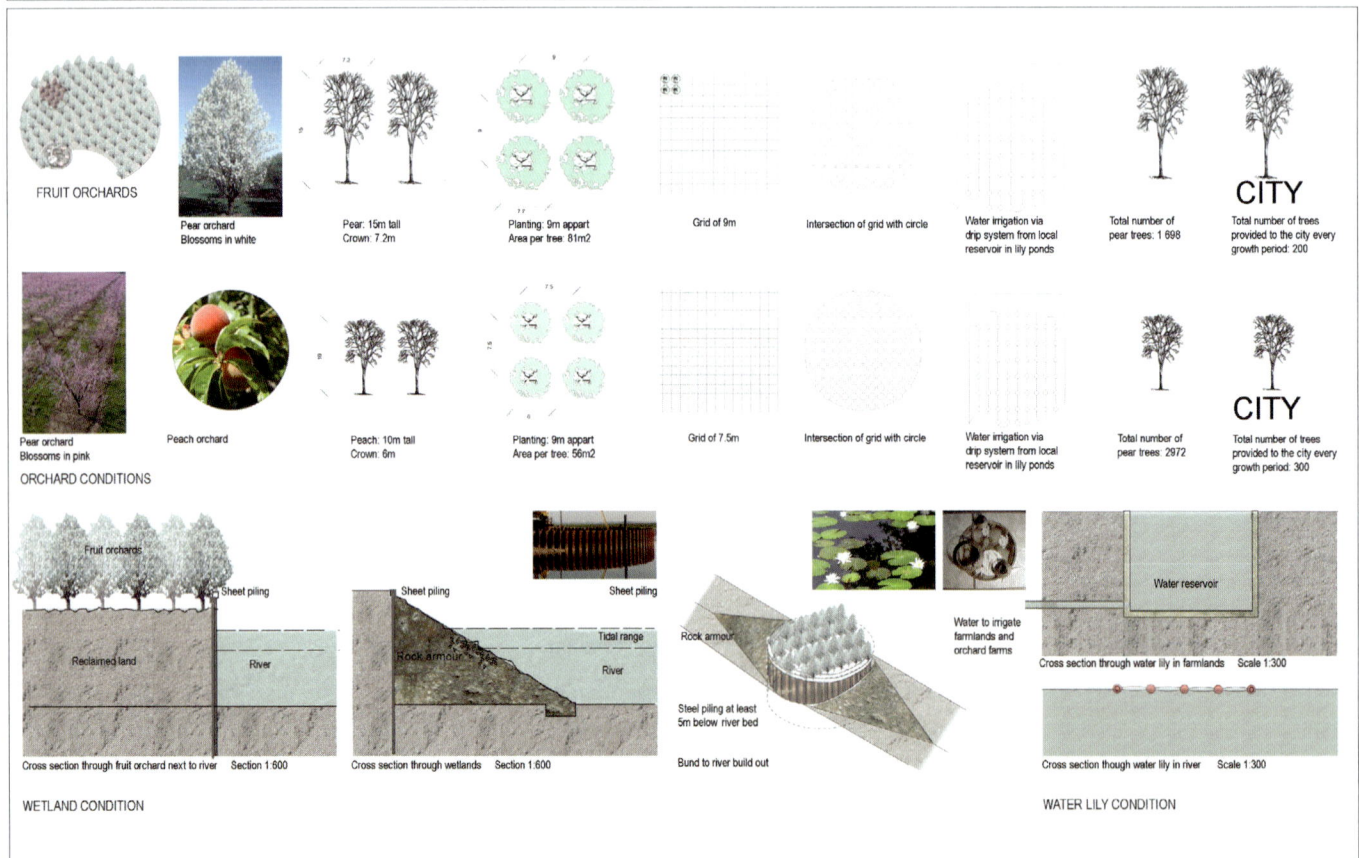

FRUIT ORCHARDS

Pear orchard
Blossoms in white

Pear: 15m tall
Crown: 7.2m

Planting: 9m appart
Area per tree: 81m2

Grid of 9m

Intersection of grid with circle

Water irrigation via
drip system from local
reservoir in lily ponds

Total number of
pear trees: 1 698

Total number of trees
provided to the city every
growth period: 200

CITY

Pear orchard
Blossoms in pink

Peach orchard

Peach: 10m tall
Crown: 6m

Planting: 9m appart
Area per tree: 56m2

Grid of 7.5m

Intersection of grid with circle

Water irrigation via
drip system from local
reservoir in lily ponds

Total number of
pear trees: 2972

Total number of trees
provided to the city every
growth period: 300

CITY

ORCHARD CONDITIONS

Fruit orchards

Sheet piling

Reclaimed land

River

Cross section through fruit orchard next to river   Section 1:600

Sheet piling

Sheet piling

Rock armour

Tidal range

River

Cross section through wetlands   Section 1:600

WETLAND CONDITION

Rock armour

Steel piling at least
5m below river bed

Bund to river build out

Water to irrigate
farmlands and
orchard farms

Water reservoir

Cross section through water lily in farmlands   Scale 1:300

Cross section though water lily in river   Scale 1:300

WATER LILY CONDITION

Data + Communications

Relatively basic web-based software is used to collect data from digital metres and to manage the MUSCO accounts online, resulting in reduced management costs. The MUSCO will also serve as the local data network provider, making the upgrade of data cabling to energy metres cost efficient.

A community intranet offering free access between residents and the MUSCO will create new community connections by offering information on local training and employment opportunities, become part of the local education system and provide all civic data. It will also manage a car-sharing scheme, provide real time public transport information and the delivery of guaranteed locally produced food to the doorstep of subscribers. A similar intranet arrangement will allow the COS and other local farmers to offer their produce to the community directly, reducing costs and improving local competitive advantage over larger national- or multi-national-scale retail suppliers.

Structural Sequencing

The very large elevated decks, freestanding pavilions and auditoria are designed for longevity, low maintenance and minimal disruption to the existing landscape. The decks are economic concrete structures with lateral stability of the structural tables provided by the columns acting as sway frames in bending. Additional elements, adjacent pavilions and suspended structures within the deck are framed in steel to reduce weight and to accelerate construction times. Thermal movement is allowed for with movement joints every third bay.

Work commences with the levelling and paving of access roads. From these primary routes, pre-cast concrete frames are assembled on short driven piles to provide support for timber boardwalks. The principal deck beams are formed in place with sliding shutters. Pre-cast deck elements are then added sequentially. The assembly will be completed with grouted joints to ensure robustness.

An innovative adaptation of the French engineer Jean Muller's bridge construction methodology is adopted to avoid disrupting the ground plane. A temporary steel 'jetting frame' is erected at deck level and cantilevered forward as work proceeds so that new foundation bases, columns and cross-beams can all be placed from above without direct access to the support points.

157

facing page top: Construction sequence of a lawn pier.

facing page bottom: Planting strategy for the orchards; construction of the water's edges.

following page left (top): Contemplation in the sound gardens.

following page left (bottom): Gathering on the lawn piers.

following page right (top): Demarcation of the fruit orchards through colour and smell.

following page right (bottom): Nurseries forming the facades of agritourism villas.

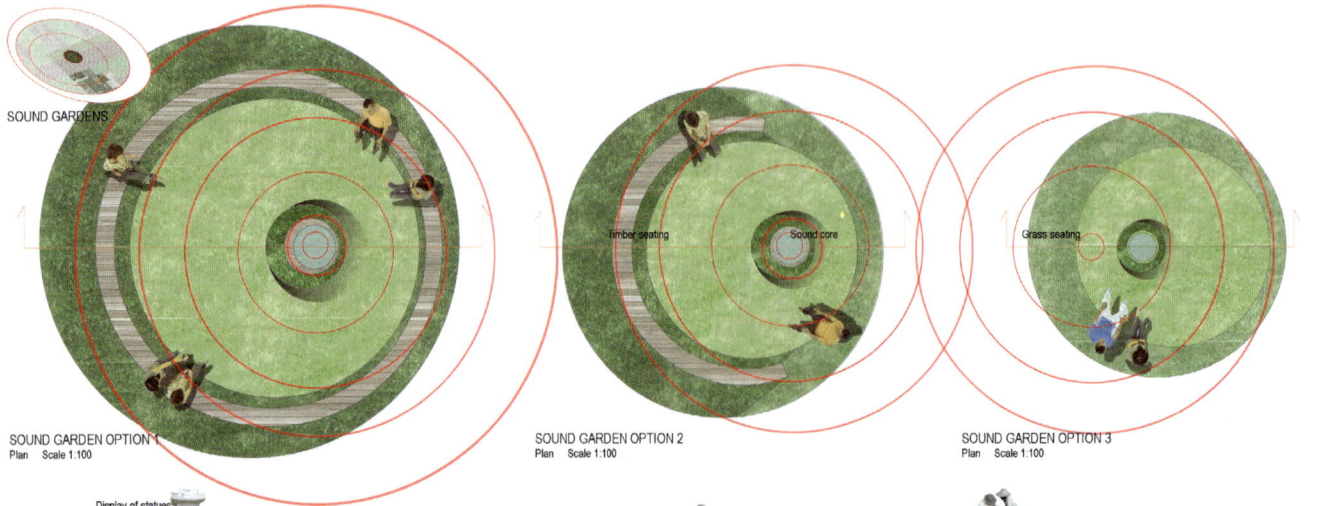

SOUND GARDENS

SOUND GARDEN OPTION 1
Plan   Scale 1:100

SOUND GARDEN OPTION 2
Plan   Scale 1:100

SOUND GARDEN OPTION 3
Plan   Scale 1:100

Timber seating   Sound core

Grass seating

Display of statues

Sound core

Sound Garden section, option 1   Scale 1:100

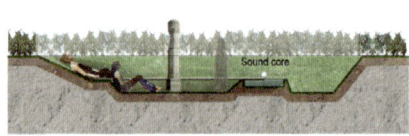

Option 2
Scale 1:100

Option 3
Scale 1:100

CONTEMPLATION
Scattered through the fields, the sound gardens
create moments of contemplation and quiet
observation of some of the country's important
statues and monuments. Each sound garden has
the flexibility to vary in size, furniture detailing
and sound type depending on the site typography
and surrounding programmatic conditions.

Tombstone of Chia, Si-chaek Tombstone of Im Nan-su

Stone Figure of Im, Seo   Tombstone of O.Kang-pyu's Stone Figure

Bonggi-ri Dolmen 2   Tombstone of Kang, Sun-young Stone Figure

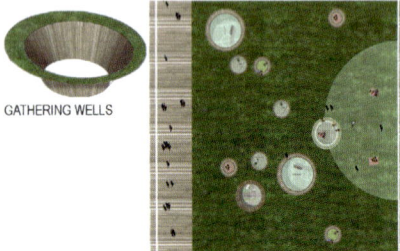

GATHERING WELLS

INHABITABLE
GRASS WELL
Plan   Scale 1:100

Clusters of gathering
wells on the piers

Football pitch below

Football pitch below

FLAT GLASS WELL
Plan   Scale 1:100

INHABITABLE
TIMBER WELL
Plan   Scale 1:100

INHABITABLE GLASS WELL
Plan   Scale 1:100

Section   Scale 1:100

Section   Scale 1:100

Section   Scale 1:100

Light fitting underneath

GATHERING MOMENTS
The piers are introduced with punctures of different textures and sizes,
ranging from grass, glass, timber and water surface. Strategically positioned
in clusters on top of the pier, they are a valuable element of the piers, not just
by bringing valuable natural sunlight to the farmlands underneath the pier, but
also encouraging public gathering moments for picnic areas, watching the
sport games below, public festivities and cinema viewing during the night.

Stepped entrance to inhabitable gathering wells

Contemplation zones

Green wall

Watching the football game below

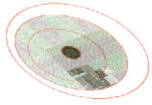

SOUND GARDENS

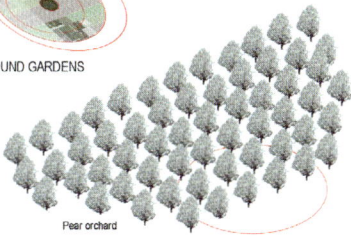

Pear orchard

Pear tree

Extent of sound represented with peach tree plantations

Peach tree

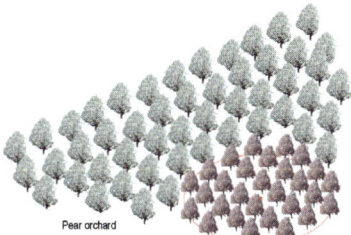

Pear orchard

Extent of sound demarcates division of trees

PHYSICAL DEMARCATION OF THE EPHEMERAL
The ephemeral sound zone from the sound garden is physically demarcated and suggested within the orchard farms through changes in height of surrounding trees.

Pier above

Farmlands below

Creeping Bent - Agrostis stolonifera

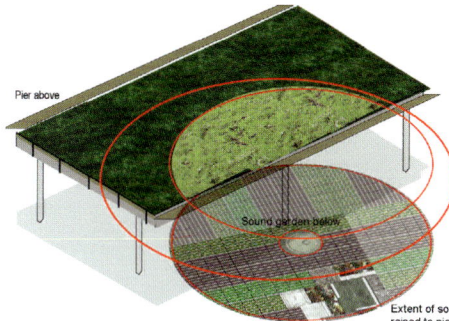

Pier above

Sound garden below

Extent of sound raised to pier level

Perennial Ryegrass - Lolium perenne

PHYSICAL DEMARCATION OF THE EPHEMERAL
The ephemeral sound zone from the sound garden is physically demarcated within the pier through changes of grass texture and colour. This further allows for a more fluid interface between the piers above and the fields below.

AGRITOURISM VILLAS

December

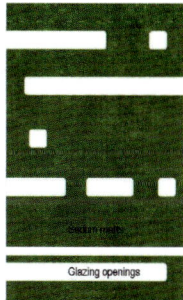

Sedum matts

Glazing openings

Circulation core

Appartment 1   Appartment 2

Sedum matts

Light wells

Light wells   Seedling deck

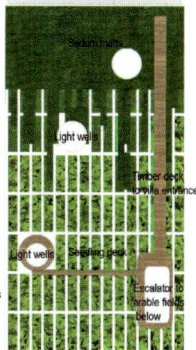

Timber deck to villa entrance

Escalator to arable fields below

Villa plan and elevation   Scale 1:600

SEEDLING DECK
The villa deck is used for growing vegetable seedlings, primarily for supplying the city through the MUSCO system. Although the deck is managed by professional farmers, its proximity to the villas means that the park visitors can participate in the growing as an educational experience. The concrete deck is divided into a grid of rectangular beds of 6m by 1.5m. These are in turn divided into trays of 1m by 1.5m wide that can accommodate 150 to 450 seeds depending on the vegetable requirements. The efficient rotational systems of growth used in the deck can accommodate winter and summer vegetables so that the deck is constantly used throughout the year, thus boosting supply to the city.

Seedling trays   Scale 1:100

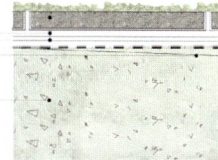

Seedling baskets
Drainage board
15mm protection board
Damp proof membrane
Screed to falls 1:60

Pre-cast concrete slab 600mm thick

Sectional detail through deck   Scale 1:50

Seedling period or soil period

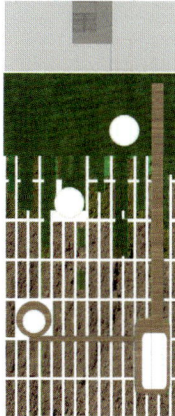

Winter
(December–March)

Summer vegetables seedling period

Spring
(April–May)

Re-planting of winter vegetables period

SEASONAL COLOUR CHART   Scale 1:600

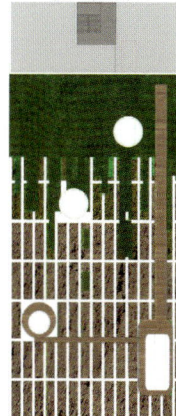

Summer
(June–September)

Winter vegetables seedling period

Autumn
(October–November)

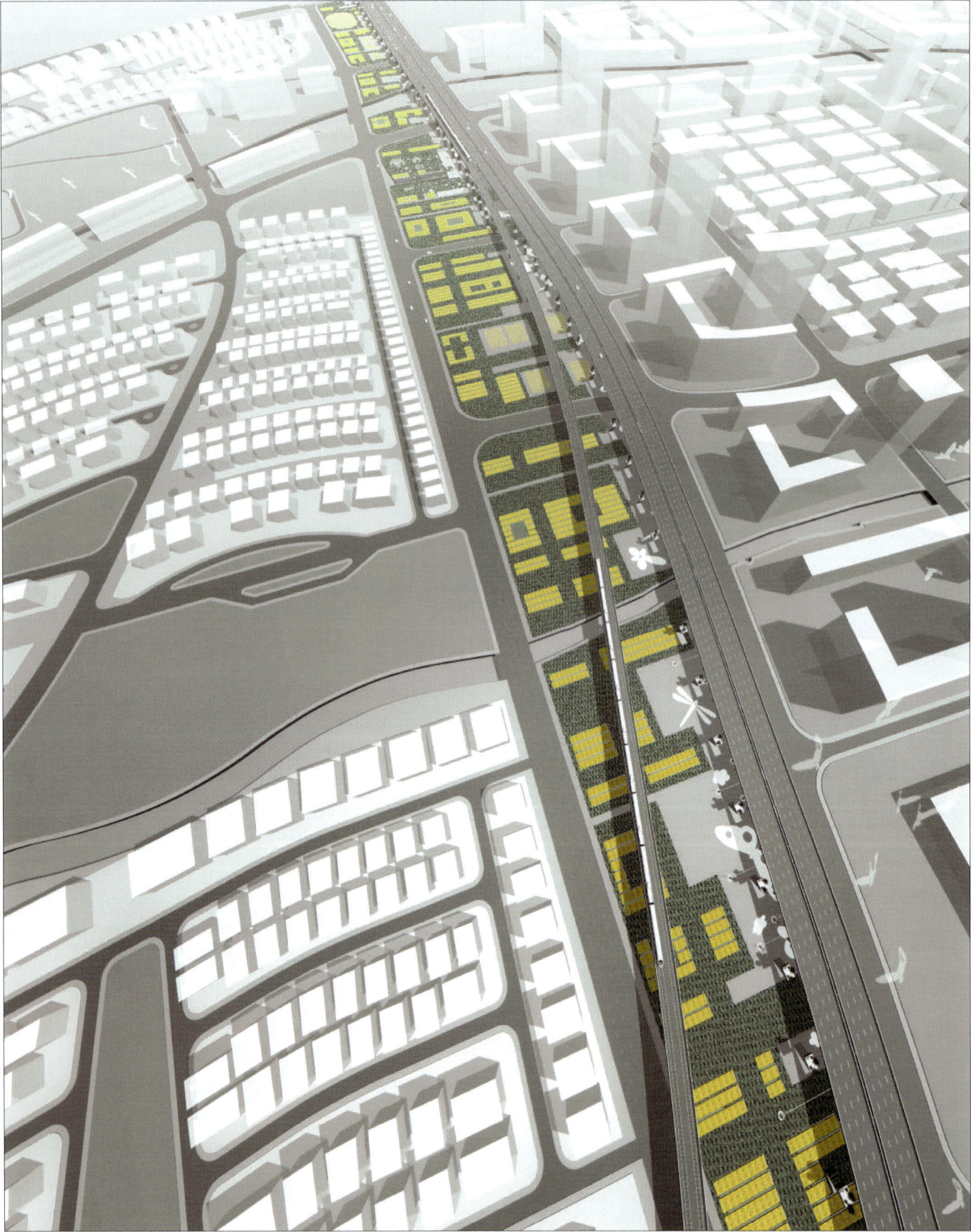

The Linear Park China

'I am interested in the ideal typical approximation of everyday phenomena – in creating the essence of reality.'

– Andreas Gursky, Hayward Gallery Exhibition, 2018

163

Many of Gursky's epic photographs offer commentaries on the various ways the landscape has been altered or controlled by human inhabitation. In 'Rhine II' (1999), for example, bands of grey and green present the river as it might have been, and as it can no longer be. The narrow horizontal bands of sky, water and manicured grass so palpably real belie the absent coal power station at the opposite bank, which Gursky digitally effaced among other markers of industry. Despite the view being entirely fictional, the neatly controlled interpretation of the real mediates complex feelings about the postmodern relationship between human and nature: loss and betrayal, culpability and longing. By approximation, Gursky draws the viewer closer to the real – there is no natural landscape in which man has not already been. Gursky is 'not addressing a 19th century society warming up for industrialisation, but a post-industrial society resolutely directed toward consumerism, individual narcissism, and leisure'.[1] Gursky's photographs are never simply descriptions of terrain.

'We don't live in the Biosphere... we live in cities.'[2] Landscape architect Martha Schwartz similarly rejects the homogenising notion of landscape as a perpetual and uncritical greening of urban space. Famous for her ironic use of cultural objects in so-called Pop-gardens – from bagels to gold painted frogs, from blue plastic palm trees to concrete tires – Schwartz's landscapes seek to contest a complacent imitative slathering of nature and regurgitation of landscape in recognisable terms, mainly naturalistic. Wallpapering a building's surface with grass squares to disrupt the mindless predilection for the 'anachronistic lawn' or repurposing familiar Central Park garden elements, like park benches into an installation that mocked New Yorkers' incapacity to read landscape beyond Frederick Law Olmstead's 19th century design, Schwartz blurs the boundaries between the natural and human-made to better diagnose and ultimately heal our imaginative paralysis about the entanglement of culture and ecology. This interiorisation of the exterior uses the vocabulary of the domestic to make evident the power of landscape to both form and critique culture. When she painted pilasters in a Harvard dormitory and its adjacent lawn after a new sunken garden disrupted the vertical and horizontal dimensions of the space, Schwartz termed her painted and temporary integration of the two as a 'gesture of healing'.[3]

The Linear Park in DongYi Wan embraces Schwartz's notion of healing and Gursky's post-industrial commentary in bringing about an awareness of ecological fragility. One cannot erase the aggressive

facing page: Aerial view of a resilient landscape that integrates nature with infrastructural scars to highlight ecological fragility.

1. S Jacobs, 'Blurring the Boundaries between City and Countryside in Photography', CLCWeb: Comparative Literature and Culture, vol.14, no.3, 2012, p.8 [https://doi:10.7771/1481-4374.2040], retrieved 16 August 2018

2. F Bernstein, 'At Home With/Martha Schwartz; Making Landscapes Pop', New York Times, 21 December 2000

3. K Campbell, Review of T Richardson (ed.) 'The Vanguard Landscapes and Gardens of Martha Schwartz', The ArtBook, vol.12, no.2, May 2005, pp.53–54

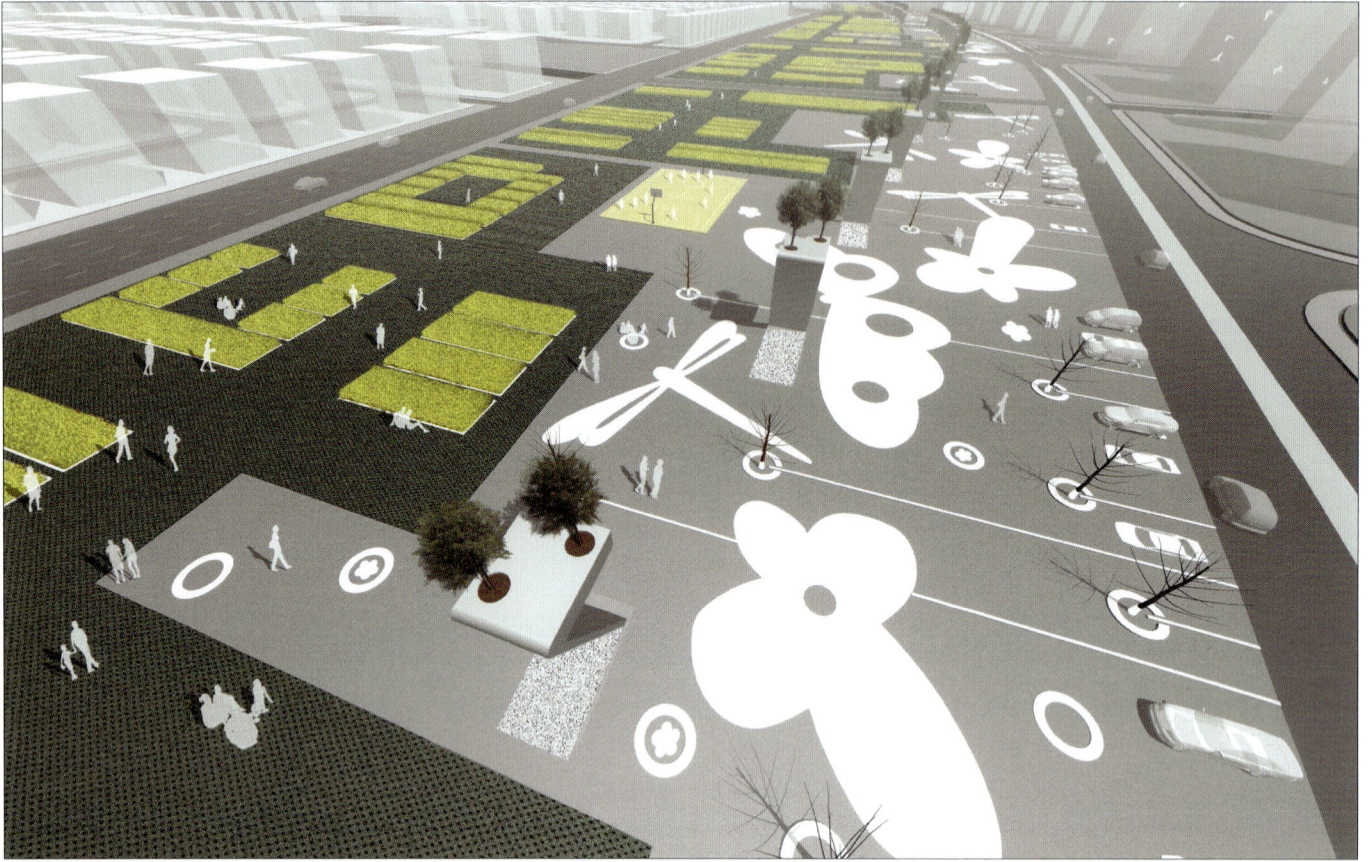
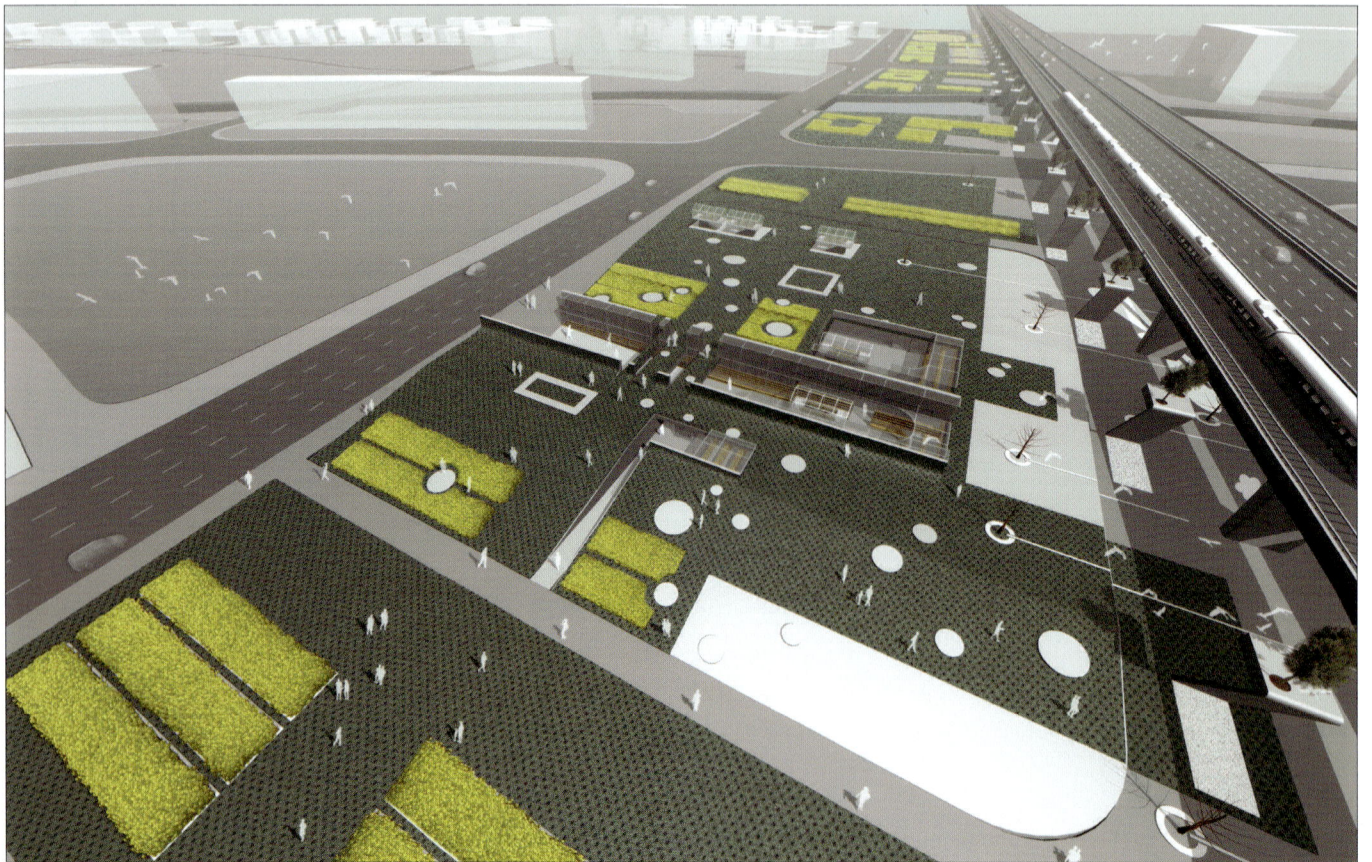

spatial and psychological scars on the landscape that the highways and the railway tracks have made across the site – three imposing strips of infrastructure float 10m above the ground, dispensing a forest of structural footings, and hence overshadowing almost 30 percent of the ground surface with little opportunity for any foliage. Physical structural alterations to the existing engineered structures are not allowed, while visual permeability in the east–west direction of the park is vital. These limitations create vast stretches of undesirable space beneath and around the infrastructure, coded with a narrative of blight, its indeterminate and unused qualities producing consequent actions of neglect and disregard as they become targeted as sites for dumping. Instead, the theorist Andrea Mubi Brighenti would identify the scarred space as a 'phenomenon on the ground', a 'happening', a 'combination' or an 'encounter'.[4] The Linear Park is an urban interstice for interplay between real nature and human-made landscape, a cultivation of the urban exterior space by transplanting the interiorised patterns of the human imaginary.

Like a piece of music score, the park complements the rhythmic design proposal for the waterfront and is divided into two zones – the 'human-made' asphalt garden and the 'real' floral garden. The Linear Park and Tangshan Earthquake Memorial Park share similar resilient philosophies and engage with the symbolisms of the natural world. From the overshadowed areas of the park, a new human-made asphalt covered garden emerges, punctuated with flower motifs, tree trunks, painted patterns and greenhouses of moss. As simulated ornamentations, these are icons of nature rather than nature itself – akin to William Morris' evocations of the organic form with his famous wallpapers, a loving artifice of nature communicates something more than the acanthus leaf or daisy can express in their original state. What is achieved in this lively efflorescence of colour and line, repeated and continuous, is how 'a beautiful piece of nature... [presses] itself on our notice so forcibly that we are quite full of it, and [how we] can, by submitting ourselves to the rules of art, express [this] pleasure to others'.[5] A radical departure from the fashion of trompe-l'œil that sought to create the illusion of the outside world, the paradoxically exuberant yet constrained motifs of Morris's wallpapers translated nature into subtle patterns that transformed the imagination into its own oasis. By creating a parallel in the outside world of the visual language of human thought, the park, like Morris's wallpapers, embraces a conceptualisation of space that is intuitively reflexive and self-governing.

Morris worked against the grain of his contemporaries' rejection of human-made forms. Deeming the human-made as ugly, their point of view turned into an unrealistic, and ultimately unsustainable, pursuit of unspoiled nature as the source of regeneration and beauty. Morris had a more integrated, and thus revolutionary, artistic language in mind. In stylised inscriptions on walls, Morris's wallpapers enchanted the quotidian everyday with glimpses of an alternative world. Anticipating the graffiti artists of our cities, or artists like Gursky and Schwartz, Morris's wallpapers insisted and demonstrated that what humans can make could, and should, be beautiful. The Arts and Crafts movement which Morris would come to lead clarified an alternative way to return to nature, by human's status as maker, creator and artist. The modern resilient imagination animating the Linear Park likewise proposes such a spirit. The Linear Park's visual logic and markings are derived in part from graffiti's transgressive property which comes from what geographer Tim Cresswell has called '(graffiti's) placement where it shouldn't be or doesn't belong'. If graffiti is a practice of 'heretical geography' that questions 'the taken-for-granted, everyday

165

facing page top: The 'human-made' asphalt garden of over-scaled floral patterns and dead tree trunks (the highways and rail tracks removed from image).

facing page bottom: The 'real' floral garden is an approximation of natural forms, highlighting the chrysanthemum plant.

4. J Ferrell, 'Graffiti, Street Art and the Dialectics of the City', in 'Graffiti and Street Art: Reading, writing and representing the city', K Avramidis & M Tsilimpounidi (eds.), Routledge, 2017, p.33

5. W Morris & M Morris, 'The Lesser Arts of Life [1882]', The Collected Works of William Morris: With introductions by his daughter May Morris, vol.22, Cambridge University Press, 2012, p.260

normative landscapes' of urban space,[6] the Linear Park manifests a similar practice in the recuperating landscape created by infrastructure. In the urban environment, graffiti works not by calling attention to any specific content, but by calling attention to what Jean Baudrillard names their self-referentiality. The organic graffiti of the Linear Park points to its own scriptive occupation of urban space 'scrambling the signals of urbania and dismantle the order of signs'.[7] Using a cool colour palette of blues, white, greys and dark brown, the Linear Park's artificial garden thus writes a narrative of resilience in the shadows of infrastructure.

167

In this approximation of a naturalistic garden, floral motifs appear via ground sunken up-lights projecting over-scaled blue floral patterns, embroidering the under-croft of the engineered structures while dead tree trunks on light stools demarcate spatial territories across the forest of structural footings. The undersides of the highway infrastructure are 'painted' by the interplay between projected light and the shadows cast upward by the tree branches. In the meantime, their dark brown branches are varnished, the glazed surfaces dispersing light by reflection. Fields of two-dimensional painted floor patterns march along the length of the park, sometimes accommodating car park provisions, while the bold white floor treatment refreshes the nocturnal version of the park. At night, ultraviolet surfaced graphics glow to enhance the tempo of the long site, with some patterns extending into the 'real' floral garden. Colonies of moss-covered greenhouse forms bring touches of green to the otherwise dark zone. These green structures are equipped with an automated water-sprinkling system. Within the hybrid 'natural' homes, interiors are timber cladded, providing users with temporary domestic comforts for the enjoyment of music. As components of the park's transgressive vocabulary of beauty, these elements '[enliven] the city's lost spaces, decorating its shadows and corners, occupying urban interstices and creating new ones by their presence'.[8]

For the other half of the Linear Park, the approximation of natural forms continues with a conscious choice to highlight the chrysanthemum plant, the most cultivated of botanical species in Chinese horticulture. Instead of masking the urban environment by a futile attempt at reintroducing a wilderness of grasses and native plants, the Linear Park continues the harmonic conversation between the natural and the human-made. Therefore, the other half of the Linear Park has a 'soft texture' of perfectly manicured grass and a never-ending tapestry of chrysanthemum. The classically calm 'real' floral garden is contrapuntal to the dynamism of the artificial asphalt garden. The glorious and poetic significance of the chrysanthemum 'centre of excellence' is magnified by the species' fascinating iterations.

With its late blooming, the chrysanthemum's association with longevity translates not coincidentally with its enduring symbolic power in Chinese cultural works. As the earliest flower cultivated for ornamental purposes, the varieties of the flower have likewise bloomed in the consciousness of poets and painters: from the 35 varieties described in the 12th century to the 70 or more enthusiastically illustrated in a collection of landscape paintings by the next. By 1708, almost 300 varieties are recorded.[9] While Queen Mary's Garden, at the centre of Regent's Park in London, is internationally renowned for its thousands of roses, the Linear Park aptly chooses the chrysanthemum as the symbol of its historical and aspirational identity. The association with retirement and leisure renders the chrysanthemum a fitting flora to

facing page: The park has a 'soft texture' of perfectly manicured grass and a never-ending tapestry of chrysanthemum.

following pages: Lighting plays an important role in the compositional poetics that extend the scenic position to complement a resilient imagination.

6. T Cresswell, 'The Crucial "Where" of Graffiti: A Geographical Analysis of Reactions to Graffiti in New York', in 'Graffiti and Street Art: Reading, writing and representing the city', K Avramidis & M Tsilimpounidi (eds.), Routledge, 2017, p.8

7. J Baudrillard, 'Kool Killer, or the Insurrection of Signs', in 'Symbolic Exchange and Death', IH Grant (trans.), Sage, 2017, p.101

8. J Ferrell, 'Graffiti, Street Art and the Dialectics of the City', in 'Graffiti and Street Art: Reading, writing and representing the city', K Avramidis & M Tsilimpounidi (eds.), Routledge, 2017, p.33

9+10. C Thacker, 'The History of Gardens', University of California Press, 1997, p.55

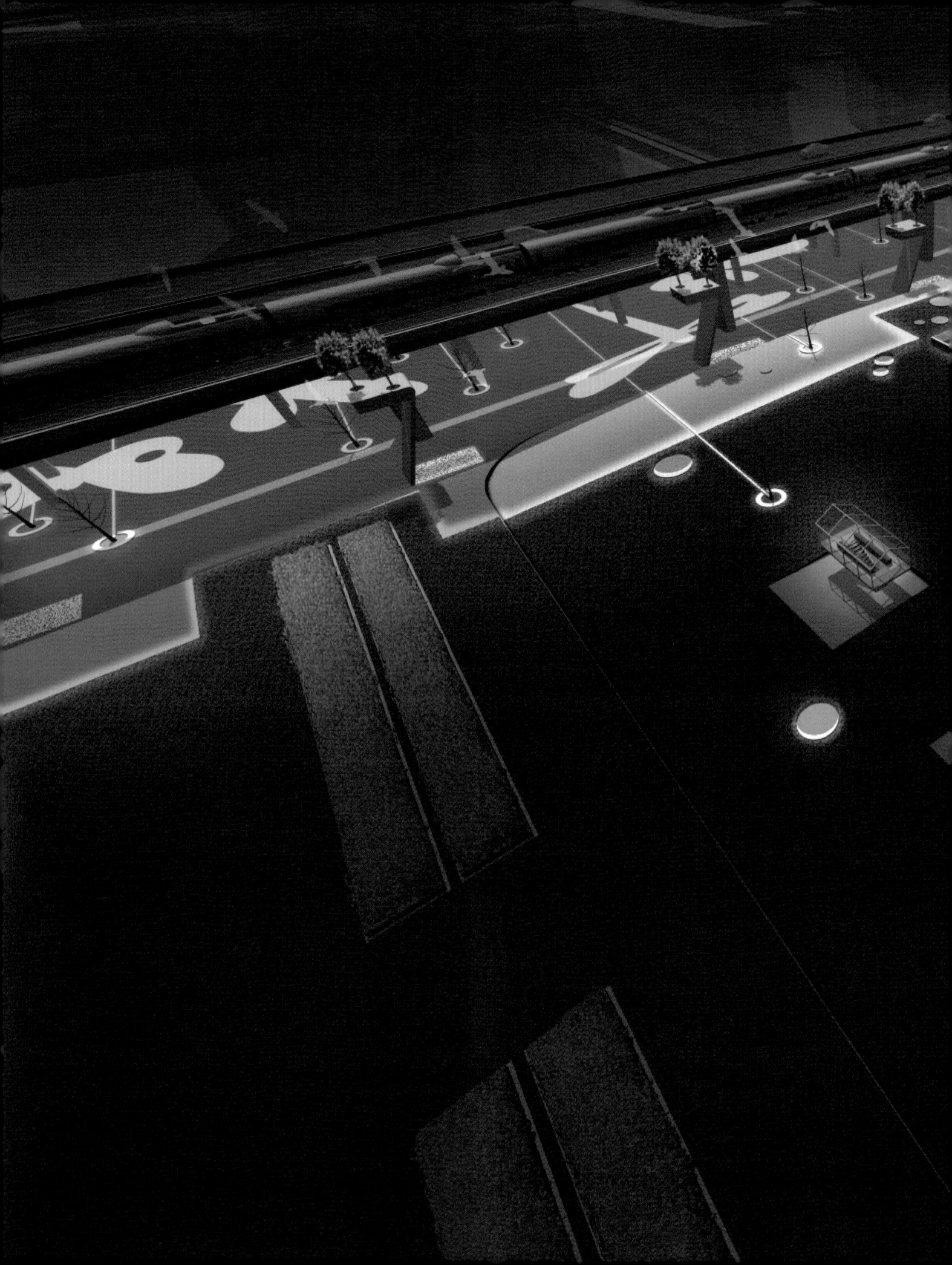

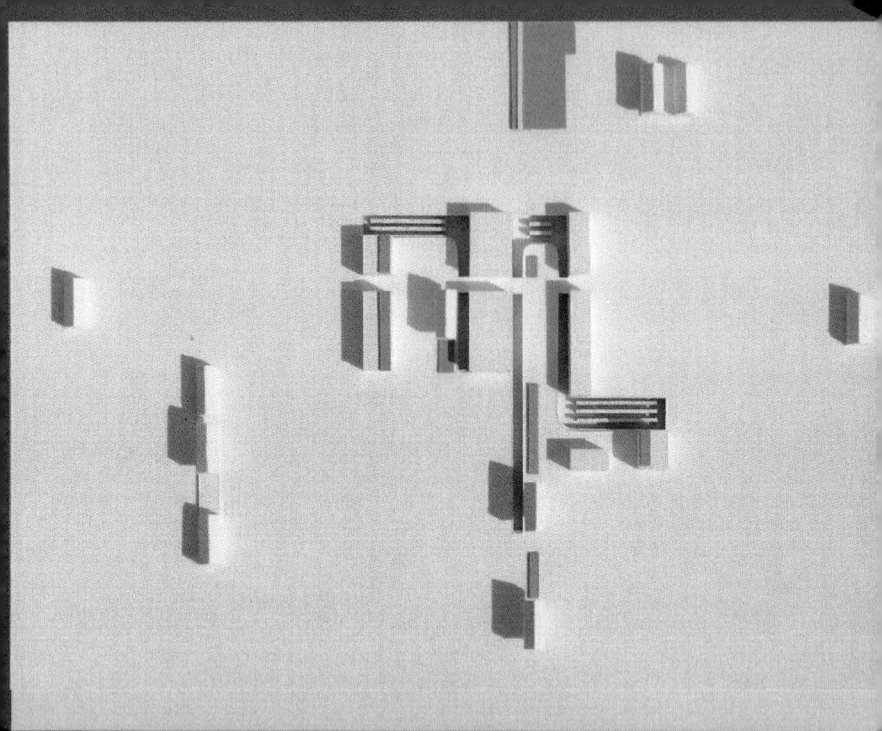
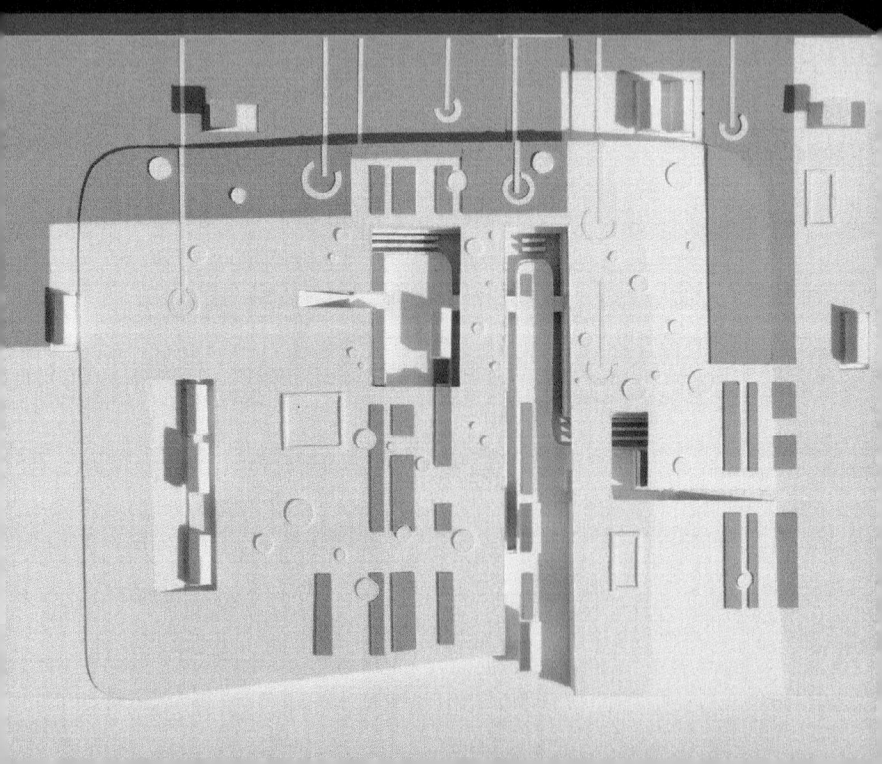

preside over the aggressively blighted space. Thus, apart from the local community, horticulturists, floral experts, and chrysanthemum enthusiasts would frequent the park. The park is home to an annual chrysanthemum festival, a gathering that is a contemporary echo of the gatherings Tao Yuan Ming, the Chinese flower lover famous for rejecting court life precisely to cultivate the cherished blossom, would have hosted in his own garden, in which the chrysanthemum becomes the centrepiece of communal assembly. As wine is made from its petals, the flower in fact unites the pleasures of the garden with the delight of kindred spirit and poetic inspiration: 'I plucked the chrysanthemums beside the hedge. In calm, I found the southern hills.'[10]

171

Other elements in the soft zone play upon the motif of the interiorised exterior at multiple key locations. Sport courts and running tracks are woven into the park. The library is tucked beneath the chrysanthemum and grass carpet; users can choose from everyday reading materials or perusing a curation of horticultural books whose pages resonate with the dramatic real floral displays. In addition, 'floating trees' clouds hover nine metres above the ground, and visually engage with passengers of cars and trains on the highway. At the level of the park, the floating trees mark the areas of rest and seating. The trees are of local species and have an independent self-regulated watering system. The main lighting system follows the threshold between the floral garden, and the 'human-made' asphalt garden. The lighting system complements its counterpart at the waterfront. Finally, the vertical fluorescent park lighting is provided within the slender columns of the floating trees.

Just as Gursky reclaimed the Rhine through a visually domesticated intelligence of an industrial landscape, the Linear Park is likewise fully invested in compositional poetics that extend the scenic position to complement a resilient imagination. The park's synthesis of multiple registers of meaning and function, creates a fascinating opportunity for the spatial practice that Henri Lefebvre characterised as rhythmanalysis – 'when rhythms are lived and blend into another, they are difficult to make out. Noise, when chaotic, has no rhythm. Yet, the alert ear begins to separate, to identify sources, bringing them together, perceiving interactions.'[11] Such an attention is vital for the integration between self and the resilient landscape. 'If we don't listen to sounds and noises and instead listen to our body... usually we do not understand the rhythms and associations which none the less comprise us.'[12] The Linear Park conducts an arrangement that sutures the seemingly discordant notes of the site. It imagines a user who can 'go deeper, dig below the surface [and] listen closely... Continue and you will see this garden and the objects polyrhythmically, or if you prefer, symphonically.'[13]

facing page: Models showing the spatial poetics of the chrysanthemum 'centre of excellence'.

11, 12+13. H Lefebvre, 'Writing on Cities', K Kofman & E Lebas (eds.), Blackwell Publishers, 1999, p.222

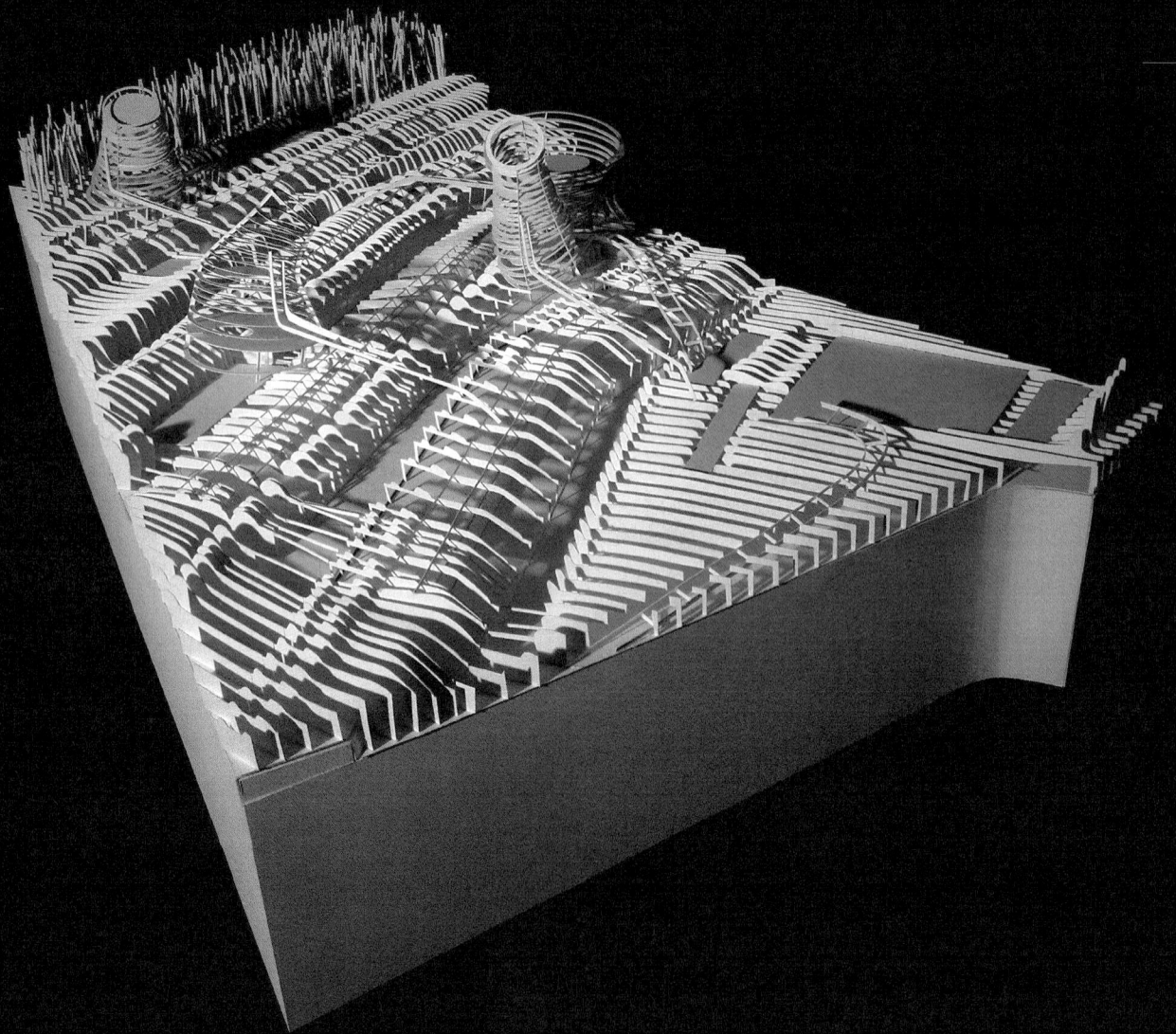

A Workplace in a Garden Ireland

'...our factory stands amidst gardens as beautiful (climate apart) as those of Alcinous, since there is no need of stinting it of ground, profit rents being a thing of the past and the labour on such gardens is like enough to be purely voluntary, as it is not easy to see the day when 75 out every 100 people will not take delight in the pleasantest and most innocent of all occupations; and our working people will assuredly want open air relaxation from their factory work.'
– William Morris, 'A Factory as It Might Be', Justice Magazine, 1884

The Utopian socialist William Morris advocated the idealism of 'factory gardens'. Within a decade of his article being published, pleasure landscapes for workers to partake in leisurely strolls, rest and eat lunch outdoors came into being at larger factories. Wholesome community activities including exercise were compulsory for the youngest workers and allotments were made available for employees' children. Cadbury, Rowntree and Leverhulme pioneered such landscapes, and were passionate about gardens for workers' wellbeing and nature as an instrument of social reform. To attract and retain a high quality and stable workforce, especially female workers, the company towns offered pension and welfare provisions, education opportunities, and most notably well-designed homes in beautiful gardens and parks with creeks. At Port Sunlight, the manicured lawns, flowerbeds and trees in the 120-acre pleasure ground was cared for by the landscaping department of the company to create a uniform look – the picturesque order of the place would not be disturbed.[1]

The idealism might have diminished after the Second World War; however, integrating nature within the workplace is tremendously beneficial in the pursuit of the Smartcity in the 21st century. Apple's futuristic headquarters in California, for example, has a ratio of 20:80 built area to landscape.[2] The ring-shaped structure by Foster + Partners encircles a park with more than 9000 trees, all of which are drought-resilient to survive a climate crisis, with a sequentially ripening orchard of cherries, plums, apricots, persimmons, and of course, apples. The ecological preservation of soil, bark and flowers 'will be dotted with not only green oaks but also hundreds of fruit trees that will provide both magnificent blooms at different times of year and fresh food for the four-story Caffé Macs on campus', described David Muffly, senior arborist at Apple.[3] Not to be outshone, Google's campus at London's King's Cross has a 300m-long rooftop terrace with fields, wildflower meadows and a 200m running track. Other corporate proponents of the factory garden have bowling greens, woodlands, deer parks, lakes and vegetable rooftop gardens. Some might perceive these initiatives as cynical ploys to offset corporate guilt; for the workers and their families access to green spaces nevertheless is a lifeline to supplement their food budget, ensuring fresh produce and family outdoors time.

facing page: Model showing contours of Wexford are retained to construct the inhabitable landscape.

1. H Chance, 'The Factory in a Garden: A history of corporate landscapes from the industrial to the digital age', Manchester University Press, Manchester, 2017, pp.28–30

2. E Tyler, 'Workers of the World Unite in Green Spaces', Financial Times: Gardens, 11 May 2018

3. S Levy, 'Apple Park's Tree Whisperer', Wired: Backchannel, [https://www.wired.com/story/apple-parks-tree-whisperer], retrieved 1 June 2017

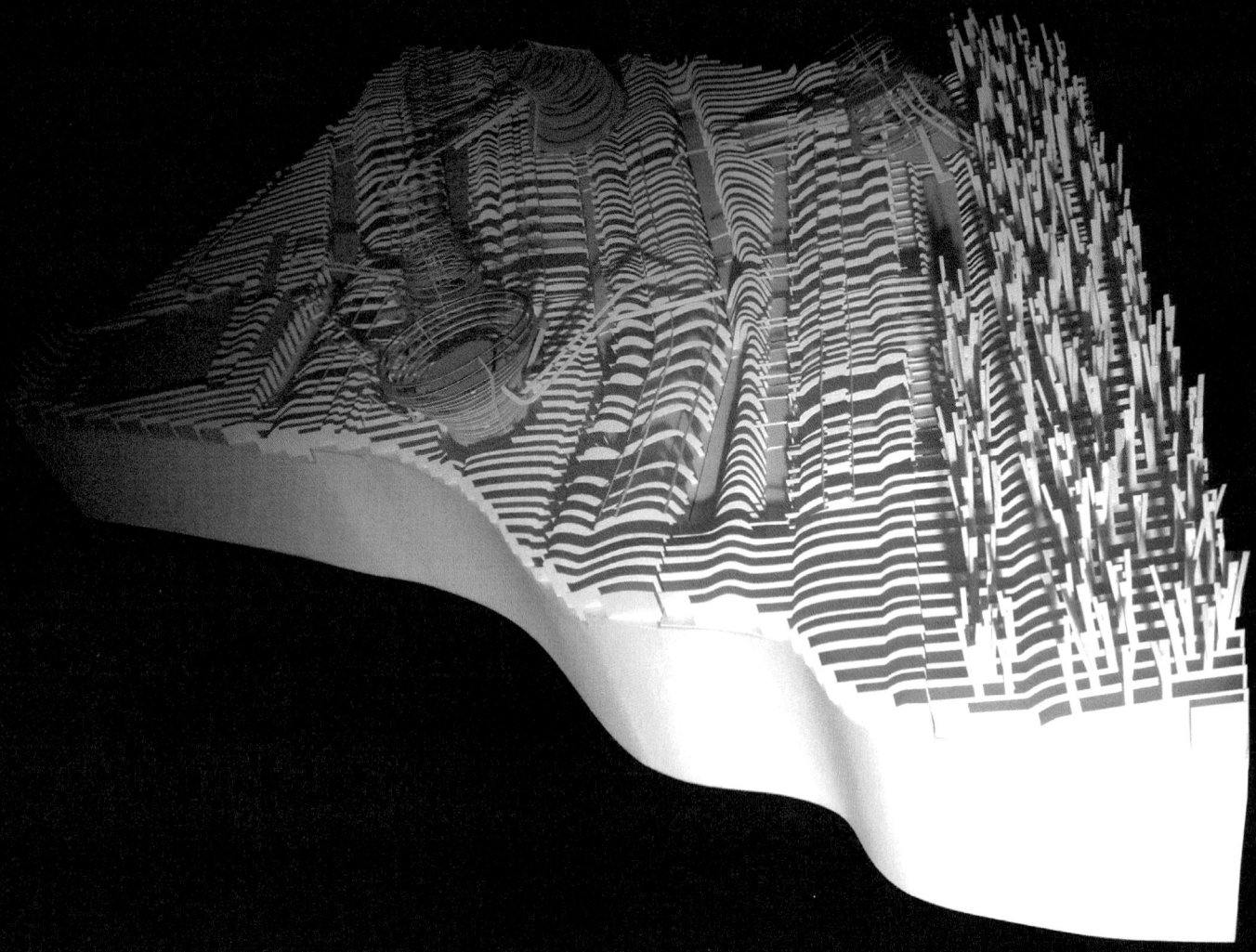

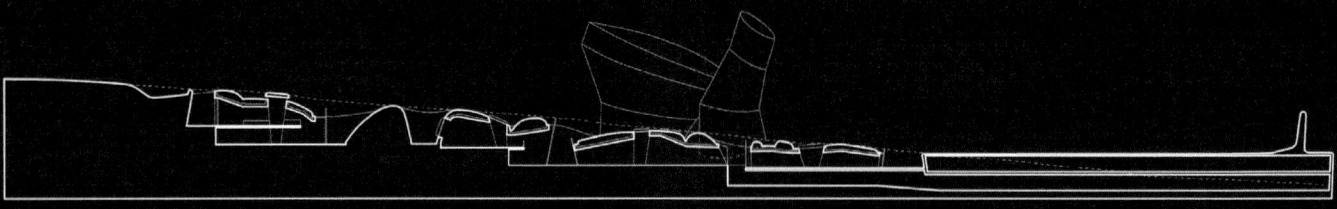

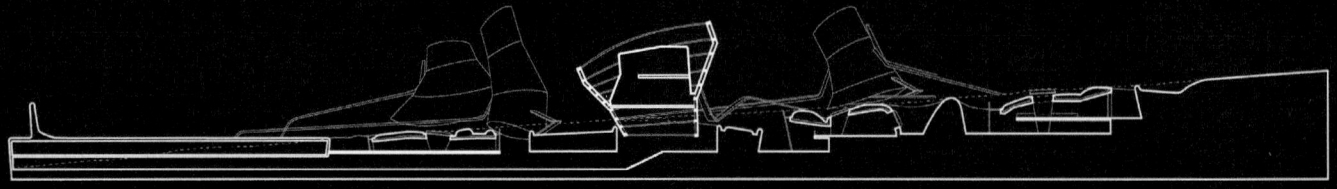

These resilient landscapes are about keeping employees happy and healthy, and consequently boosting productivity. Recounted in the Harvard Business Review in 2017, scientific research has found that 'exposure to nature might result in lower risk of depression, obesity, diabetes, and cancer. The immune system certainly receives a boost from stress-reduction, and even just the sounds of nature trigger a relaxation response in the brain, and strengthen memory and better foresight in decision-making.'[4] Charles Darwin, Friedrich Nietzsche, Ludwig van Beethoven and Mark Zuckerberg knew walking improves creative thinking.[5] Steve Jobs did his best thinking during long walks in nature, which inspired the tree filled campus. Today, many employees share similar views and like those of Morris's time, consider pleasure and wellbeing as important factors in where they choose to work.

Nature is fundamental in shaping the headquarters for Wexford County Council – the landscape is envisaged to be as much an attraction as the work spaces, and is positioned to take advantage of the natural characteristics of the grounds to enable a pleasant relationship between each department and the public. The six departmental offices peer across the green from beneath the contoured terraces, offering the most dramatic panoramic views across the site down to the River Slaney and Wexford Harbour. An array of troughs punctures the inhabitable landscape to choreograph a tempo of natural light and ventilation to allow for human occupation.

The 18th century poet Alexander Pope believed that landscapes should be designed like paintings, in particular the ideal Arcadian landscapes adorned with beautiful figures, classical artefacts and imagined buildings romantically painted by Nicolas Poussin and Claude Lorraine. In 'The Follies and Gardens Buildings of Ireland', James Howley elaborated 'the effect is both picturesque and nostalgic, evoking memories of a glorious and civilised past. The natural style sought not only to let nature take her true course but also to adorn her with buildings, to embellish and thus underline her beauty. The building simply demonstrates man in harmony with nature'.[6] There is no doubt the influence of Pope's thinking dramatically changed much of the topography of Ireland and England.

Such belief is applied to the inhabitable landscape of Wexford County Council – the beautiful ground is further enhanced by the addition of four follies judiciously placed to represent the main towns of the county – Wexford, Enniscorthy, Gorey and New Ross – whilst promoting the four civic virtues: honour, authority, culture and future. Each folly houses a specific function, respectively executive area and main lobby; council members' area; restaurant and outdoors smoking area; crèche and training rooms. Unlike the departmental offices which are deliberately simple, low-lying and nestled below ground, the follies act as 'horizon makers' and offer diverse functions and rich spatial experiences. Traditionally, the folly is merely an ornamental building of no useful purpose erected for amusement, diversion and for its picturesque extravagance to stimulate melancholic contemplation. Ireland is particularly rich in follies: sham ruins and grottoes, curious obelisks and columns, miniature temples and Gothic fantasies were inspirations brought back by wealthy patrons from their European tours. During times of unrest and, in particular, famine in Ireland, follies were mechanisms for maintaining social order. The construction of Conolly's Folly in County Kildare was to provide employment for starving farmers stricken by successive cold and wet winters, without robbing them of their dignity by issuing unconditional handouts.[7]

facing page top: The landscape breaks down the formality generally associated with municipal or institutional building typologies.

facing page bottom: Sections of the inhabitable resilient landscape.

4. E Seppala & J Berlin, 'Why You Should Tell Your Team to Take a Break and Go Outside', Harvard Business Review: Workspaces, 26 June 2017

5. C Dowden, 'Steve Jobs Was Right About Walking', Financial Post [https://business.financialpost.com/executive/c-suite/steve-jobs-was-right-about-walking], retrieved 21 May 2015

6. J Howley, 'The Follies and Gardens Buildings of Ireland', Yale University Press, New Haven and London, 2004, pp.4–6

For Wexford County Council, the landscape facilitates breaking down the formality generally associated with municipal or institutional building typologies. The main circulation route runs 'north–south'; there are four further transverse routes providing internal connections to all work spaces. A series of paths and bridges laces throughout the ground to encourage cycling, jogging and outdoors 'walking meeting' for workers, and offers an open spatial relationship amongst public, council staff and nature. There are, in addition, small 'beach' areas for sunbathing, and outdoor meetings, theatre and picnic zones with panoramic views.

177

Akin to Pope's guiding principle for garden designs, the layout of the sustainable and energy efficient council offices is based on the importance of discovering its genius loci. The environmental management together with long-term low impact maintenance are achieved by the following combinations: (1) The buried external concrete frames offer good 'U' values, and are filled with soil to facilitate the growth of grass and wildflowers; they also carry all services and heat exchange capillary tubes. The heat exchange tubes create a rise in soil temperature and a microclimate that allows an extended growing season. The park roof offers long-term external maintenance while providing efficient insulation to reduce heat loss. (2) All spaces in the building have natural lighting and ventilation. The partially buried nature of the spaces reduces direct solar gain while providing plenty of indirect light. Over the summer months, efficient cross ventilation ensures a cool and fresh temperature in the work and gathering spaces. Natural environments and greener workplaces can offer superior air quality which in turn leads to better social connection and harmony.

facing page: Wholesome community activities in a community landscape; nature is the instrument of social reform.

The natural contours of Wexford are retained almost in their entirety from constructing the inhabitable landscape – a cut and fill exercise ensures minimum amounts of material are removed from site. The excavated soil and weathered rock are valuable topsoil and aggregate. Terraces are cut just into the weathered rock head and retaining walls are formed economically with proprietary pre-cast reinforced concrete retaining elements. Earth mounds are built up as part of the landscape using geo-textile supports. Foundations are generally spread footing on rock. Work spaces are waterproofed with drained cavities and a drainage system carrying ground water around each terrace. The work spaces are constructed using proprietary formwork and pre-fabricated reinforcement cages.

The city and nature, both ecological systems are under-threat. Environmentalist Dave Foreman of the Rewilding Institute believes that for humans to survive, we need a re-equilibrated ecosystem where we commit to integrate with nature. He argued that 'although we often go to considerable lengths to protect them, the danger usually is us. The odds of us all living together, let alone soon, are slight but within the realm of possibility.'[8] Not only are Wexford County Council and other municipalities and corporations promoting employee wellbeing, they are also aiming to reduce their ecological footprint. The economy and poetics of using on-site resources is in keeping with the Smartcity's resilient strategy, and planting selection becomes an exercise to showcase biodiversity not only in native flora, but also in the fauna populations with which they are connected.

7. M Wilson, 'Dark Origins of the Garden Folly', Financial Times: House & Home, 9 September 2016

8. A Weisman, 'The World Without Us', Virgin Books, London, 2007, pp.235–239

N

0m 100 200 500 1000

	Arable organic gardening fields
	Lychee hills
	Sound gardens
	Lakes
	Guangming Flower River
	Lawn patch
	Solar gardens + Virtual gardening
	Plazas + Entrance to park
	Art displays
	Light-wells for car park

Bicycle stations + parking

Circulation for pedestrians
+ cyclists

Stairs connecting vegetable terraces

(R)   Proposed local subway station

Local subway + Express rail

Guangming Energy Park China

'Cities have the capability of providing something for everybody, only because, and only when, they are created by everybody.'

– Jane Jacobs, 'The Death and Life of Great American Cities', 1961

Hunger, or fear of hunger, remains a powerful driving force of the Chinese mentality. Many of China's inhabitants still remember the 'Three Years of Natural Disasters', the world's largest famine causing an estimated 30 million 'excess' deaths between 1958 and 1961. China's cities were not left unaffected, leading to the government issuing food coupons in urban areas and a subsequent insistence on food self-sufficiency. Farming continues to be at the forefront of national policy with Wen Jiabao, the former Chinese premier, declaring, not unjustifiably, that feeding 1.3 billion people would be China's biggest contribution to the world.[1]

Guangming Energy + Art Park stretches from the centre to the north of New Guangming Radiant City, Shenzhen, and covers 2.37km2 of agricultural land. The government brief called for an exploration of 'new relationships between (a) city and green belt, (b) urban life and park life and, (c) city development and ecological effects under rapid urbanisation'.

facing page: Guangming Energy Park masterplan.

left: The park maintains a synergistic relationship with the city through its food and renewable energy production.

1. J Wen, 'Speech at the World Food Summit: Five years later', 2002 [http://www.fao.org/docrep/005/Y4172M/rep2/china.htm], retrieved 15 May 2018

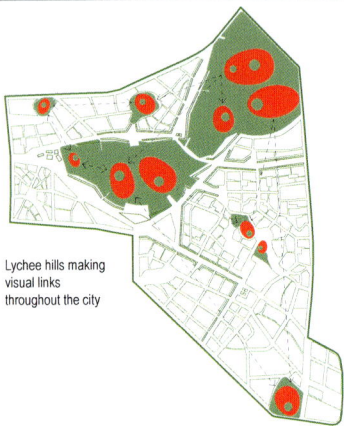

Lychee hills making
visual links
throughout the city

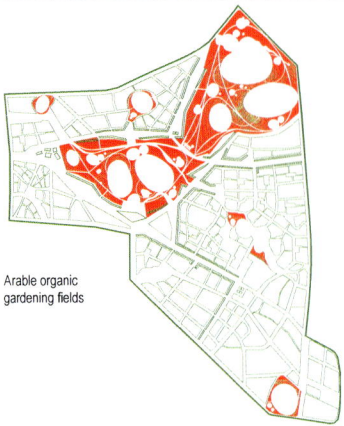

Arable organic
gardening fields

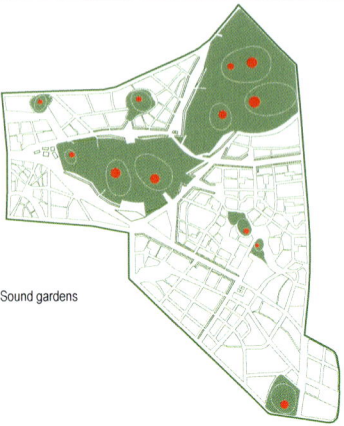

Sound gardens

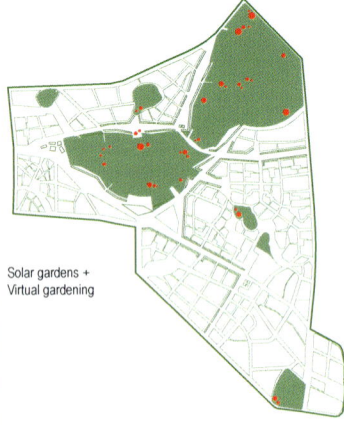

Solar gardens +
Virtual gardening

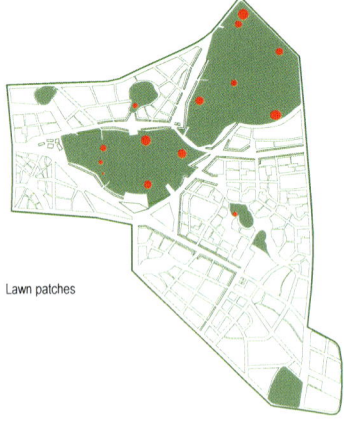

Lawn patches

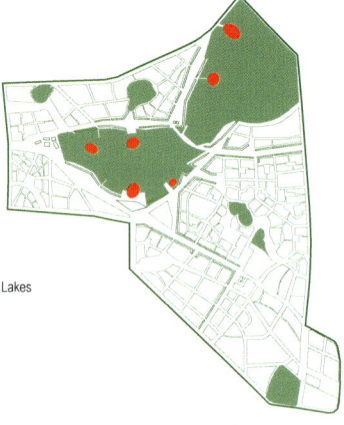

Lakes

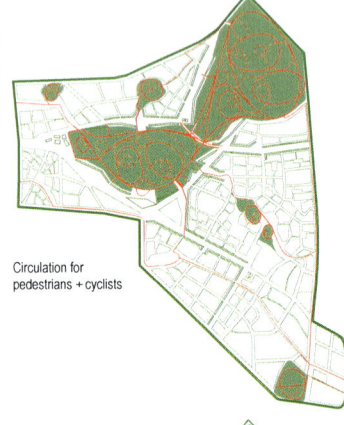

Circulation for
pedestrians + cyclists

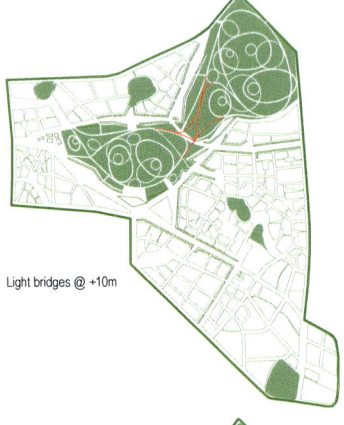

Light bridges @ +10m

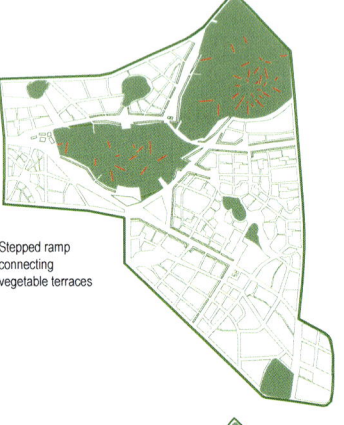

Stepped ramp
connecting
vegetable terraces

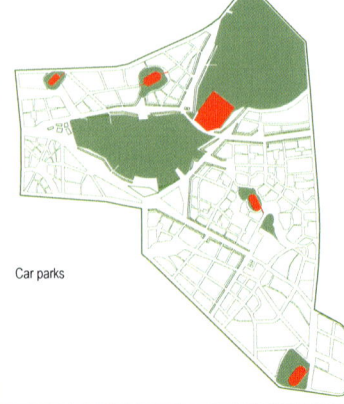

Car parks

Skylights for car park

Proposed local
subway station

The park is a site for energy creation of many types, one of which is the production of human fuel: food. The agricultural heritage of the site, along with the local farming skills and livelihoods of the populace, are therefore preserved whilst providing neighbourhood sustenance for Guangming's residents. Seventy per cent of the park is covered by an 'arable carpet' similar to that found in Nordhavnen, but without the marine-based vegetation. The remainder of the site is taken up by hills of lychees, plazas, lawns, photovoltaic gardens, and paths for cyclists and pedestrians. Existing elements of the green zone are preserved, adopting a 'do-more-by-building-less' approach. Natural resources are enhanced by redistributing and reorganising the park into landscaped clusters of flexible programme, and making the land more accessible, welcoming and ecologically sound.

The main park is linked to five satellite gardens at higher altitude. The existing topography is amplified by mounded non-biodegradable landfill forming a farming and leisure network that spreads through the city to propagate the park's environmental and resilient landscape strategies, and contributing to the character and iconography of the new city. These 'green-passive-ripples' seed new civic, recreational, agricultural and tourist facilities into the wider region. The park also maintains a diverse ecosystem and a synergistic relationship with the city through its food and solar power production.

The Radiant City is not car-free but discourages car use. Located beneath the central and most constricted part of the site are a subway station and the city's principal car park. Commuters emerge directly into the park from the subterranean areas past a cycle station and pass through the park to the north or south either on foot or by bicycle, decelerating the frenetic pace of the Chinese city.

The brief from the Shenzhen Municipal Planning Bureau called for innovative administration strategies for the park, suggesting a possible public–private partnership renting green space to individuals or companies to construct and maintain. Such partnerships are all the more essential in a resilient and productive landscape, with the arable carpet divided into 300 plots and tenured by the town's resident farmers. Every two years, the park hosts an Arable Garden Festival to mark the start of a new gardening calendar. A lottery is drawn to select new gardeners for plots, with the ten winners from the previous year's competition given the option to extend their occupancy, mirroring the land reform of the 1980s during which rural plots were leased to individual households for 15-year periods; farmers were permitted to sell excess produce once they had fulfilled an agreed quota.

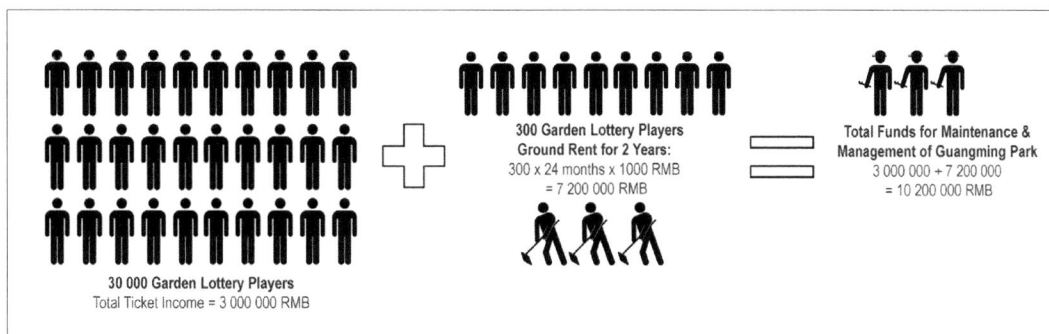

30 000 Garden Lottery Players
Total Ticket Income = 3 000 000 RMB

300 Garden Lottery Players
Ground Rent for 2 Years:
300 x 24 months x 1000 RMB
= 7 200 000 RMB

Total Funds for Maintenance &
Management of Guangming Park
3 000 000 + 7 200 000
= 10 200 000 RMB

facing page: Infrastructure plans of Guangming Energy Park.

left: Administration strategy for the park.

Economic, Social + Cultural Viability

The construction of Guangming Energy + Art Park is to be phased in conjunction with the development of the city. The topography is sculpted into contoured zones using inert recycled landfill and excavated soil from the city's construction. The existing farming community will be employed to construct the new landscape and will play an important role in the long-term park gardening workforce.

The existing farming strip within the valley is conserved and extended by employing terraces on inclined ground, reclaiming more land for agricultural production. Fresh produce will be sold directly to the community and the gardeners' markets will become vital social spaces where produce and news can be exchanged. The park will also organise an annual lychee-picking festival, drawing in tourists from neighbouring cities and beyond.

As one of the world's first large-scale urban agriculture initiatives, the Energy + Art Park will generate income and inter-city exchange through agritourism, offering visitors the opportunity to pick their own produce and have it prepared for them in the on-site kitchens or local restaurants. The Park is also the city's cultural centre, running a year-round programme of art and music. Urban topiaries, giant sculptures and digital art sit alongside fields of local produce, equating the importance of, and connection between, culture and agriculture. A series of public flora sound gardens and manicured lawns punctuate the vegetable and lychee landscape for relaxation and contemplation.

Landscape Infrastructure

Topographical adjustments are managed to minimise disturbance to the existing wildlife and natural habitats by employing non-invasive construction methods and intelligent manipulation of the ground. The revitalised lakes and new topography, sculpted from inert landfill and retained by gabions and adobe walls, provide controlled natural spaces and choreograph the naturally prevailing winds over the water to passively cool the Energy Park. The lower lying areas that constitute the organic farming carpet are a 'Centre of Excellence' for education and research, simultaneously offering leisure facilities and preserving tradition. Educational activities such as natural habitat exploration, bird watching, art exhibitions and nature walks work in conjunction with the local training, schools and tourism board. These activities encourage inhabitants to engage with and learn from their inner-city park, and help preserve the natural ecosystems financially.

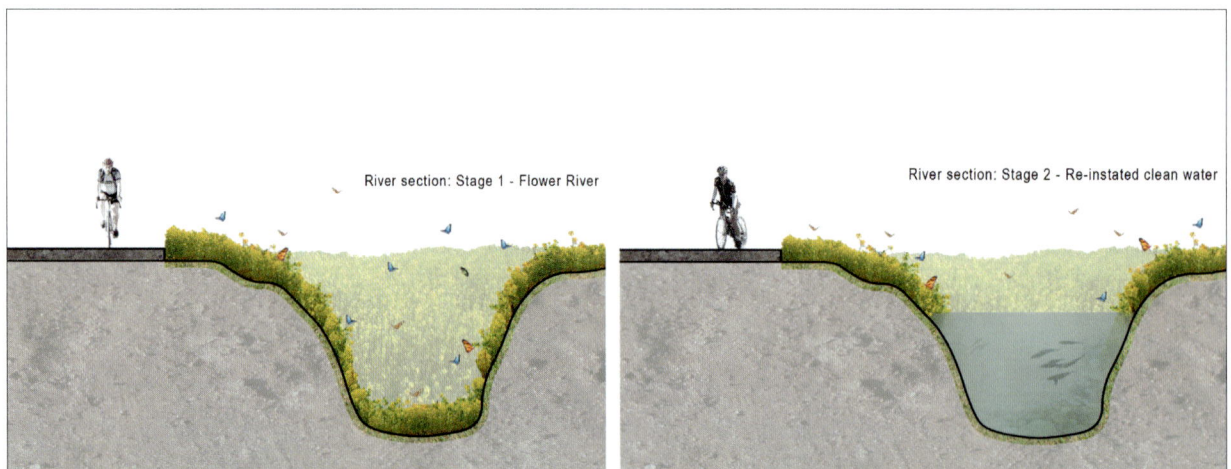

River section: Stage 1 - Flower River

River section: Stage 2 - Re-instated clean water

Irrigation of rural crops accounts for more than half of China's total water demand, and where inefficient delivery is high, excessive amounts of groundwater are being directed to agriculture. In the Energy Park, existing watercourses on the site are enlarged to form lakes and reservoirs to retain rainwater, and the canal is restored to reinforce the hydrology and water-based ecosystems present. The increased expanse of water encourages displacement cooling of the surrounding areas and fresh water fish farming using the mulberry dyke fishpond system.

The municipal authorities have determined that cleaning the entire Maozhou River that flows through and terminates within the park is unfeasible. The smaller arteries of the watercourse are therefore to be locked off and planted with flowers. A floral river will flow through the Energy Park to spread colour and biodiversity throughout the new town. Once the Maozhou River has been cleaned, the locks will be deactivated to allow nature to run its established course.

Public art in China has in the past been highly politicised, whether in the form of official works under the sponsorship of the Communist Party, or guerilla art in reaction to an autocratic regime. Guangming Park will provide a forum for new forms of art free of political affiliations, and will be curated to play off the surrounding landscape. Thematically, the installations will experiment with scale in the manner of Claes Oldenburg and Jeff Koons, and involve the community in their creation. Urban-scale topiaries will be scattered through the park and change throughout the calendar year, reflecting the farming culture and heritage of Guangming.

facing page: Guangming Flower River – mustard flowers are planted in drained polluted canals to encourage biodiversity.

left: 'Green-passive-ripples' seed civic, cultural and agricultural facilities into the wider region.

184

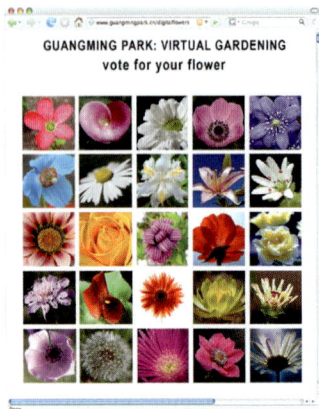

GUANGMING PARK: VIRTUAL GARDENING
vote for your flower

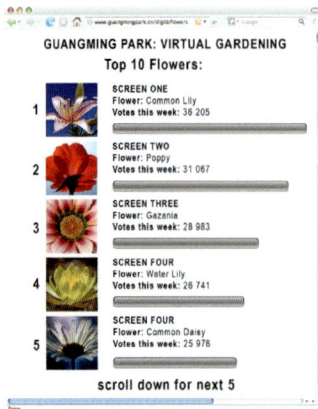

GUANGMING PARK: VIRTUAL GARDENING
Top 10 Flowers:

SCREEN ONE
Flower: Common Lily
Votes this week: 38 205

SCREEN TWO
Flower: Poppy
Votes this week: 31 067

SCREEN THREE
Flower: Gazania
Votes this week: 28 983

SCREEN FOUR
Flower: Water Lily
Votes this week: 26 741

SCREEN FOUR
Flower: Common Daisy
Votes this week: 25 976

scroll down for next 5

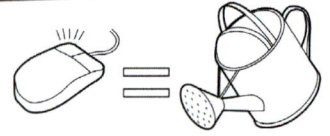

1 Click   1 Virtual Watering

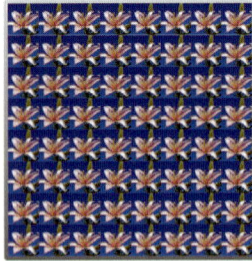

3000 votes

10 000 votes

20 000 votes

30 000 votes

Virtual gardens will contrast with their real-world counterparts. As part of the new dynamic cultural landscape, the inhabitants of Guangming are encouraged to select floral arrangements online which are abstracted and displayed on large-scale digital screens. During the day, the screens fold into a horizontal aspect to become shading devices and to collect solar energy from an array of photovoltaic cells. When dusk falls, the screens rotate into a vertical position to screen a luminescent floral spectacle that lights the boardwalk and discourages crime. Once voting is complete, the public can 'virtually water' their flowers to watch them grow. After two weeks, the images are reset and voting recommences.

The addition of sound gardens dispersed through the park creates a multi-sensory environment; accessed off the boardwalk, the sound gardens provide moments of contemplation and quiet observation of some of the park's more distinguished flora. Each sound garden has its own identity defined by the colour and smell of its unique plants. The gardens vary in size and music genre depending on the site typography and surrounding conditions. Sound is generated from both organic and mechanical sources.

Access

Plazas are located at entrance points on the perimeter of the park. They serve as meeting hubs, information points and spaces to inhabit with creative and recreational activities. The stone floor is an important theme for these spaces, providing a stage for traditional pastimes that occupy urban space such as the practice of tai chi, water calligraphy, public poetry and chess. The city's car parks are housed beneath the green landscape of the Energy + Art Park and its satellites. Visitors and residents are encouraged to leave their car and utilise the extensive network of cycle paths to promote an environmentally greener city. Glazed towers are stationed at key locations from which bicycles can be rented with a refundable deposit. The towers operate on a rotary mechanism to take up a small footprint and appear as illuminated beacons at night.

facing page top: Virtual Gardening.

facing page bottom: Digital screens attached to the solar structures are used for outdoor screenings as well as virtual gardening.

Park Management

The photovoltaic gardens take advantage of the region's abundant sunlight to harvest energy during the day to contribute to the national grid, and to power a low energy lighting system in the park. Maintenance is minimised by the disposition of agricultural land that is tended by the farmers. Additionally, there is a park keeper's lodge located in the centre of the park next to the subway station. Upkeep of the park is paid for through lottery money and ground rent, used for dredging and lake maintenance, hedge and topiary pruning, mowing, leaf collection, electronic maintenance and general cleaning.

Just as China has been unwavering in its support for policies on limiting the effects of climate change, the country still needs to take significant steps to coordinate sustainable development of its people, resources and the environment. In 2018, the Chinese government assigned three sustainable development zones to implement the United Nations 2030 Sustainable Development Goals – Taiyuan, Guilin and Shenzhen are to tackle issues on air and water pollution, desertification, and ecological and resource management respectively.[2] Only by securing all forms of sustainable energy security, will it be possible to maintain a resilient society and secure a foundation for the development of a Smartcity.

2. S Song, 'Here's How China is Going Green', World Economic Forum, 26 April 2018 [https://www.weforum.org/agenda/2018/04/china-is-going-green-here-s-how], retrieved 15 May 2018

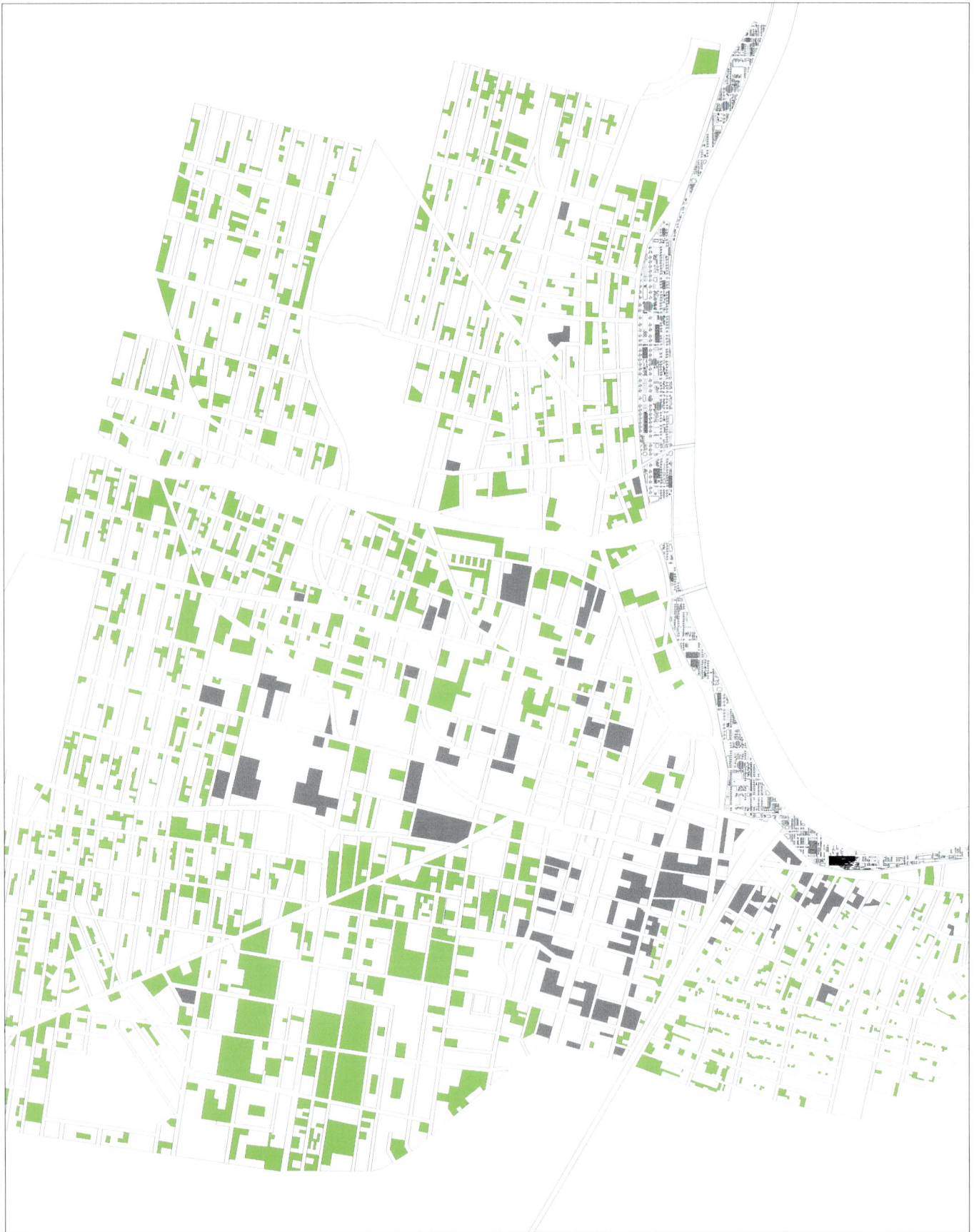

Newark Gateway Project USA

'One swallow does not make a summer, but one skein of geese, cleaving the murk of March thaw, is the Spring.'

– Aldo Leopold, 'A Sand County Almanac and Sketches Here and There', 1949

In the field of crystallography, a small crystal is dipped in a supersaturated solution as a nucleus that accumulates material on its surfaces to grow into a large lattice structure. Compounds in solution are free to move and interact inter-molecularly, and the seed provides a pre-formed pattern for colliding molecules to follow, a process known as seeding. Similarly, cloud seeding disperses particles of silver iodide that have a similar crystal structure to ice into the air as nuclei for raindrop formation.

The new visitors' centre for Newark, located five minutes walk away from Pennsylvania Station, is a Smartcity nucleus that attracts residents interested in a sustainable lifestyle to sow seeds into the wider community. Whilst the centre is a single building, the scale of its ambit is urban. In the same way that a museum spreads culture and a football ground esprit de corps, Newark's new gateway building propagates a Smartcity lifestyle through self-regulating and promoting car-free commuting, fresh produce and the distribution of seeds to expand growth of the extant urban agriculture movement.

Located five miles west of Manhattan and two miles north of Staten Island, Newark is the largest city in the state of New Jersey. The make-up of its residents is diverse, with five political wards that have significant African American, Italian, Jamaican, Hispanic and Latino populations. Despite an upturn in recent years, the city has suffered heavily from problems of poverty and crime, resulting in a continual population loss since the late 1960s.

Newark Penn Station is a major transportation hub served by Newark Light Rail, New Jersey Transit commuter rail and Amtrak, as well as local, regional and national bus services. Due to its proximity, the visitor centre is ideally located as a gateway to the city with the potential to transform how people commute into and out of Newark by providing cycle parking, storage, shower, laundry and changing facilities. Since the 1950s, Newark has been an auto-centric environment with minimal concession to cyclists, but there is a growing lobby for the implementation of cycling infrastructure in a city once known as the nation's bike racing capital.

 A less orthodox but equally transformative service provided by the gateway building is the provision of fresh locally grown produce available to commuters on their way home. The South and West Wards of

facing page: Newark Gateway Project masterplan – Vacant and derelict sites (green) are transformed into community food production spaces; A new visitors' centre within the ribbon of community gardens along the Passaic Waterfront.

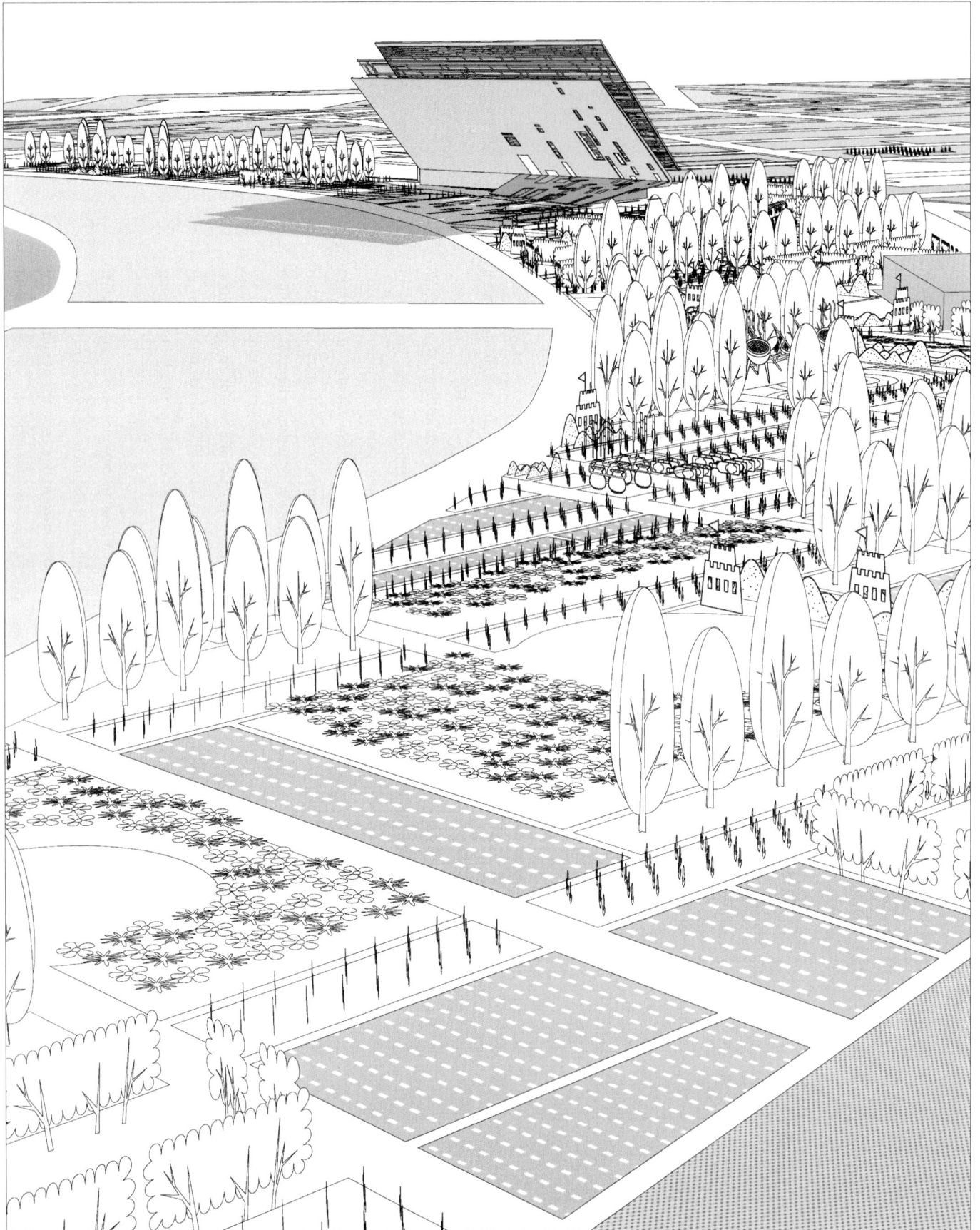

189

facing page: The new Passaic
Waterfront and Riverfront Park, and
visitors' centre.

left top + bottom: The Centre is
located along the Passaic Waterfront,
minutes away from Penn Station;
fresh local produce is made available
to commuters.

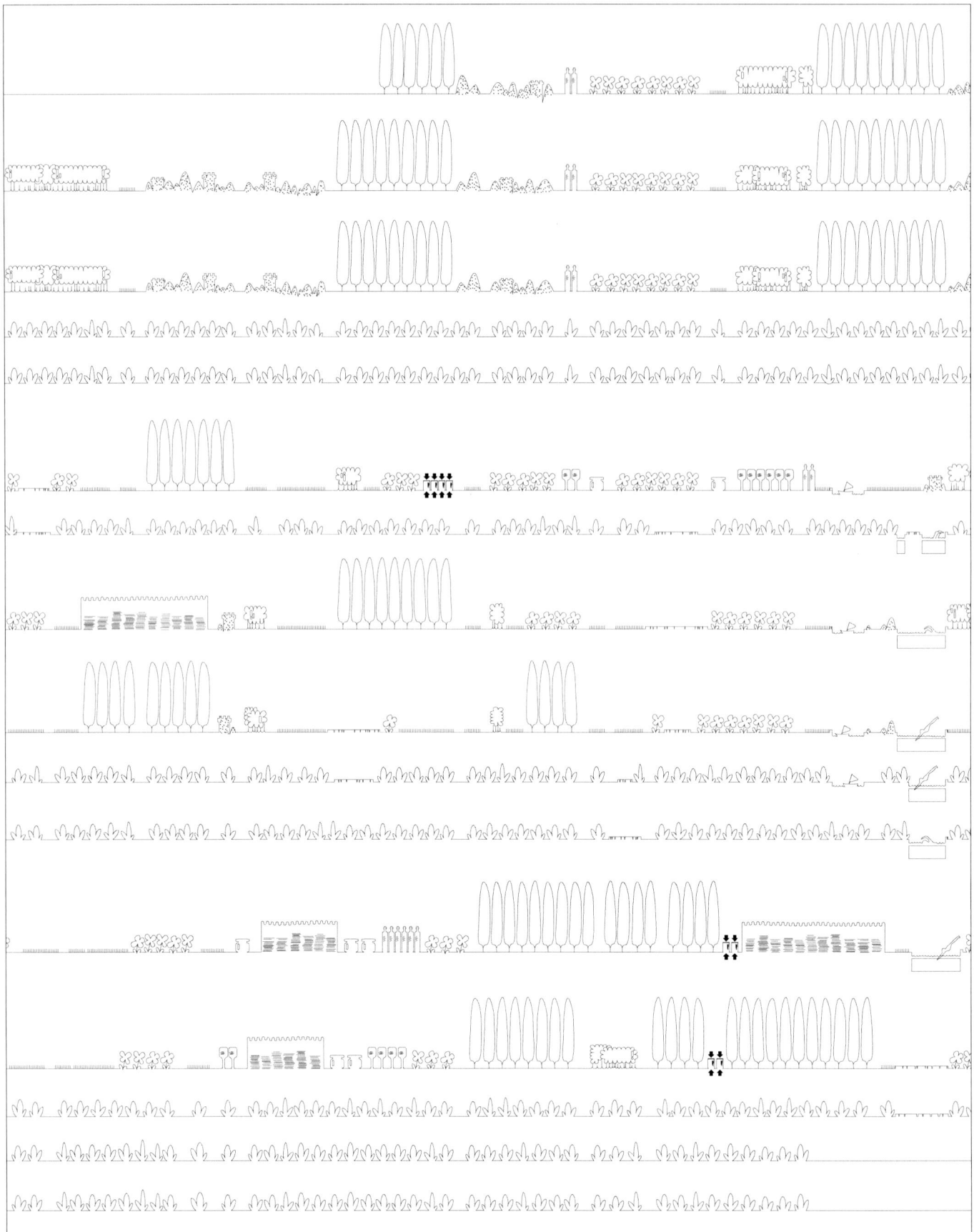

Newark, along with the dense deprived neighbourhoods of cities such as Detroit and Memphis, have been subject to a phenomenon known as urban 'food desertification' – large supermarket chains, particularly in recent times of economic uncertainty, consider these neighbourhoods as high-risk ventures and have been slow to enter less affluent markets. Unless residents are prepared to travel beyond the city, the default option for the residents of Newark is fast food rather than fresh food.

The small-plot intensive farms of the Brick City Farms initiative represent small oases in the urban food desert, employing portable EarthBoxes to grow fruit and vegetables on a number of unused sites. The positioning of these resilient landscapes in the midst of their customer base means that the urban farmers can direct market, removing the middlemen and consequently reducing the cost of fresh nutrition paid by the general populace. Swiss chard, aubergines, peppers, cucumbers, rocket, collards, spinach, turnips and herbs have been successfully cultivated and sold to the community as well as local restaurants. The sub-irrigated planters, raised above ground contaminated by former industrial activities, are able to offer extremely high yields per square foot, and are efficient in their consumption of water and fertiliser. Encouragingly, the number of individuals willing to invest their time, funds and energy into the programme is showing little signs of abatement.

It is apparent from cursory inspection that there remain a great number of vacant lots that could be brought back into use, transforming derelict sites contributing to anti-social behaviour into community food production spaces and maximising the spatial efficiency of precious urban real estate. To date, the Mayor has provided legislative assistance to the Brick City Farm programme, and private landowners have been sufficiently enlightened and community-minded to lease unemployed space for food production rather than as parking lots. Expansion is nonetheless necessary to significantly change the lifestyles of the Wards' residents and the appearance of the city.

The gateway building takes the form of a striking inclined fin along the riverbank that is half billboard, half greenhouse. This fin is partially covered in photovoltaic panels orientated towards the south for the collection of solar energy, and also contains a seed nursery supplying the community farms.

Within the shadow of the glass billboard, a gently inclined timber platform accommodating storage units is the venue for the farmers' market and a public plaza forming part of the planned Passaic Waterfront and Riverbank Park. Within a limited building footprint, the centre also provides exhibition spaces, a lecture theatre and an information suite for education on sustainable living.

The visitors' centre, while being the nucleus for the city's new green network of community gardens and an icon for healthy living, is just the start of the Passaic Waterfront development. Currently home to abandoned chemical factories and derelict warehouses, there are plans for a green promenade with a collection of thriving mixed-use developments to revitalise the neighbourhood by 2025. In the spirit of the temporal small plot intensive farms, there is scope for further expanding the resilient landscape of vegetable and floral gardens, bottle and paper recycling banks, sandboxes and bathing pools that could offer a lasting legacy of social and sustainable principles for the future park.

facing page: Sections through the waterfront park.

following page: The Soil Remediation Plan – 'Phytoremediation' employs the natural biological, microbial or physical activities and processes of the plants to detoxify or immobilise environmental contaminants in a soil, water or sediments.

yarrow | alfalfa | field chickweed | winter rye | winter yarrow | red clover | bermuda grass | big bluestem | hydrangea | garden lupin | golden rod | golden rod

192

JANUARY

FEBRUARY

MARCH

APRIL

MAY

JUNE

JULY

AUGUST

SEPTEMBER

OCTOBER

NOVEMBER

DECEMBER

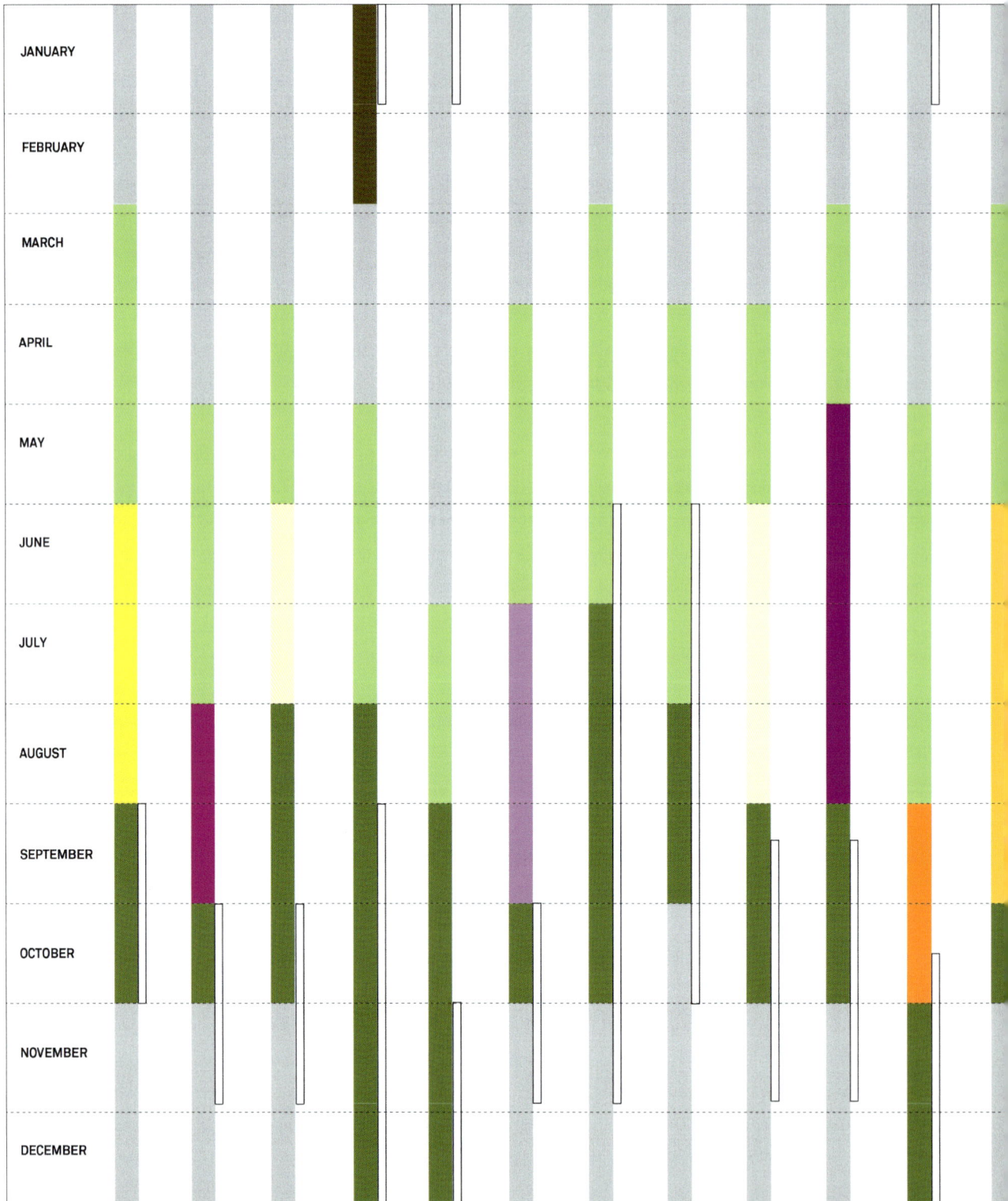

harvest/ digestion   blooming   blossoming

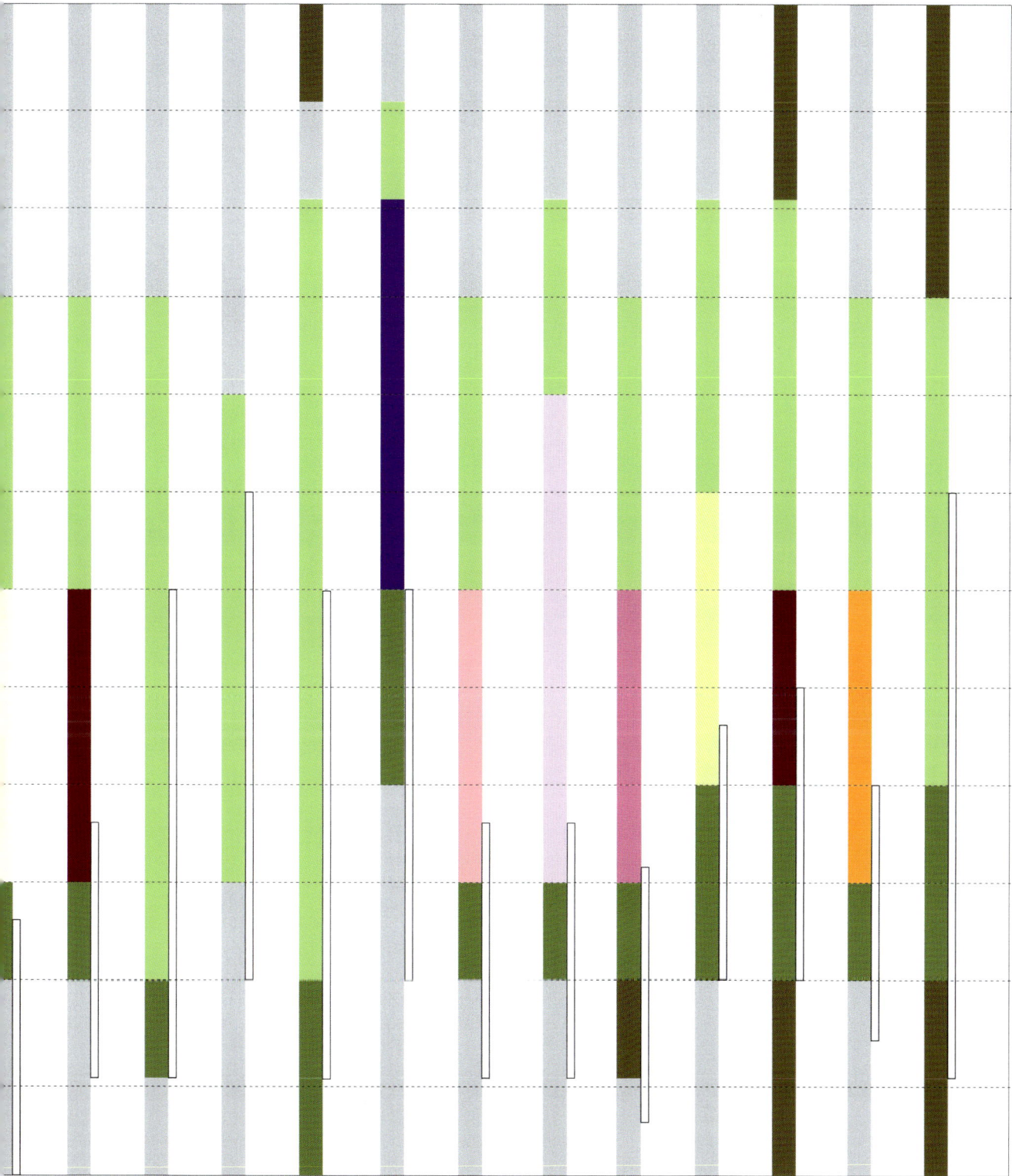

garden orache

kale

minner's lettuce

bermuda grass

violet

fox glove

marsh mellow

vetch

dog wood

mulberry

sunflower

willow

193

drying stage

The City of a Thousand Lakes China

'The rural landscape became a lost Eden, a place of one's childhood, where the good air and water, the open spaces and hard and honest work of farm labour created a moral open space that contrasted sharply with the perceived evils of modern urban life. Constable's art functions as an expression of the increasing importance of rural life. "The Hay Wain" is a celebration of a simpler time, a precious and moral place.'

– Steven Zucker, 'Constable and the English Landscape', 2015

One sees John Constable's 'one brief moment caught from fleeting time a lasting and sober existence'[1] as oneness with nature while gazing at his iconic six-foot paintings set in the countryside of his childhood years, or at the quick impressionistic brushstrokes in his 'plein air' sketches. In contrast to the fashionable historical paintings of the era, his landscapes speak to the viewer about an attachment to a disappearing countryside. The paintings of pastoral scenes with elms, hedgerows and cornfields where time has not been allowed to move on reveal Constable's disdain towards a dystopian industrialised England. Robert Hughes wrote in his review that Constable 'did not so much idealise stability as worship it. Peace, security, the untroubled enjoyment of unproblematic Nature: such is the main motif of his work.'[2]

What is it that makes Constable's landscapes so apt for discussing the changes of our fast disappearing countryside two centuries later, and perhaps their contribution towards cultivating the Smartcity so important? His indelible signature in the pictorial tradition of British landscape is more than just his passionate observation of nature and an idyllic view of the English country life. For Constable, nature is the arbiter of resilience – the death of his landscapes serves as a cautionary tale of indiscriminate urbanisation and that not all change is for the better. Urbanisation is 'one of the most irreversible human impacts on nature, and poses direct threats to productive ecosystems, affects energy demand, alters the climate, modifies hydrologic and biogeochemical cycles, fragments habitats, and reduces biodiversity'.[3] As a result, such idyll landscapes are rare in modern Britain today.

Not surprisingly, China's rapid economic growth through urbanisation has changed much of the country's landscape, particularly in the loss of agriculture land. The policy is most prevalent in rural provinces of low and middle incomes. Adverse climate conditions, low yields and incomes from farming, and the enticement of blue-collar work opportunities have accelerated rural to urban migration, causing further abandonment of agricultural livelihoods. In 2015, total cultivated land area in China was about 1.827 billion mu (121.8 million hectares), or 1.39 mu (0.09 hectares) per capita, about a third of the

facing page: A city whose physicality, social wellbeing, culture and economy revolve around water and its natural landscape.

1. M Craske, 'Art in Europe 1700–1830', Oxford University Press, New York, 1997, pp.37–38

2. R Hughes, 'The Wordsworth of Landscape', Time Magazine: Art, 25 April 1983 [http://www.time.com/time/magazine/article/0,9171,923575-2,00.html], retrieved 28 August 2018

3. KC Seto, M Fragkias, B Güneralp & MK Reill, 'A Meta-Analysis of Global Urban Land Expansion', PLoS ONE 6(8): e23777, 2011 [https://doi.org/10.1371/journal.pone.0023777], retrieved 28 August 2018

global average.[4] The government's 'red line' was 1.8 billion mu (120 million hectares) to achieve 95 per cent self-sufficiency – the Ministry of Land and Resources has since improved the goal for farmland conservation to at least 1.865 billion mu (124.33 million hectares) by 2020.[5] In an unprecedented move, China, which has long been preoccupied with guaranteeing self-sufficiency in food, has allowed some farmland to lie fallow to reduce huge stockpiles of grain and restore depleted soil.[6] These long overdue actions are key for the alternative form of economic growth in China to succeed: one that cultivates sustainability and mutually beneficial realignment to rural communities and the natural landscape.

197

The rural province of Gaochun in Jiangsu Province is an ecological hinterland of Nanjing, situated between two large lakes – the 196-square-kilometre Lake Shijiu to the north and the 35-square-kilometre Lake Gucheng to the south. The total fresh water surface occupies one-third of the province's physical area and is actively protected by the local government. Within the vast fertile water plain is a picturesque landscape of 'a simpler time, a precious and moral place', that echoes that of Constable's landscapes. The tranquillity of the water-fields, ponds, rivers and lakes is immensely inviting. Gaochun is the city of a thousand lakes whose physicality, social wellbeing, culture and economy revolves around the awareness of water and its natural landscape. The city's constitution for self-regulation is focused on the quality and quantity of its fresh water: climate change is expected to threaten water resources and food security, and increase the vulnerability of rural communities.

facing page: Gaochun is reimagined as an aquaculture inhabitable landscape that respects heritage and traditions.

following pages: Plans of the resilient landscape.

Currently, the economic draw of neighbouring cities has rendered Gaochun a surreal environment – the aquaculture villages are almost completely inhabited by the third age and the very young, while those of working age have migrated for job opportunities, only to reappear transitorily in droves at key moments during the crab-harvesting calendar. Gaochun needs to establish an ecological symbiosis between nature and built form to protect local sustainability and resist migration out of rural communities. It is, therefore, vital for local policymakers to develop resilient landscape strategies to not only safeguard the idyllic rural environment, but also promote economic growth by keeping productive land in use.[7] High-speed digital networks together with employment opportunities, education and health facilities will incentivise rural wellbeing, agricultural livelihoods and facilitate an equilibrium between rural and urban integration – the need to relocate to highly dense urbanised areas will be no longer necessary.

The new development of Gaochun has one eye on an urban future and one eye looking warily at the state of unsustainable ghost cities in China. From high-rise apartment complexes to shopping malls, hundreds of grand scale cities in China are largely empty, obsolete before they were ever inhabited. Employing a similar social-economic policy as in Guangming Smartcity, Gaochun recognises the roles and land rights of its agricultural communities. The resilience of the city rests on community cohesion, works with nature, and understanding the significance of its ecosystems. The director of the planning department of China's National Development and Reform Commission, Xiaodong Ming, acknowledges that 'the next task is to improve the layout patterns of urbanisation, the goal is to balance the spatial layout, optimise the scale of cities and towns. It requires promoting urbanisation strictly in accordance with plans for managing land, water and ecology.'[8] The rural and urban integration is reimagined as aquaculture inhabitable landscapes that respect heritage and traditions rather than civil engineering

4. Z Yan, 'China Seeks a Balance Between Food Security and the Urbanisation', www.china-embassy.org.uk, 15 October 2015 [http://www.chinese-embassy.org.uk/eng//xnyfgk/t517904.htm], retrieved 29 August 2018

5. XinHua News, 'China Sets New Goal for Farmland Conservation', www.xinhaunet.com, 24 June 2016 [http://www.xinhuanet.com/english/2016-06/24/c_135463245.htm], retrieved 30 August 2018

6. D Patton, 'China to Map 70.5 Million Hectares of Critical Arable Land: Xinhua', Reuters, 11 April 2017 [https://www.reuters.com/article/us-china-grains/china-to-map-70-5-million-hectares-of-critical-arable-land-xinhua-idUSKBN17D0Z6], retrieved 30 August 2018

7. X Deng, 'Urbanisation and Impacts on Agricultural Land in China', in 'The Routledge Handbook of Urbanisation and Global Environmental Change', KC Seto et al. (eds.), 2015, p.45

8. X Ming, 'Rethinking Infrastructure: Voices from the Global Infrastructure Initiative', McKinsey & Company, 2014, p.52

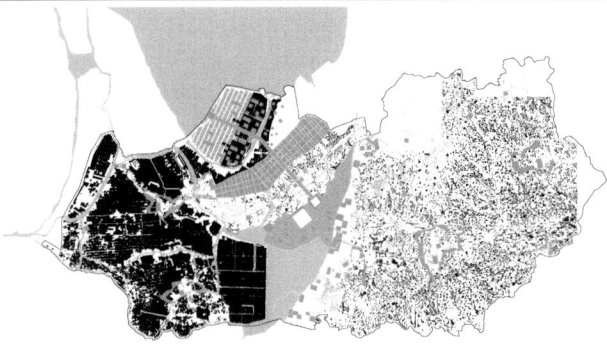

Aquatic Agriculture

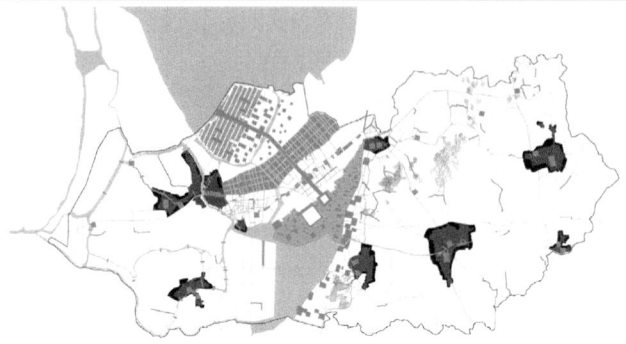

Bamboo Forest Rings

Orchards

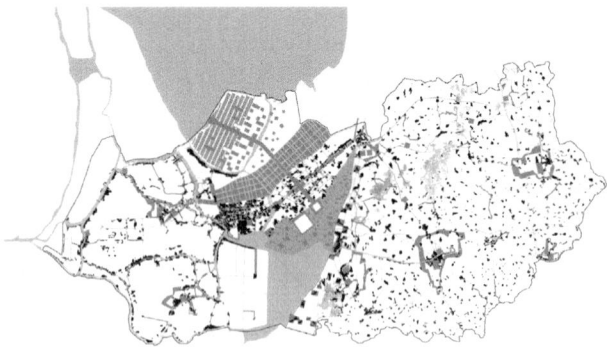

Retained Buildings & Villages

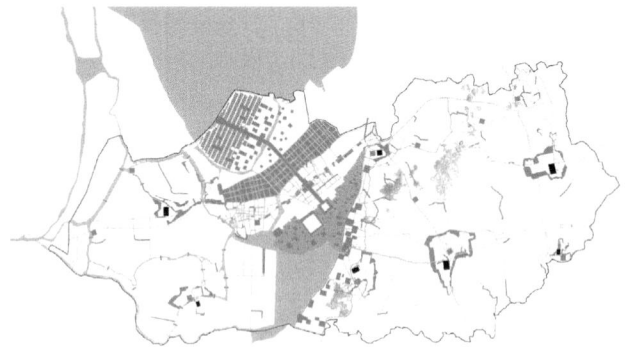

Anaerobic Digester Parks

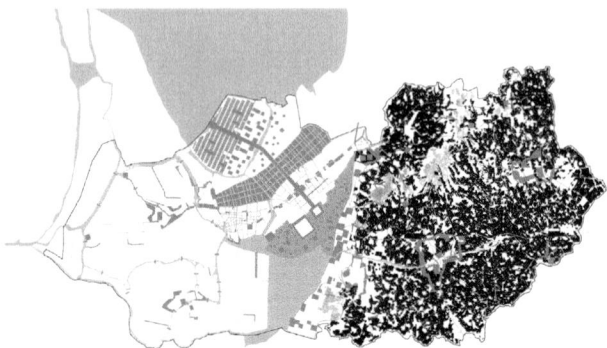

Retained Fields

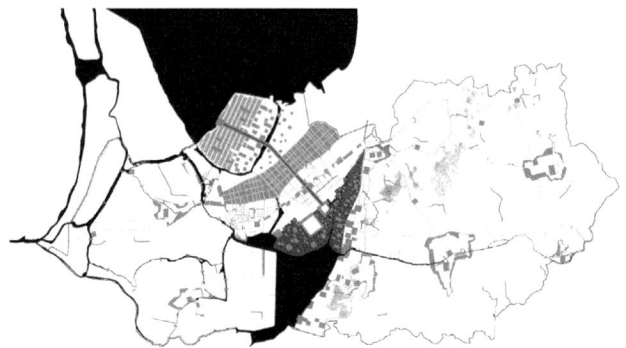

Existing Lakes & Rivers

Constructed Wetlands

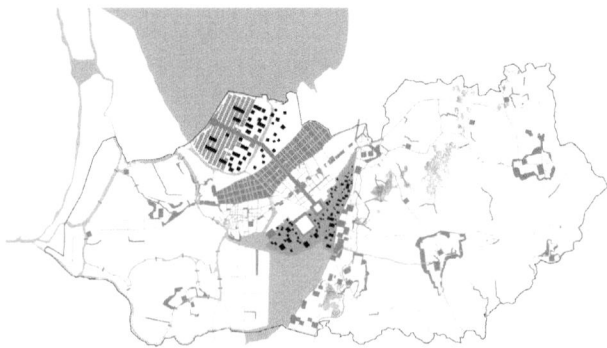

Orchard Islands

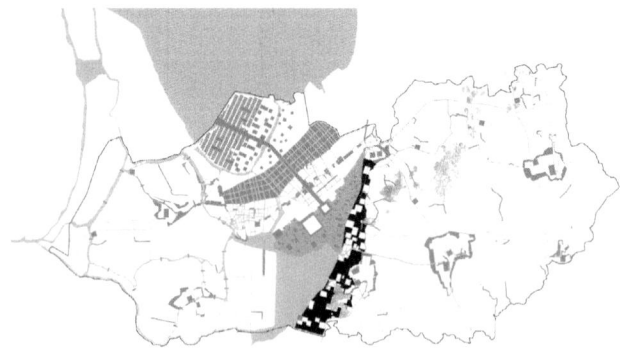

Urban Agricultural Strip

self-regulate: The City of a Thousand Lakes

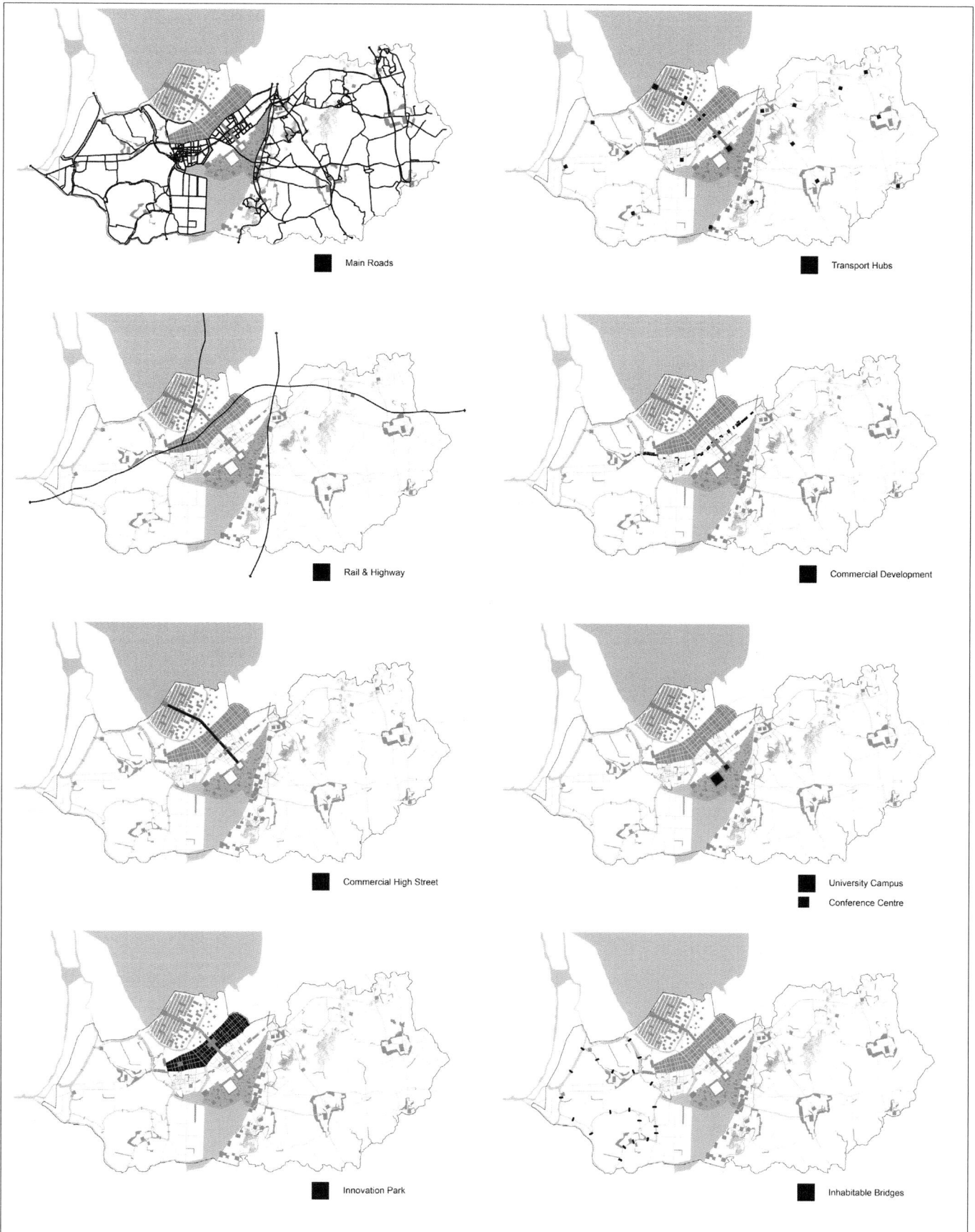

Main Roads

Transport Hubs

Rail & Highway

Commercial Development

Commercial High Street

University Campus

Conference Centre

Innovation Park

Inhabitable Bridges

CONSTRUCTED WETLANDS

REED BEDS / LOTUS / FISHING

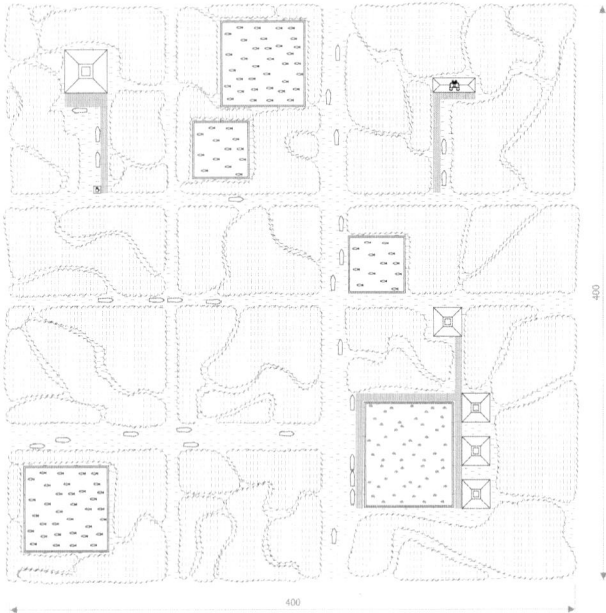

0   50m   100m   200m

REED BEDS WATER PURIFICATION | LOTUS FIELD FLOATING BAMBOO BORDER | LOTUS FIELD FLOATING BAMBOO BORDER | FISH NURSERY FLOATING BAMBOO BORDER + NET | TICKET OFFICE 20m x 20m 400m² | RESTAURANT 20m x 20m 400m² | OBSERVATION HIDE | TELEPHONE 5m x 5m | CAFÉ 5m x 5m 25m² | WC (FEMALE) 10m x 10m | WC (MALE) 10m x10m

400

400

BIOGAS ENERGY PARK

SCHEME 1 (SINGLE)

STANDARD: 80
DISABLED: 16
BICYCLE: 512

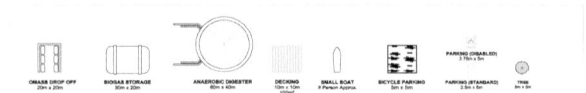

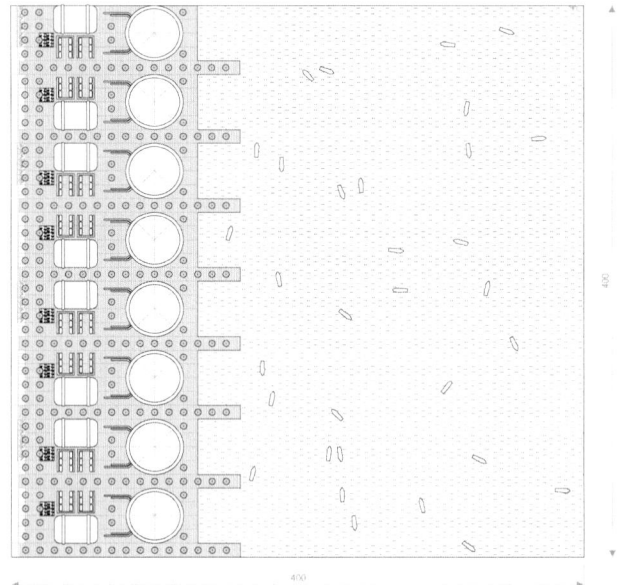

0   50m   100m   200m

GRASS DROP OFF 20m x 20m | BIOGAS STORAGE 30m x 10m | ANAEROBIC DIGESTER 60m x 40m | DECKING 10m x 10m 100m² | SMALL BOAT 6 Person Approx | BICYCLE PARKING 5m x 5m | PARKING (DISABLED) 3.75m x 5m | PARKING (STANDARD) 2.5m x 5m | TREE 5m x 5m

400

400

ORCHARD ISLAND

CINEMA ISLAND

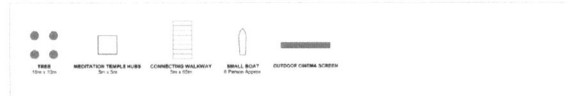

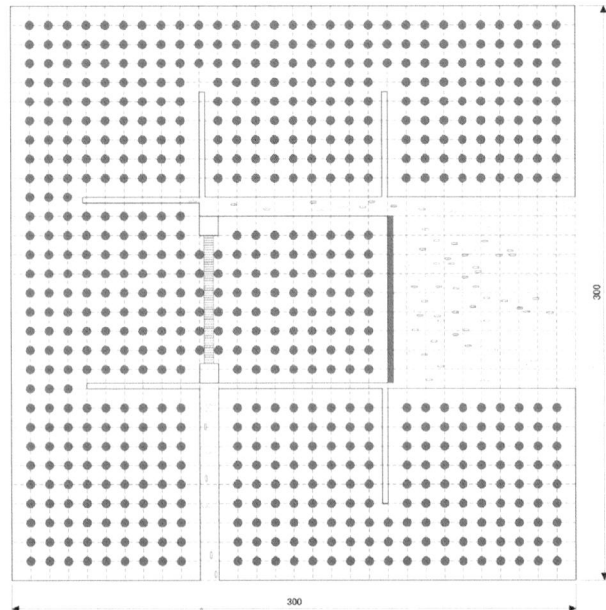

FRUIT TREES: 658
UTILITY HUBS: 2

0   50m   100m

TREES 10m x 10m | MEDITATION TEMPLE HUBS 5m x 5m | CONNECTING WALKWAY 5m x 50m | SMALL BOAT 6 Person Approx | OUTDOOR CINEMA SCREEN

300

300

ORCHARD ISLAND

MEDITATION GARDEN ISLAND

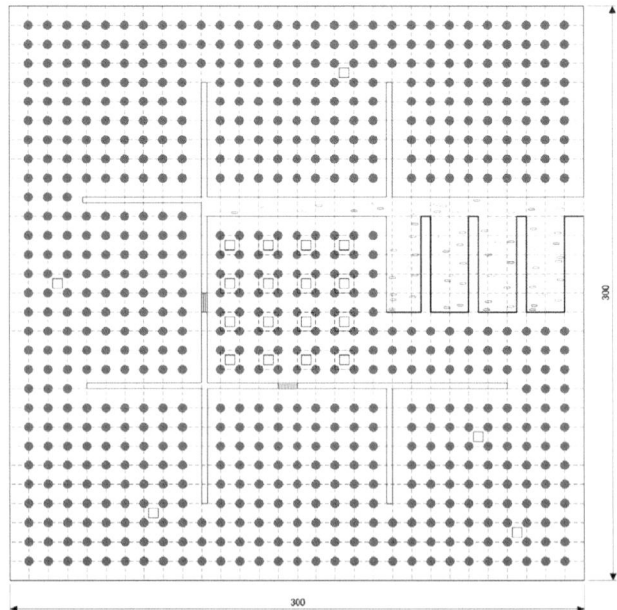

FRUIT TREES: 691
TEMPLES: 21

0   50m   100m

TREE 10m x 10m | MEDITATION TEMPLE HUBS 5m x 5m | CONNECTING WALKWAY 5m x 50m | SMALL BOAT 6 Person Approx

300

300

constructs, alluding to the nation's foundation in agricultural practice and the development of vernacular Chinese agricultural cities. The community understands the benefits of employing gentle yet strategic adaptations of the district's low-lying land and vast complex jigsaw of lakes, ponds and paddies that stretch out as far as the horizon.

Pastoral life in the city of a thousand lakes is idealised and creates purposefully constructed versions of nature. The landscape adapts to form aquatic habitats for the province's celebrated mitten crab, fish, and ducks, and together with bamboo, lotus and wild grasses softens the clusters of human inhabitation. The water quality of Gaochun's water-fields and lakes, with abundant insects and aquatic weeds, provides the ideal environment for the crab farming. There are farming outposts scattered on mud embankments to store and distribute animal feed, and to collect duck droppings for fertiliser. The serenity of the vision and its waters is only interrupted at specific times of the year to harvest crab, fish and lotus. In July and August, the water is overwhelmed by a superb display of jade-green circular lotus leaves, and white and pink flowers. The aquaculture has long maintained an empowering poetic presence; there are no vacuous developments, and all sectors of the city are animated and made accessible by water.

Satellite villages are dotted in amongst the multi-use landscapes of expansive water. Behind perimeters of bamboo forests, the villages function as energy hubs and actively convert biomass and human waste into biogas, and recycle the province's grey water. Children are kept healthy and play in never-ending water and bamboo covered landscapes, whereas the elderly wander at their leisure through gardens and orchards devoted to specific indigenous flora. Learning from its weather and seasons has validated the capacity for self-regulation of Gaochun – the new inhabitable landscape with housing displays a metabolic relationship with nature and is a celebration of the idyll landscape. From the dwellings, the inhabitants can contently take in the panoramic views of lakes and ponds, and the unrivalled fall foliage of flaming-red maple trees and shimmering golden gingko leaves.

The lakes are a social-economic resource as well as threat in the form of potential floods. The southern lake is a water filtration wetland, which hosts numerous artificial floating islands. This phase of the development aims to prepare the communities for a potential rise in water level, and the possibility of living on water. The multi-use islands provide additional space for orchards, bamboo forests and community activities. A new university campus and stadium are moored at the end of the axis that runs through the city. They are linked to an existing innovation park and into the new residential area in the north. The high-street axis connects the aquaculture developments to commercial activities, prioritising the services industry. Instead of a high street dominated by cars, Gaochun has vessels gliding along at a glacial pace in the plum and cherry tree-lined canal – the pink, magenta and orange coloured blossoms draw large crowds of visitors. Tourism forms a large part of the local economy, especially during the Crab Festival in September. The mitten crab has been an economic contributor to the city; crab connoisseurs from within China and abroad travel to Gaochun to experience the seasonal local delicacy.

The subtle aromas of the landscape arouse a new interpretation of the senses and memories of a traditional Chinese garden. The melancholy shimmer and flatness of the waterlogged lakes preserve and reflect

previous pages: The celebrated crab, fish, ducks and lotus are invaluable assets for local rejuvenation.

facing page: Modules of resilient landscape of lucid waters and lush flora.

following pages: The orchard island modules offer programmatic and spatial flexibility.

ORCHARD ISLAND

TYPE 1 GREENHOUSE

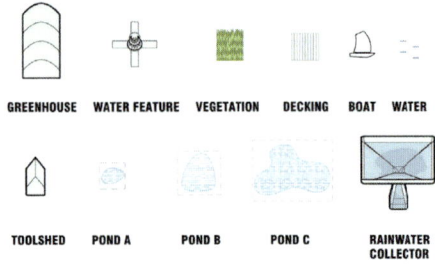

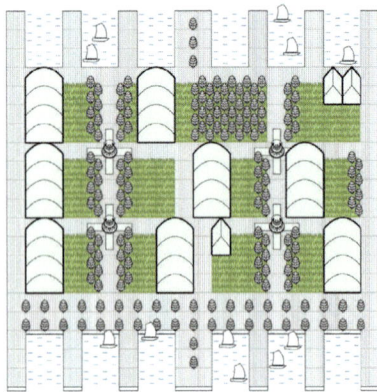

GREENHOUSE   WATER FEATURE   VEGETATION   DECKING   BOAT   WATER

TOOLSHED   POND A   POND B   POND C   RAINWATER COLLECTOR

ORCHARD ISLAND

TYPE 2 POND / WETLAND

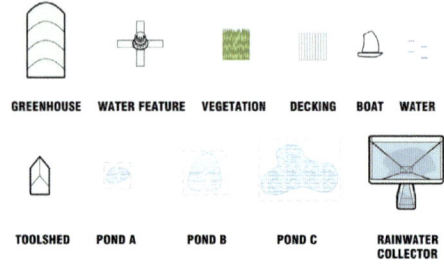

GREENHOUSE   WATER FEATURE   VEGETATION   DECKING   BOAT   WATER

TOOLSHED   POND A   POND B   POND C   RAINWATER COLLECTOR

ORCHARD ISLAND

TYPE 4 RAISED PLANTERS

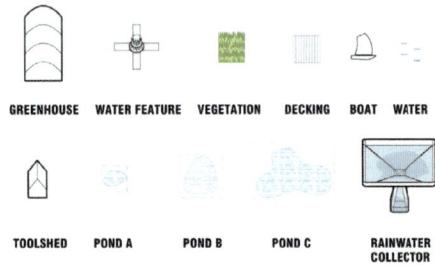

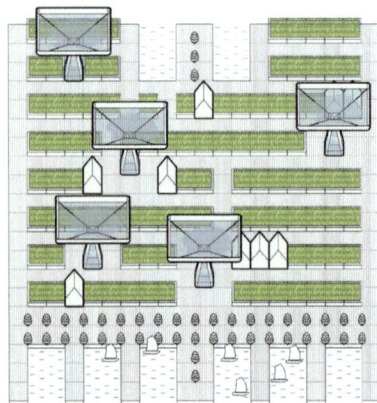

GREENHOUSE   WATER FEATURE   VEGETATION   DECKING   BOAT   WATER

TOOLSHED   POND A   POND B   POND C   RAINWATER COLLECTOR

ORCHARD ISLAND

TYPE 3 ARTIFICIAL FISHING LAKE

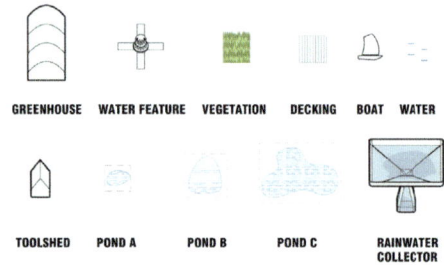

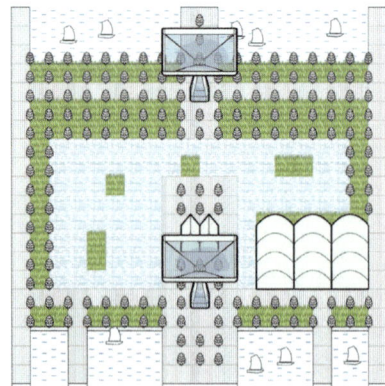

GREENHOUSE   WATER FEATURE   VEGETATION   DECKING   BOAT   WATER

TOOLSHED   POND A   POND B   POND C   RAINWATER COLLECTOR

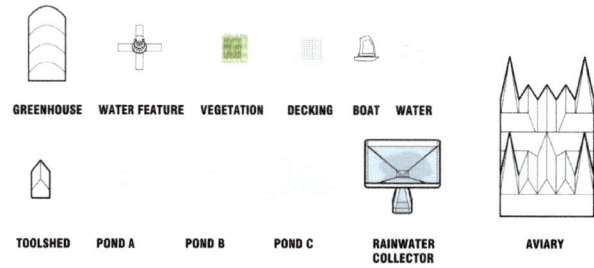

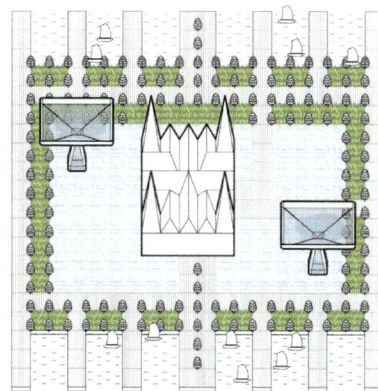

ORCHARD ISLAND
TYPE 5 BEEHIVES

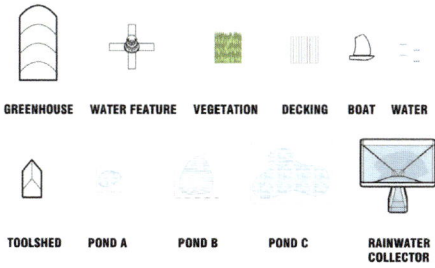

GREENHOUSE WATER FEATURE VEGETATION DECKING BOAT WATER

TOOLSHED POND A POND B POND C RAINWATER COLLECTOR

ORCHARD ISLAND
TYPE 6 AVIARY

GREENHOUSE WATER FEATURE VEGETATION DECKING BOAT WATER

TOOLSHED POND A POND B POND C RAINWATER COLLECTOR AVIARY

ORCHARD ISLAND
FOLDED LANDSCAPE TYPE 1

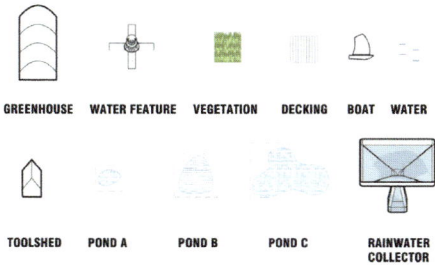

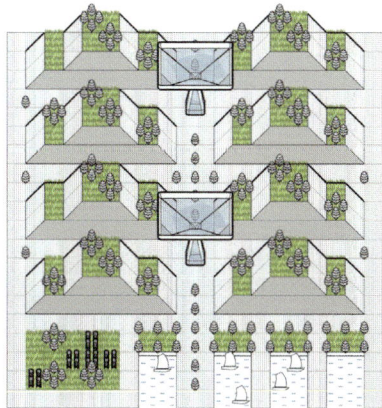

GREENHOUSE WATER FEATURE VEGETATION DECKING BOAT WATER

TOOLSHED POND A POND B POND C RAINWATER COLLECTOR

ORCHARD ISLAND
FOLDED LANDSCAPE TYPE 3

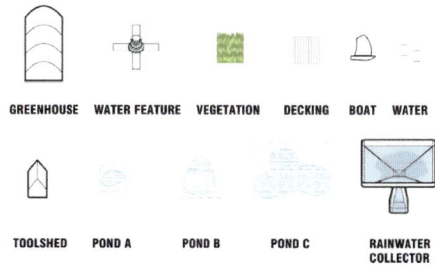

GREENHOUSE WATER FEATURE VEGETATION DECKING BOAT WATER

TOOLSHED POND A POND B POND C RAINWATER COLLECTOR

a peculiar yet poetic and beautiful aspect of the province. The water somehow holds a sense of time, a sense of the sublime in its reflective surface and its inability to be stopped. Gaochun does not retreat to the 'traditional' nor does it shy away from strategic density – it has achieved multiple interpretations of the Smartcity manifesto to allow the province to localise its economic growth, rediscover its identity and contribute to long-term national policies. At the 18th National Congress of the Communist Party of China, President Xi Jinping listed 'ecological civilisation' as one of the five goals in the country's overall development plan, along with economic, political, cultural and social progress, and that ecological civilisation included the need to respect, protect and adapt to nature.[9] He later underscored, 'a good ecological environment is the fairest public product, and the most accessible welfare for the people'.[10] The thousand lakes of Gaochun serve to remind the communities of their faith, and it is nature that unites and provides the capacity for self-regulation. In a country as large as China, pockets of resilient landscape with lucid waters and lush flora are invaluable assets for national rejuvenation.

207

facing page: The melancholy flatness of the waterlogged lakes and nature afford Gaochun its ability to self-regulate.

left: The resilient landscape allows the province to localise its economic growth, rediscover its identity and contribute to long-term national policies.

9. L Li & Y Huang, 'President Xi Sets The Pace on a Better Environment', Xinhua Insight, www.xinhuanet.com, 21 March 2017 [http://www.xinhuanet.com/english/2017-03/21/c_136144907.htm], retrieved 30 August 2018

10. China Daily, 'Lucid Waters and Lush Mountains are Invaluable Assets', www.chinadaily.com, 9 October 2017 [http://www.chinadaily.com.cn/china/19thcpcnationalcongress/2017-10/09/content_33032118.html], retrieved 29 August 2018

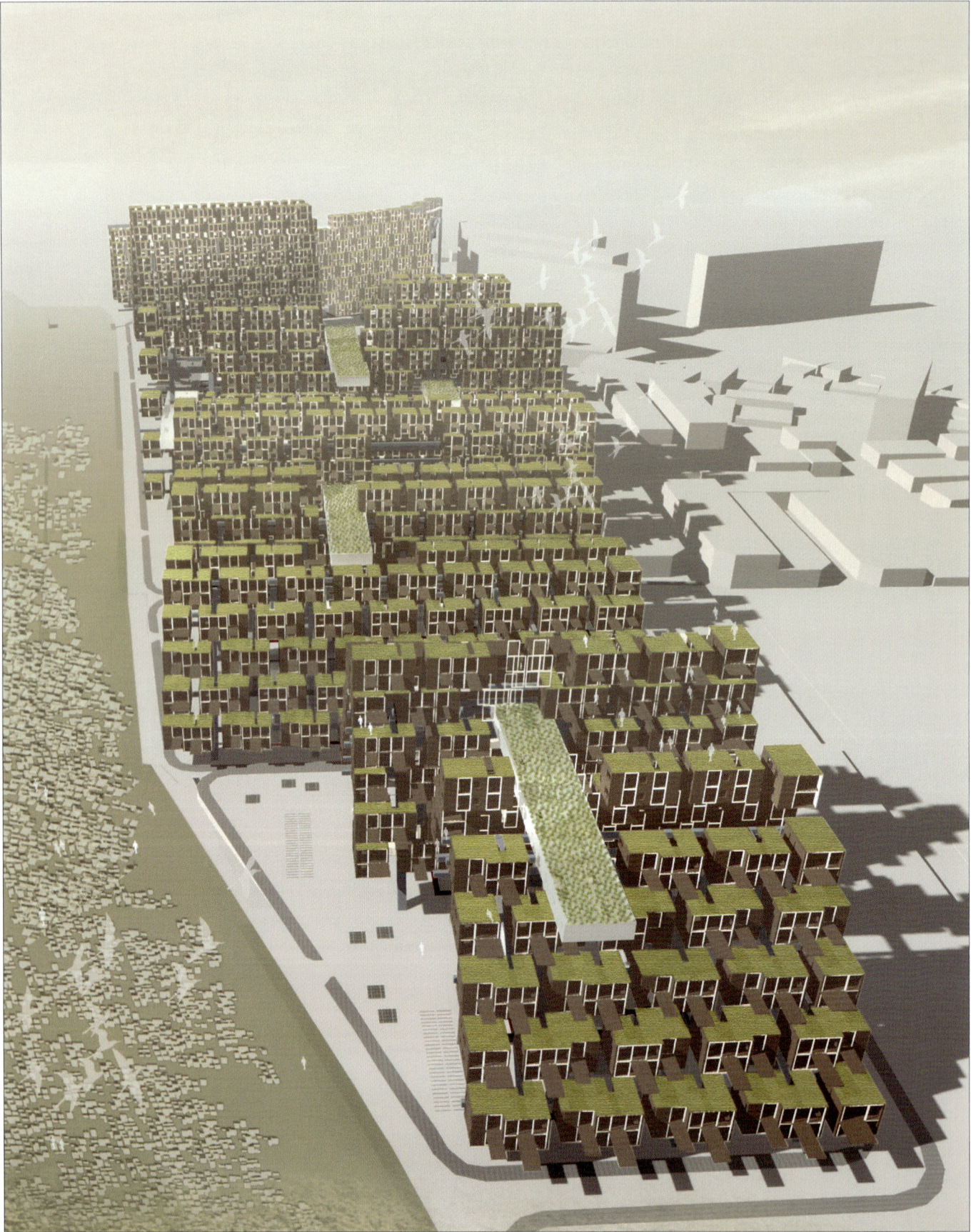

Rifle Range Regeneration Malaysia

'The reader can hardly conceive my astonishment, to behold an island in the air, inhabited by men, who were able...to raise or sink, or put it into progressive motion, as they pleased... But the greatest curiosity, upon which the fate of the island depends, is a loadstone of a prodigious size, in shape resembling a weaver's shuttle.'

– Jonathan Swift's Laputa, 'Gulliver's Travels', 1726

209

In Swift's satirical travelogue, the fictional worlds of Gulliver mirror the conditions of the protagonist's own 18th century England. On his third voyage, Gulliver is shipwrecked, then saved by the wondrous floating island of Laputa. The inhabitants of Laputa are technologically advanced although inept at everyday life, and thus illustrate the blindness afflicting obsessive scientific pursuits. Swift's Laputa is provocative, inspiring Hayao Miyazaki over 200 years later to imagine his own 'Castle in the Sky'.

Miyazaki's film continues Swift's critique of a distorted worship of technological progress, but imbues it with an ecological message urging on a vision of a self-regulating resilient ecosystem. In the film, Laputa is a floating wonderland where technology, in the form of robots and levitating cubes, and nature co-exist in an 'Atlantis-in-the-air', a lost civilisation whose mysterious origins we later discover.[1] We learn that Laputa's royal family abandoned the planet long ago, disavowing Laputa's darker identity as a floating weapon of war in a gesture of repentance and atonement for human abuse of technology for destructive purposes. Without human intervention, Laputa reverts to a state of harmony between robots and living creatures. Miyazaki's point then, according to critic Thomas Lamarre, is that 'received technologies themselves (are not)... troublesome. It is the way they are used and imagined, and how they are handed down and received.'[2] When the protagonist and the heiress to Laputa's secrets decide to cut the island off forever from human reach, robots and animals return to these symbiotic, self-regulatory relations. In fact, at the core of the planet, Miyazaki has imagined a giant central tree whose roots are intertwined with his version of Swift's lodestone.

The urban regeneration of Rifle Range in Penang takes up this provocative narrative by adopting nature to create a self-regulating, energy efficient community of prosperity and wellbeing. The re-evaluation of domestic and community place making of the development explores vertical densities to afford space for cohesion that attract residents from outside the area, preventing an isolated culture and stimulating wider social capital. In his analysis of Miyazaki's film, Anthony Lioi writes: 'Swift's version contained no arboreal core... [so] it is a sign of Laputa's cultural difference that the most advanced technology is forested, as if the Laputians recognise that their power is ultimately based in nature.'[3]

Facing page: A green cooling oasis under the equatorial sun.

1. R Harrington, 'Movies: "Laputa": Adventures in sky-fi.', The Washington Post, 2 September 1989

2. T Lamarre, 'From Animation to Anime: Drawing movements and moving drawings', Japan Forum, vol 14, no 2, 2002, p 356 [doi:10.108 0/09555800220136400], retrieved 20 September 2018

3. A Lioi, 'The City Ascends: Laputa: Castle in the Sky as Critical Ecotopia', ImageTexT: Interdisciplinary Comics Studies. 5.2 (2010), Dept of English, University of Florida [http://www.english.ufl.edu/imagetext/archives/v5_2/lioi], retrieved 23 September 2018

In a Smartcity, technology is less a signifier of progress than a means for enabling. Therefore, Rifle Range responds to context, programme and climate change with two landscape resilience strategies that shelter and heal: (1) the Green Vernacular Canopy – the formal continuation of the adjacent mountains as the new inhabitable landscape embedded with residential units, inspired by traditional Malay vernacular architecture, as well as (2) the Atrium Park – a continuous ground level concourse bringing nature into the development, while providing public facilities for work, rest and play. The two landscapes are of urban scale, establishing a series of microclimatic atriums to form complex spatial relationships that engage in rich and unpredictable social and environmental synergies.

In the first strategy, the Green Vernacular Canopy comprises nine green 'mountains' ranging from 12 storeys to 20 storeys, each with terraced residential units inclined towards each other. The inhabitable mega-structures are draped with a landscape of urban agriculture. The green roof has the key role of providing shade and nutrition, cultivating social cohesion, reducing heat penetrating the living spaces, and collecting rainwater for recycling. A two-metre gap between each residential unit allows a rich amount of sunlight to permeate, as well as prevailing wind to cross-ventilate.

Cities as well as smaller communities are vulnerable ecologies – rising frequencies of heatwaves exacerbate the well documented urban heat island (UHI) effect, creating 'marked heat sensible fluxes' that distinguish an urban space from its surrounding areas.[4] Combined with pollutants, urban heat fluxes 'aggravate events of discomfort and may cause direct impacts on human health such as respiratory difficulties, fatal and non-fatal strokes, and alteration of sleep cycle'.[5] The deathly membrane of hot air effectively encloses the city, suffocating the most vulnerable within, and sealing the urban landscape from living organisms – for example spores, insects and small animals that contribute to biodiversity. Such a barrier can be seen as a type of micro-climatological catastrophe as fragile ecological chains within urban ecosystems are broken, depriving the city of its capacity to protect itself and its resources.[6]

Geoengineering surfaces to impact climate change gained traction in recent times, notably with Steven Chu, the US Secretary of Energy for President Barack Obama, and a Nobel prize-winning scientist's memorable scenario of painting buildings and roofs white, and switching pavement colour from black to a more reflective light grey. According to Chu, the reduction of carbon emissions would be synonymous to a near apocalyptic magical trick of withdrawing the world's entire stock of automobiles from the streets for over a decade.[7] The Green Vernacular Canopy provides similar benefits, but with augmented cooling effects, pollution absorption, biodiversity and a landscape which resilient cities hold precious.

The second self-regulating landscape, the Atrium Park, is located beneath the Green Vernacular Canopy and the residential units. The park extends the green strategy into the public realm, and is an adaptable landscape for cultural and sporting activities. Within the microclimatic pockets, trees and floral patches shape social hubs and sound gardens for public relaxation and recreation, and offer total visual connections across the site. The atrium incorporates a new outdoors sport facility situated adjacent to the existing school. The public facility, which includes swimming pools, and courts for basketball and badminton, is for residents as well as for locals and neighbourhood school children to enjoy.

The lack of urban canopies has resulted in economic deprivation in neighbourhoods, from a three-decade-long study showing spatial correlation between income and green space in Phoenix, to an analysis of the decreased numbers of street trees in rights of way of poor neighbourhoods in Tampa.[8] Introducing green solutions to address spatial justice would commit the urban landscape itself to form oases of both environmental and social inclusion for the city's inhabitants.[9] The

Cool Air Out

Hot Air In   Cool Air Out   Cool Air Out   Hot Air In

Uninterrupted Views

Shaded Zones

Cross Ventilation Through Flats

Solar-Balustrade
Structure + Services

Ventilation by Stack Effect

Green Balconies
Inclined Circulation
Fish Ponds

Evaporative cooling from fish ponds

Guest Flats
Residents Flats
Shopping

Cross Ventilation

Car Park
Atrium Park

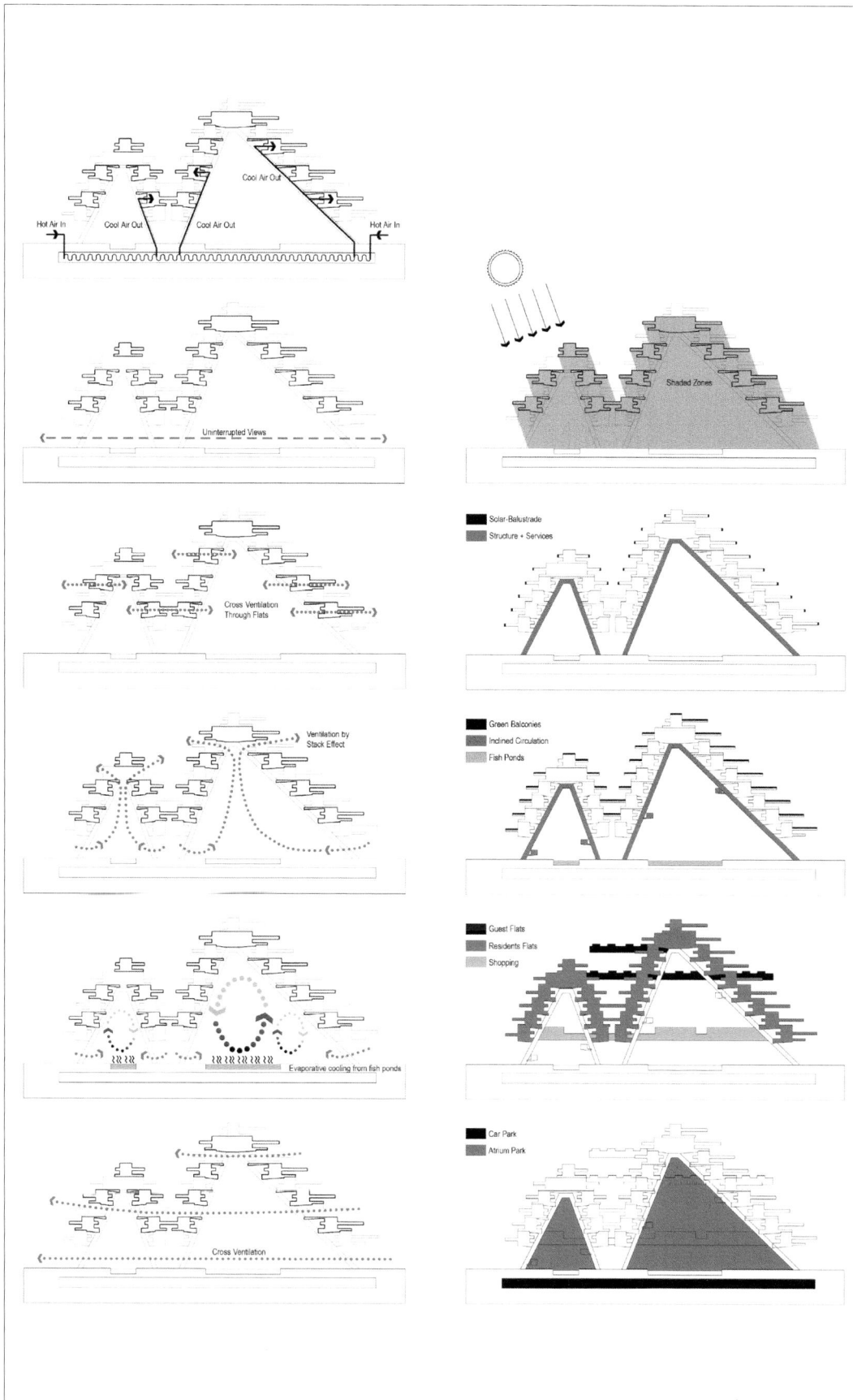

previous pages: The Green Canopy creates a microclimate for the regeneration.

left: Sectional diagrams showing environmental strategy.

following page: Aerial views of the Green Vernacular Canopy.

213

4,5+6. T Georgiadis et al., 'City Resilience to Climate Change', in 'Rooftop Urban Agriculture', F Orsini et al. (eds.), Springer International Publishing, 2017, p.255

7. S Connor, 'Obama's Climate Guru: Paint Your Roof White!', The Independent, Independent Digital News and Media, 23 October 2011 [www.independent.co.uk/environment/climate-change/obamas-climate-guru-paint-your-roof-white-1691209.html], retrieved 22 August 2018

8+9 CS Greene et al., 'Canopy of Advantage: Who Benefits Most from City Trees?', Journal of Environmental Management, vol.208, 2018, p.25, [doi:10.1016/j.jenvman.2017.12.015], retrieved 18 September 2018

10. M Kober, 'A Travel Through Oases in French and Arabic Literature', in 'Oases and Globalisation: Ruptures and continuities', E Lavie & A Marshall (eds.), Springer, 2017, p.18

11. M Kober, 'A Travel Through Oases in French and Arabic Literature', in 'Oases and Globalisation: Ruptures and continuities', E Lavie & A Marshall (eds.), Springer, 2017, p.19

12+13. A Lioi, 'The City Ascends: Laputa: Castle in the Sky as Critical Ecotopia', ImageTexT: Interdisciplinary Comics Studies. 5.2 (2010), Dept of English, University of Florida [http://www.english.ufl.edu/imagetext/archives/v5_2/lioi], retrieved 23 September 2018

symbiotic relationship between landscape and human changes the social politics of circulation and spatial programming – a network of pedestrian elevated 'sky' streets at five metres above ground connects commercial units with residential units, relieving human from vehicular traffic. The arrangement of the commercial arcades in the atrium encourages maximum permeability and social flow. The Atrium Park offers a public territory in the third dimension and hybridises aims for urban renewal with social resilience.

The passive environmental design strategies aim to resist heat gain and allow excess heat dissipation through the landscape to reduce the cooling demand. To minimise ongoing cost and to maximise the positive impact for a sustainable community, thermal comfort in the region's equatorial climate can be achieved at most times of the year without the intervention of mechanical system. This can be achieved by extending the logic of self-regulation to the following:

Solar Control: An essential strategy to control heat gain for the development and the interior spaces comprises three distinct elements: (a) control of aperture geometry, (b) consideration of solar-optical properties of transparent and opaque surfaces, and (c) provision of shading elements in the form of the Green Vernacular Canopy – this creates unique spaces in the Atrium Park demonstrating the power of natural forces to enhance comfort.

Natural Ventilation/Air Permeability: Apart from providing good indoor air quality, ventilation plays a major role in maintaining acceptable thermal comfort and improving energy performance. The ventilation strategies include a non-compact layout of the residential banks providing effective cross ventilation; window size and orientation to wind direction; coupling with outdoor spaces; and the provision of sky courts and transitional spaces. Natural ventilation into indoor spaces is enhanced by cross ventilation, wind assisted ventilation, and stack effect. Balanced wind-stack ventilation is applied in continuous atriums or core spaces in buildings. The A-frame configuration of the mega-structure of the Atrium Park promotes buoyancy effects and the apex of each cluster has an adjustable baffle to control flow rates.

Heat Sinks: The development uses natural heat sinks such as air, water, ground and built mass. Increased use of vegetation in the form of the Green Vernacular Canopy and the use of building elements – ventilated double façade and double roof – allows heat loss and prevents overheating. Indirect cooling can be achieved through roof ponds or spaces around water bodies in the Atrium Park. The open water surfaces of the watercourse, ponds and fish pond encourage evaporative cooling to the communal spaces and the tempered air is confined by the enclosure of each square.

How might urban renewal be reimagined instead as dreaming up a self-regulatory ecosystem? Rifle Range is a climatological adaptation alongside the logic of nature itself that is aligned with another self-sufficient refuge – the oasis. In Middle Eastern literature, the oasis is both mythical and locatable: 'they exist in contrast to deserts as a divine miracle as well as a fragile human creation... and opens up a holy dimension in space.'[10] The ecosystem has become associated with the fantastical and the exotic, akin to an 'island in the human imagination'.[11] Oases have in cultural imaginaries served to characterise spaces of exception, from Robinson Crusoe's island to the countless floating islands referenced in the speculative world-making genre of videogames. What might it mean if we instead see future urban renewal developments as oases, not of exception, but of possibility? As Lioi commented about Miyazaki's floating island, such oases are not 'Neverland(s) in which children are invited to stay children forever'.[12] Instead, this Laputa of terraced green roofs and refreshing atria of shade and water-cooled air is thinking of design as 'take(ing) responsibility for planetary flourishing, (by) invit(ing) ecotopia to come down to earth'.[13]

self-regulate: Rifle Range Regeneration

Car Park

Concrete Pavings

Existing Sand Fields

Timber Decking

Tree Gardens

Grass Lawn

Agriculture

Agriculture

Agriculture

Rock Field (Medium Polished Rocks)

Timber Decking

Sand Beach with Trees

Marina Clubhouse

Shallow pools

Swimming Pools

Marina

0m   50   100   200   300

Dongyi Wan East Waterfront China

'It is really beautiful. It is flood water. I see opportunity and productivity. But we have forgotten the art
of survival and ancient wisdom of living with water.'
— Kongjian Yu, 'Sponge City and Symbiotic Solutions to Urban Water Issues', 2018

The waterfront development at Dongyi Wan East in the Ronggui subdistrict of Shunde in Southern China
is an amphibious landscape bordering the DeSheng River, providing recreational and urban agriculture
resources to a new residential quarter that occupies a greenfield site of approximately five square
kilometres.

Following China's era of 'opening up' to the outside world that began in 1978, Shunde has taken advantage
of its cultural position and geographical proximity to Hong Kong, Macau and Taiwan to develop into a
major manufacturing centre from traditional agricultural origins and was designated as a pilot city for
the comprehensive reform of Guangdong in the 1990s.

Located in the centre of the Pearl River Delta, the area is characterised by intersecting rivers that flow
around and through the new residential and school district, resulting in a special relationship between
the water and its residents. The waterfront site, which occupies approximately half a square kilometre,
lies three metres below the perimeter road that separates the main development from the DeSheng
River and is subject to annual flooding. Although the floods are relatively predictable, arriving during
the summer monsoon period, the waterfront is unsuitable for conventional development. This condition,
however, frees up welcome opportunities to adapt into an innovative hybrid landscape of marina,
artificial beach, cultivated land and wetland wildlife reserve.

The Pearl River Delta is a vast alluvial plain containing possibly the highest density of intersecting rivers
in the world, and is one of the most fertile regions in China. The climate is moderate and rainfall plentiful,
providing ideal conditions for growing crops such as pak choi and choi sum. By establishing the majority
of the low-lying plain as agricultural land, the floods are transformed from a nuisance and potential
catastrophe into a benefit by using the soil recharged with nutrients from alluvial deposits. Additionally,
the floodplains support rich biodiversity – the river water supplies an instant rush of nutrients in the
form of decomposing organic matter on which microorganisms flourish. The microorganisms in turn
attract a sequence of predators on the food chain ruled by migratory waterfowl such as winter swans
and cranes that reward human study.

facing page: Dongyi Wan East
Waterfront masterplan.

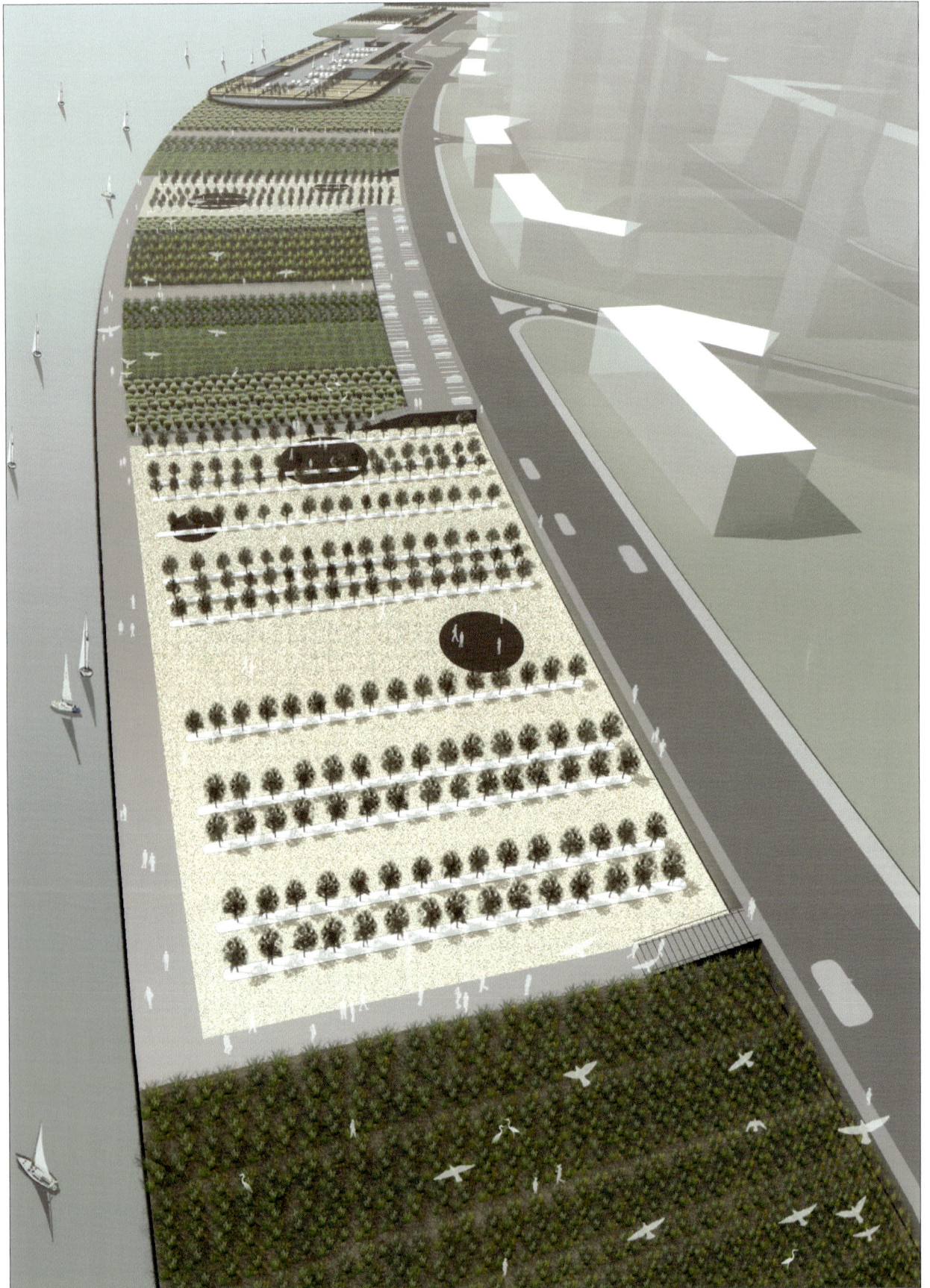

The wetland wildlife reserves and agricultural land are just two of the textures that make up an amphibious community landscape. A variety of hydrophilic ornamental grasses, concrete walkways, riparian tree gardens, sand fields, grass lawns and hard standing for seasonal parking combine to form a heterogeneous and unstructured recreational ground made of flood resistant or sacrificial organic material where traditional pursuits of calligraphy, painting and wildlife watching are encouraged.

More formal and flood-sensitive community spaces are accommodated around the marina to the southeast of the site where the activity plane is raised level with the adjacent developments and unaffected by the floods. Heavy river traffic and the unknown quality of the water in the DuSheng River prevent swimming activities and the water can be too cool in the winter months. To ensure year-round occupation, heated outdoor pools, steam baths, jacuzzis and diving facilities are arranged around the marina and clubhouse. The pools vary in depth and temperature and are contained by an infinity edge, extending the pools visually into the aquatic landscape. The remainder of the raised plain is landscaped with lawn grass, artificial beaches, trees, timber decking and semi-polished pebble fields.

The character and occupation of the waterfront change dramatically through the year. During the flood period, the majority of the site is submerged and the ambience contemplative with only the marina area animated. The fields and hard landscaping are barely visible beneath the surface of the water. After the

facing page + left: Amphibious community landscape mixed with wildlife reserves and farmland.

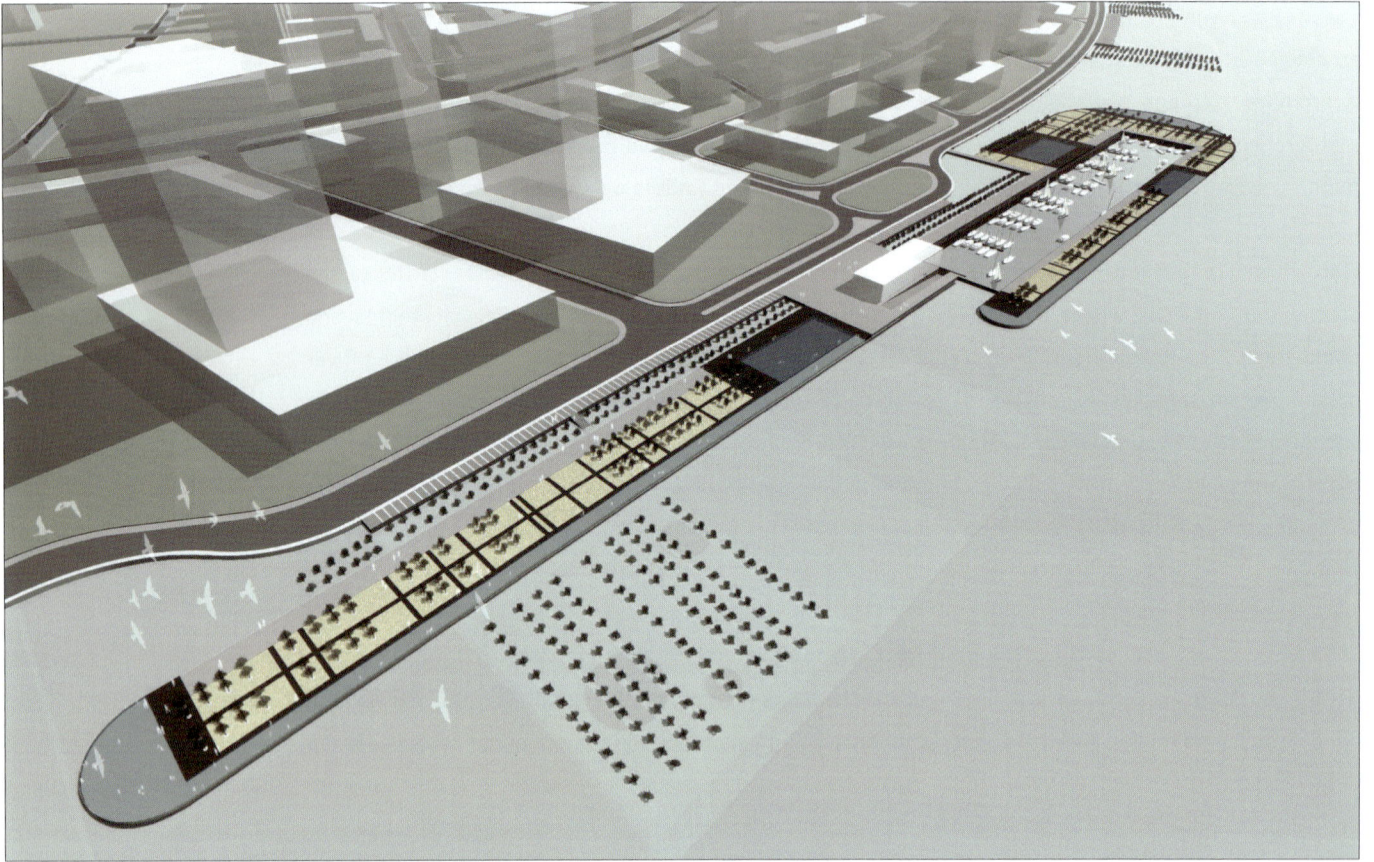
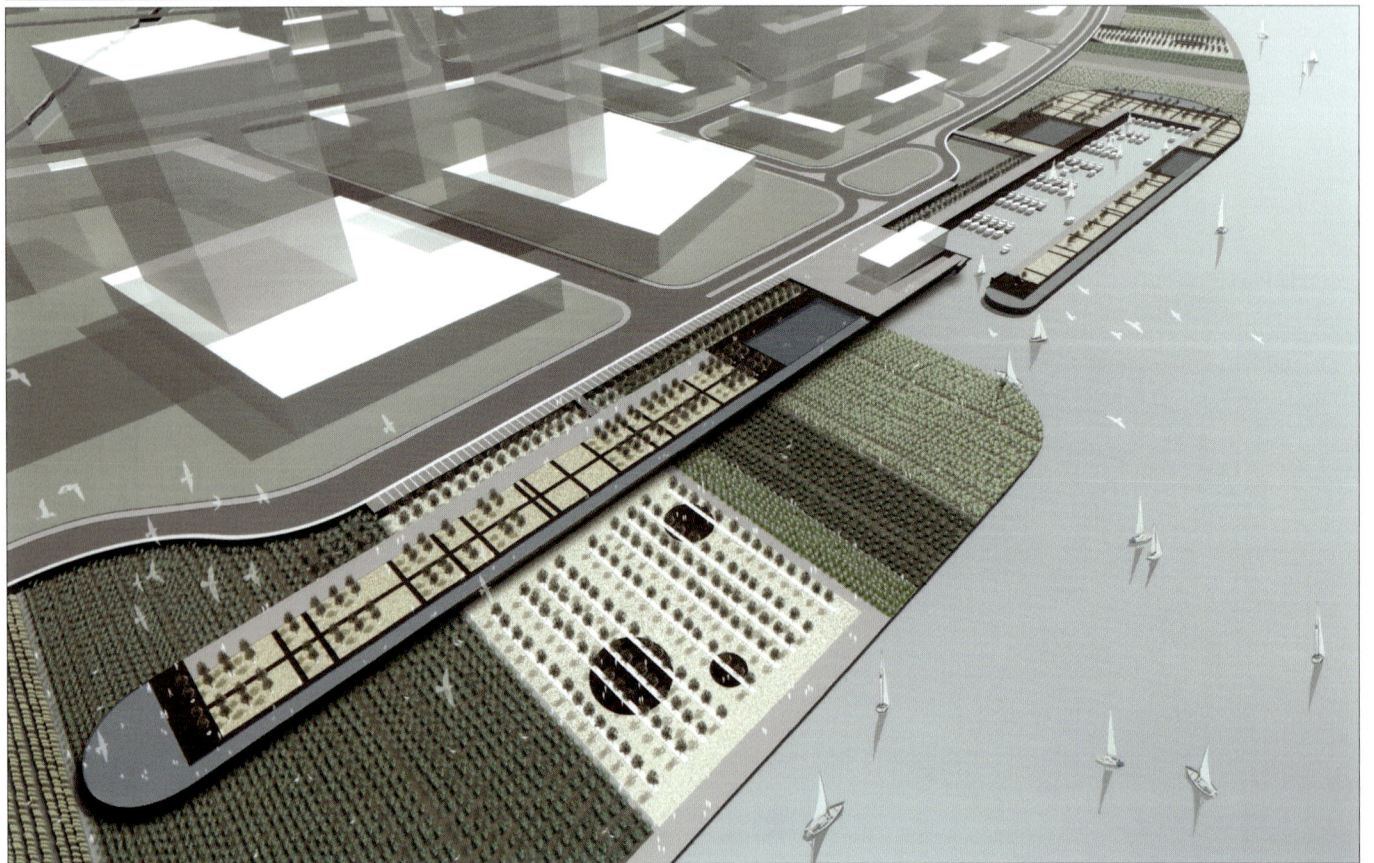

waters recede, there is a clean-up operation with the timber and concrete decks swept and washed, and rubbish removed. The farming community emerges, sowing seeds and planting crops in the revitalised ground as wildlife begins to proliferate, attracting visitors and school study groups. During the spring and summer, the waterfront will become particularly busy with visitors arriving from further inland before the farmers harvest their crops and the cycle begins anew.

Dongyi Wan East is an archetypal yet unusual Smartcity trope in its application of a new functional hybridised resilient landscape on land previously considered unoccupiable. The development is sustainable, taking advantage of natural flood cycles to grow produce for the local populace. Spatially, the architecture is created through human occupation and a plethora of recreational spaces encourages social interaction. The landscape is adapted to be inclusive, with the allocation of multiple functions on the same space depending on the season. Industrial growth in the region has in the past eroded and damaged vital wetlands in pursuit of economic expansion, a process that needs urgent reversal to renew and clean the country's watercourses and to preserve natural wildlife habitats. Dongyi Wan East belatedly provides a model for coexistence with natural systems that will benefit the community as well as the environment.

221

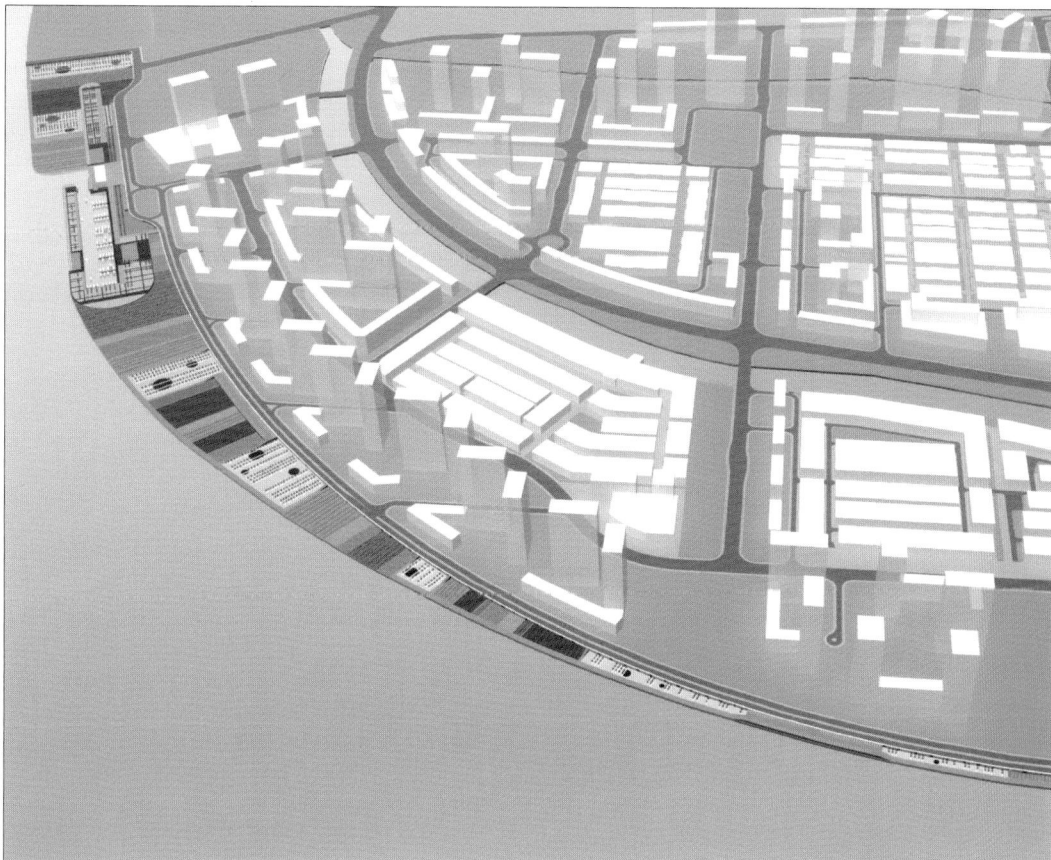

facing page top: A contemplative ambience during the flood period.

facing page bottom: Sowing, planting and cleaning after the waters recede.

left: Aerial view of the cultivated amphibious landscape.

following page: An innovative hybrid landscape that supports biodiversity.

OVERALL PLAN

SCALE: 1:500

1. Tourist Information
2. Shop (Local Crafts)
3. Covered Performance Space
4. Regional Food Outlet
5. WCs (Male, Female, Disabled, Family)
6. Restaurant Kitchen
7. Restaurant (100 Covers)
8. Restaurant Terrace
9. Gallery
10. Wildlife Exhibition Space
11. Wildlife Interactive Wall
12. Storage
13. Workshop
14. Education Classroom
15. External Education Space
16. External Playground
17. Three Flexible Meeting Rooms
18. Office For Eight Staff
19. Conference Space (200 People)
20. Plant Room
21. Caravan Site
22. Coach Parking
23. Primary Site Access Road
24. Car Park (300 Spaces)
25. Main Building Entrance

Denotes Light-well + Seating

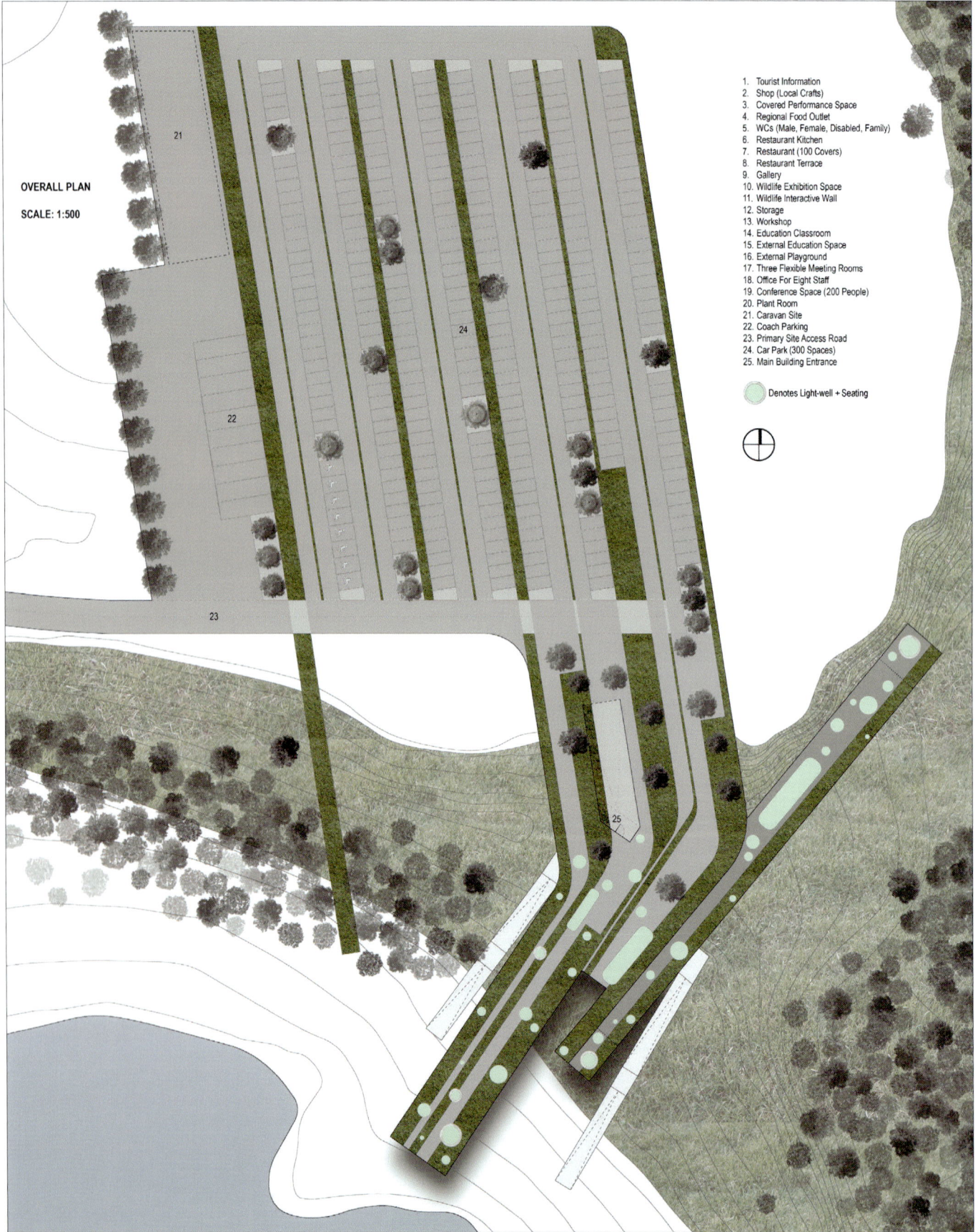

Brockholes Wetland + Woodland Reserve UK

'According to Darwin's "Origin of Species", it is not the most intellectual of the species that survives; it is not the strongest that survives; but the species that survives is the one that is able best to adapt and adjust to the changing environment in which it finds itself.'

– Leon C Megginson, 'Lessons from Europe for American Business', 1963

Polders and dykes in the Netherlands. Chinampas in the Valley of Mexico. They are poetic yet strategic resilient landscapes where humans have had to learn from nature and adapt their environments accordingly to achieve revitalisation. After drainage, the polders are planted with a succession of plants to create an increasingly fertile soil structure – starting with reeds to assist with the drying of the soil, followed by colza, and eventually grain crops. The water management landscape originated during the Middle Ages and is a product of the country's successful battles against the North Sea. With a quarter of the country below sea level and 60 percent of its people living in flood-risk areas, the tradition of Dutch 'polder-politics' cultivates community cohesion.

The Aztecs, with their technical skills and knowledge of aqua-terra, built the great city of Tenochtitlan on chinampas, 'floating gardens', in the lake. The ingenious ancient urban agriculture landscape is a collection of rectangular raised platforms built on a strong framework of sticks, wattle fencing and four willow trees to provide anchorage to the lakebed. The willow and alder trees also act as environmental shelters and habitat for birds and insects.[1] The farmers would regularly layer the chinampas with nutrient-rich mud from the bottom of the lake, and various decaying inconsumable vegetation and human waste from the city itself – the process enabled the city to treat its waste and improved fertilisation of its crops, producing three to four harvests a year. At its peak, the 750 hectares of chinampas engaged some 5000 farmers and produced approximately 80 tons a day of vegetables to feed the 21 million inhabitants of Greater Mexico City.[2] The UN Food and Agriculture Organization (FAO) selected the chinampas as a Globally Important Agricultural Heritage System (GIAHS), because they preserve agricultural biodiversity, help farmers adapt to climate change, bolster food security and reduce poverty.

Food shortages notwithstanding, energy security is another issue that will dominate the 21st century, and is a vital goal of national policy in many countries. Nuclear and wind power are not the solutions to the UK's energy security problems, particularly when many countries are deviating away from nuclear power after the crisis in Fukushima, and when wind turbines are incongruous with affordability goals and the countryside. Instead, the Smartcity seeks adaptive strategies of nature and ecological systems to create multi-layered resilience while maintaining natural beauty, as a way forward.

facing page: The masterplan of Brockholes is a study of effective management of aesthetics and function in a landscape adaptation.

1. JN Pretty, 'Regenerating Agriculture: An alternative strategy for growth', Earth Scan, New York, 1995, p.127

2. E Godoy, 'Mexico's Chinampas – Wetlands Turned into Gardens – Flight Extinction', Inter Press Service News Agency, 27 February 2016 [http://www.ipsnews.net/2016/02/mexicos-chinampas-wetlands-turned-into-gardens-fight-extinction], retrieved 14 April 2018

Brockholes is a new kind of nature reserve – a multi-layered landscape with a range of wildlife habitats added to existing woodlands and water, coupled with a renewable energy centre that actively harnesses an unpredictable instrument of nature, the sun. The one square kilometre of a former gravel quarry is shaped by integrating sensitive ecology management with effective methods of harnessing climatic resources, and smart application of historical earthwork ideas to deliver regional and national conservation gains. The ground is subtly re-profiled to manage the hydrology and camouflage the various spaces without distressing the wildlife and tranquillity of the wetlands.

227

Located in-between two existing dense sets of trees, the facilities for naturalists and bird-watchers nestle beneath a rampart carpet – the elevated earthworks lift the visitor centre above the flood zone and present a 360-degree panoramic visual engagement with the landscape and surrounding towns. The rampart carpet is asphalt covered with continuous strips of alternating wild-grasses, hedgerows of trees, and car parking. The sloped surface of hybridised textures allows continuous and unrestricted access to the roof, leading visitors directly to the main entrance of the facilities. The secondary route takes visitors along a more scenic path: through the dense trees and by the lake before entering the building from the south. The hedgerows provide semi-sheltered picnic areas, and large light-wells puncture the rampart carpet to animate the internal spaces below.

Brockholes employs a low maintenance and a fundamentally undemanding environmental strategy to minimise its energy requirements and maximise renewable energy production. The rampart carpet is heated, cooled and ventilated utilising the structural floor slabs as heat exchanger. Fresh air is supplied through the slabs to control the temperature and to allow the fabric of the landscape to function as a giant heat store and heat emitter. In summer, the system delivers passive cooling utilising night time ventilation through the slabs to remove heat from the landscape as it becomes largely self-regulating. In winter, this combines with high efficiency heat recovery air handling units recuperating heat from the return air at high peaks and transferring it into the incoming air with an efficiency of over 80 percent. The system is made up of three components: a collector, heat transfer equipment and a heat store. The collector takes the form of a horizontal pipe array incorporated into the asphalt landscape. The heat store is located beneath the visitor centre and the slopes of the rampart carpet. The heat store utilises the thermal mass of the soil under the carpet to store heat collected during the summer for re-use during the winter. The combination would deliver a close to zero carbon heating system.

In 'The Aesthetics of Wind Energy', Justin Good suggested that 'there is a deeply intuitive connection between beauty, function and purpose, especially when we are thinking about the beauty of nature'.[3] Brockholes is a study of effective management of aesthetics and function in a landscape adaptation, with nature as the energy provider. It is only a microcosm of a larger vision of sustainable energy security; the technological principle of the rampant carpet can be applied on a national level – it is estimated that the UK has a total asphalt covered road length of 245 000 miles.[4] The transition from nuclear power will not be painless, but as the Aztecs and the Dutch demonstrated, a society can successfully adapt to dramatic changes in environmental conditions and build strong resilience by making efficient use of natural resources and ecosystems.

facing page top + bottom: The planted areas of the carpet at Brockholes resemble and work on a similar principle to the thick turf insulation on the green-cloaked housing in Iceland.

3. J Good, 'The Aesthetics of Wind Energy', Human Ecology Review, vol.13, no.1, 2006, p.79

4. Department for Transport, 'Road Lengths in Great Britain: 2011', www.gov.uk, 28 June 2012 [https://assets.publishing.service.gov.uk/government/uploads/system/uploads/attachment_data/file/9072/road-lengths-2011.pdf], retrieved 14 April 2018

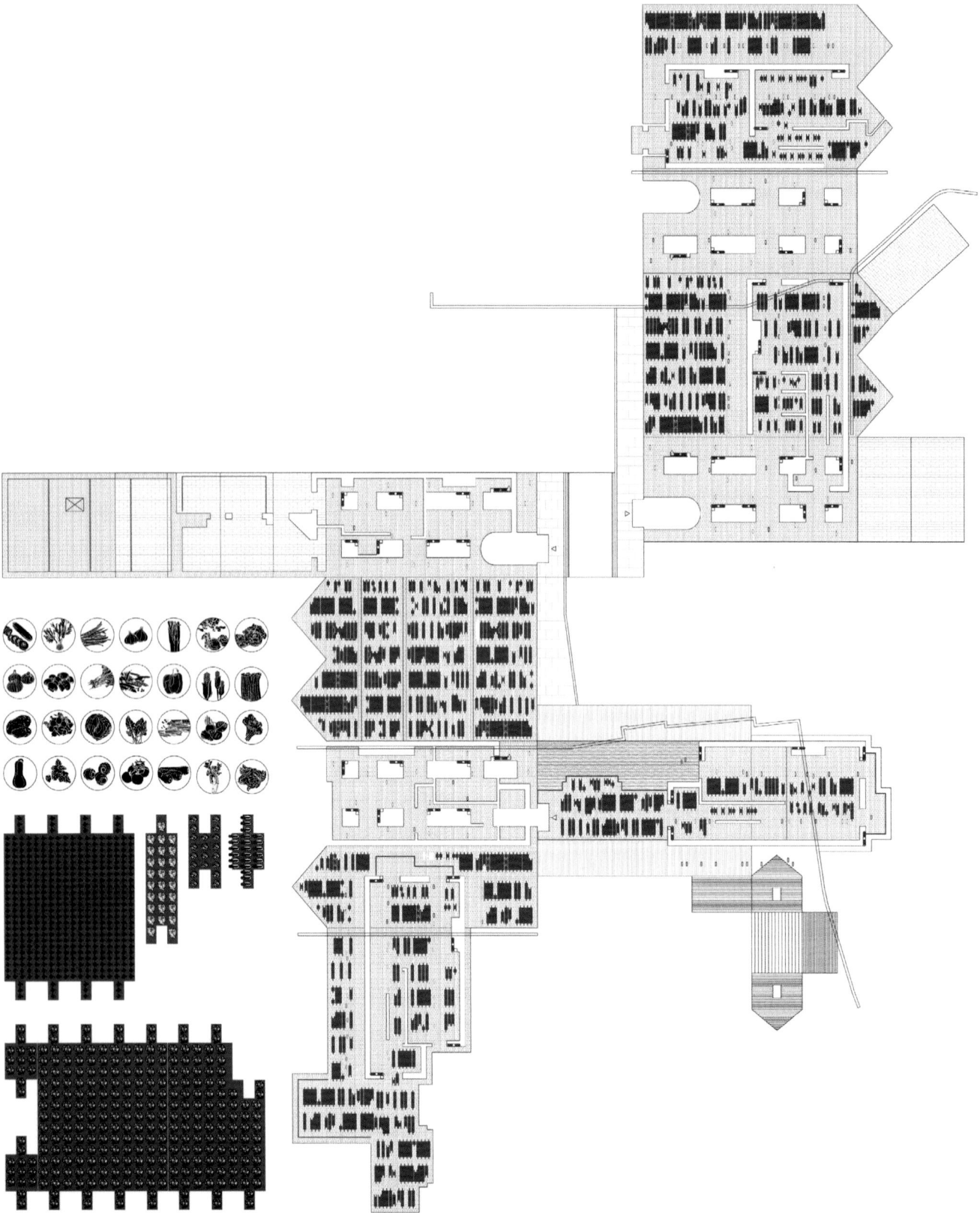

The Green Pension Plan UK

'Uju had lived to a hundred and forty-two – forty-two years older than Lea was now. It had been a good outcome for someone of her generation, someone who had been in her sixties when the Second World War began. For Lea, however, a hundred and forty-two would be a failure. Three hundred was now the number to beat.'

— Rachel Heng, 'Suicide Club', 2018

Imagine life in a near-future society where wellbeing culture has become a moral imperative and is enforced as law by the government. For the dutiful exercisers, early nighters and green juicers, living forever is the ultimate goal. Even though immortality sciences have not progressed far enough to fuel such a dystopian world, the global older population is continuing to grow at an unprecedented rate. In the UK, there were 11.8 million people aged 65 or over in 2016, projected to rise by over 40 percent – nearly one in four people will be aged 65 or over – by 2040.[1] Whilst longer life expectancies do not necessarily mean healthier living, the ageing population and increased prevalence of long-term conditions will have a significant impact on the national health and social care system. This creates a huge imbalance between earners and those likely to need care.

What sustainable strategies can cities and local governments implement to facilitate an equilibrium between younger and older generations? In a Smartcity, sustainability is defined as 'development that meets the needs of the present without compromising the ability of future generations to meet their own needs'.[2] The 'Green Pension Plan' offers the perfect opportunity for sustainability in action, centred around urban agriculture and resilient landscape interventions. The principal aim of the Plan is to develop a new notion of wealth that leverages the full potential of our ageing communities – people of the third age will physically contribute to food security, disseminate their nutritional knowledge and manual skills to future generations to tackle an accelerating obesity epidemic, benefit from mental and physical stimulation, and achieve financial independence while receiving companionship and security.

Resilient landscape interventions retro-fit communities to enable integration and diversity by adapting traditional systems into more cost-effective and practical solutions. Following the Smartcity manifesto that encourages coexistence, the Green Pension Plan has two key components. One connects communities and repairs urban scars caused by the networks of train-lines into cities and disused industrial sites. The 'urban bandage' takes its form by metaphorically bonding two unfolded configurations of a traditional Edwardian family house to create the allotments landscape. The lightweight landscape structure collects rainwater for irrigation and has an anaerobic facility to process community organic waste into

facing page: The Plan has an overarching coexistence policy that enables diverse typologies, needs and scales of green spaces.

1. Age UK, 'Later Life in the United Kingdom', 2018, p.3

2. The World Commission on Environment and Development, 'Our Common Future', Brundtland Report, Chapter 2: Towards Sustainable Development, Oxford University Press, Oxford, 1987, p.43

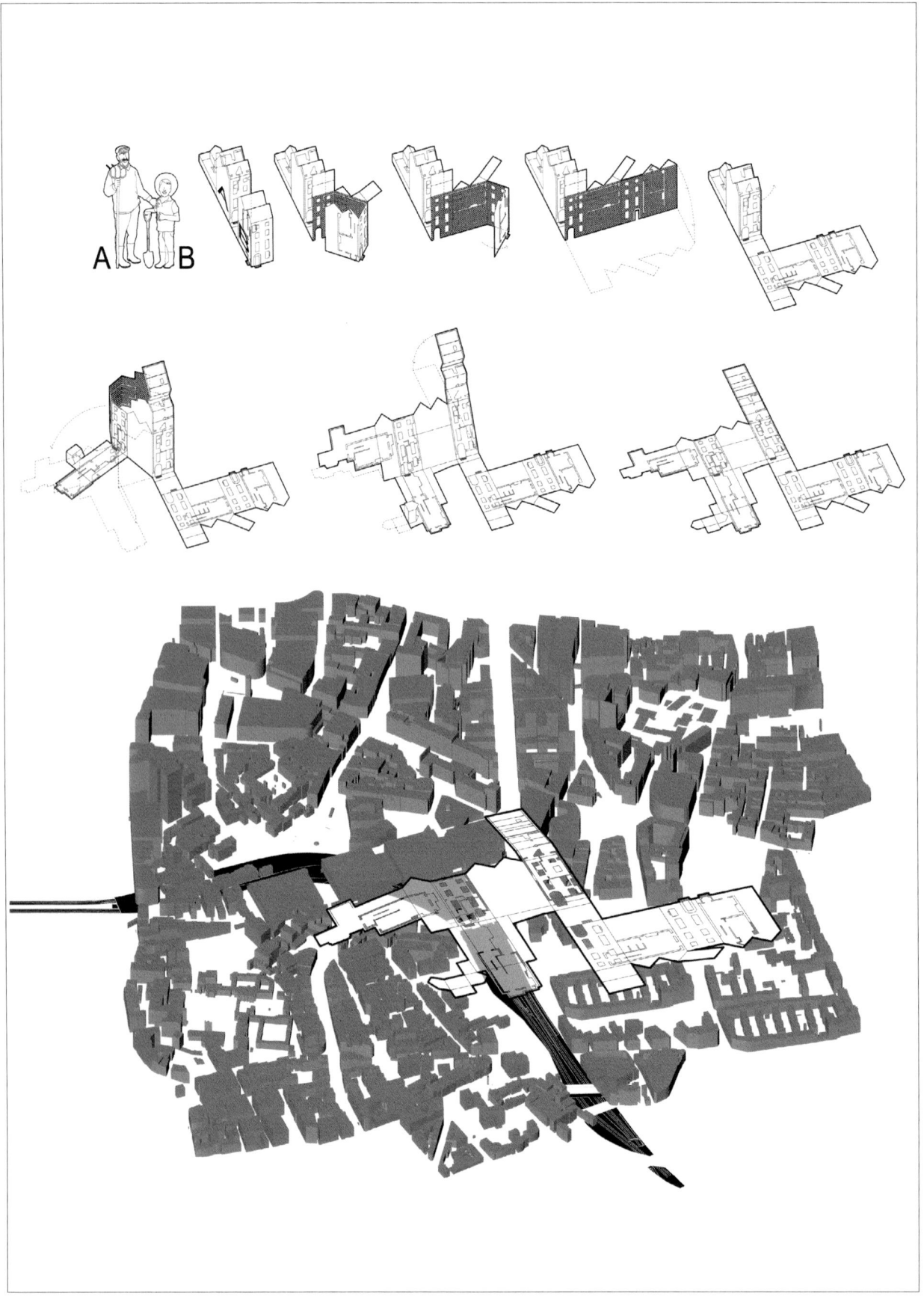

230

fertiliser for the allotments. The other half of the Plan is the 'pie houses' located in socially deprived areas of the city, which improves mobility and inequality. The pie is a metaphor for the new notion of wealth; the houses fill the city with irresistible wafts of fresh baking, establishing a social 'comfort blanket' of security, knowledge and cohesion. Each house hosts an annual pie festival after the autumn harvest. The tall chimneys of the houses transform into urban-scaled community dining tables for the celebrations of the new urban spirit.

Across the city, the matrix of urban bandages and pie houses transforms the urban pockets gradually to make changes sustainable. Jan Gehl emphasised the importance 'to give people time to adapt to physical changes, adjust their life styles, and experiment with the new ways of using the city'.[3] Change is frightening, but the life giving benefits of keeping connected with the seasons, weather and nature always outweighs any worries. Gardening is an on-going gradual act; it 'evokes tomorrow, it is eternally forward-looking, it invites plans and ambitions, creativity, and expectation'.[4] Brussel sprouts and kale adorn the allotments in winter, only to be supplanted by peas, broccoli and lettuce in summer and autumn, and the cycle starts again with a different variety. A 10 x 10 metre plot and 130-day temperate growing season will sustain a family annually with vegetables and a nutritional intake of vitamins A, C, B, complex and iron.[5] The National Institute of Health lists gardening for 30–45 minutes in its recommended activities for moderate levels of exercise to combat obesity, along with biking five miles and walking two miles in 30 minutes.[6]

Within an urban agriculture scenario, the ageing generation takes on the responsibilities of social mentors. Volunteering and citizenship give a role in life, and a sense of being needed and respected. They also decrease mortality and improve health, and provide a positive transition from work to retirement – loneliness can increase the risk of premature death.[7] The Green Pension Plan promotes a sense of belonging; it rekindles urban citizens' connection with food, nature and society, offsetting contemporary introverted pursuits in digital media with real time. Activities in the natural environment are essential ingredients of a healthy childhood, improving the coordination of children, and helping them build friendships. Together, the younger and older generations tend the allotments, help with harvesting and prepare the vegetables for the production of pies. It is crucial for communities to work collectively and move the public policy agenda to improve access to healthy food for underserved and vulnerable members of society.

The Green Pension Plan is also a provocation to raise serious questions about the priorities of governing bodies, archaic planning legislations and land ownership. City planners are often reluctant to integrate food-systems into strategies for future cities. Metropolitan governments need to identify and protect urban agricultural areas, and formalise arrangements between voluntary groups and private landowners on land that is considered 'under-used', including railway embankments, parks and peri-urban areas. Currently, such practises tend to prioritise direct economic values, rather than social values. The benefits of urban agriculture extend beyond greening and management of public spaces: they reduce refuse dumping and illegal drugs related activities. From an environmental perspective, landscape interventions support improvements of microbial diversity and activities in soils, reduction of

231

facing page: The metaphoric bonding of two unfolded traditional Edwardian family houses to configure the allotment landscape.

following pages: Life on the 'urban bandage'.

3. Project for Public Spaces, 'Jan Gehl', www.pps.org, 31 Dec 2008 [https://www.pps.org/article/jgehl], retrieved 20 May 2018

4. P Lively, 'So This Is Old Age', The Guardian, 5 October 2013 [https://www.theguardian.com/books/2013/oct/05/penelope-lively-old-age], retrieved 20 May 2018

5. P Sommers & J Smit, 'Promoting Urban Agriculture: Strategy Framework for Planners in North America, Europe and Asia', The Urban Agricultural Network, 1994, [http://community-wealth.org/sites/clone.community-wealth.org/files/downloads/report-sommers-smit.pdf], retrieved 20 May 2018

6. 'Gardening with Allergies, Arthritis and Other Problems', [http://gardening.about.com/od/allergiesarthritis/a/Garden_Fitness.html], retrieved 20 May 2018

7. Age UK, 'Later Life in the United Kingdom', 2018, p.10

Come and play at the Green Pension

Chating and chilling in the public garden...

...while their grandchildren learn to plant herbs

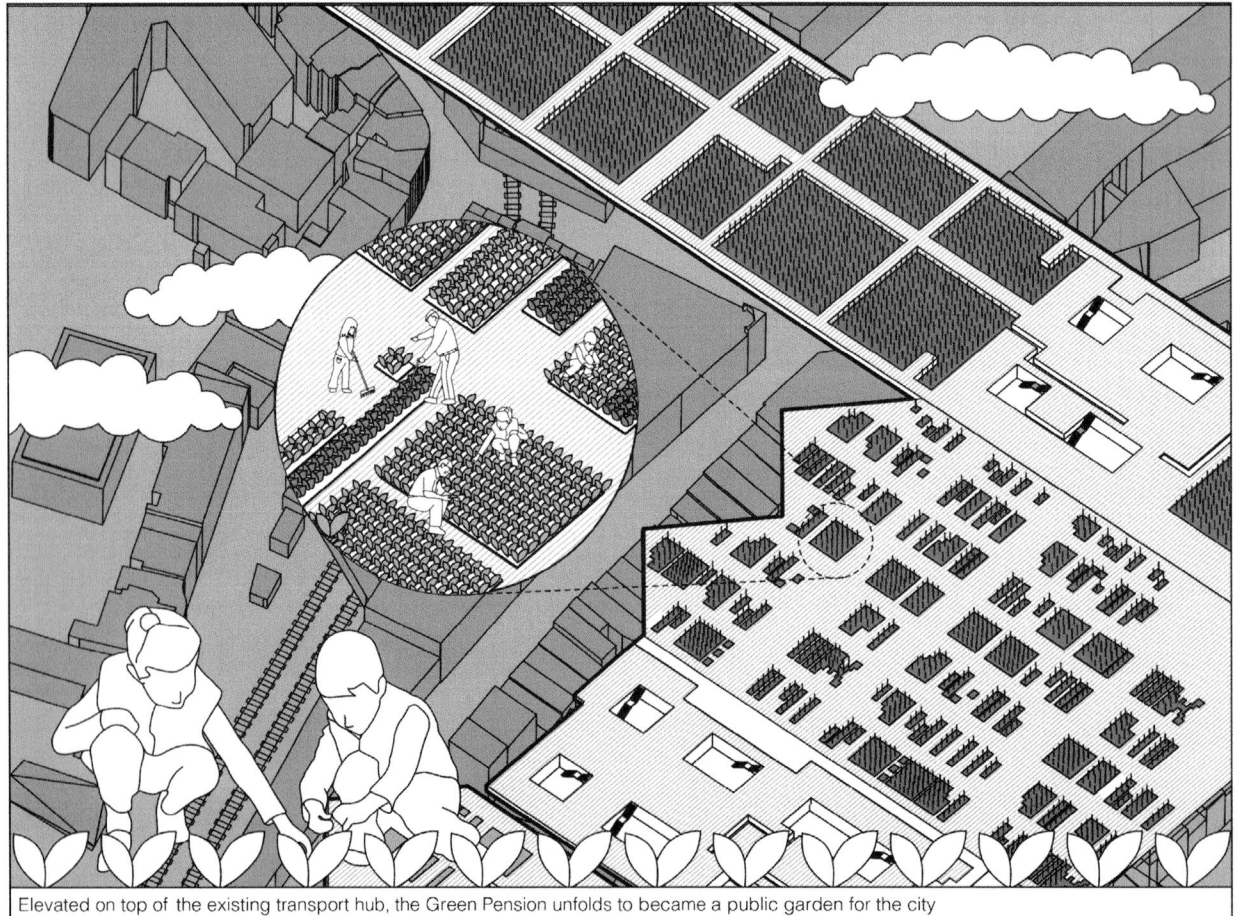
Elevated on top of the existing transport hub, the Green Pension unfolds to became a public garden for the city

Harvesting the leeks for dinner tonight

Come and grow your own food!

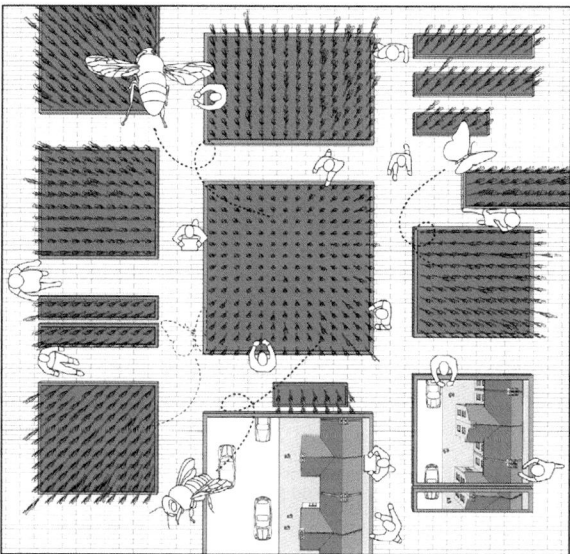

All friendly species are welcomed

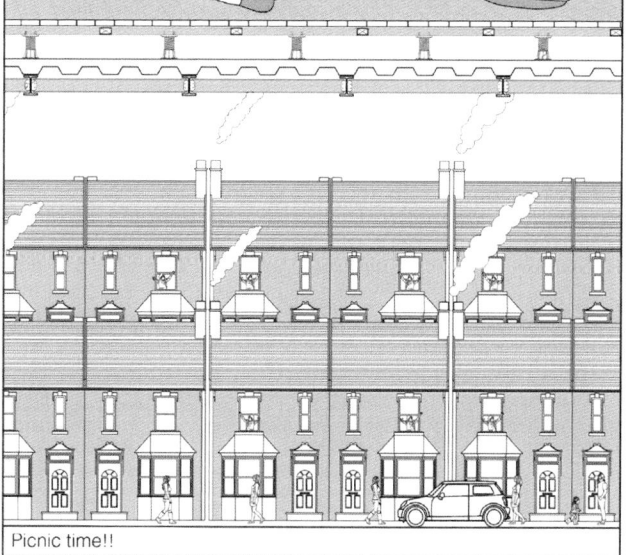

Picnic time!!

Arriving at the Green Pension!

234

Elizabeth, Alice and James welcome spring as they take their lesson in agriculture under the supervision of Grandpa Brian…

Michael and Sarah help out our Grandparents living at the Pie House by keeping the fire stoked to keep the pie oven baking hot.

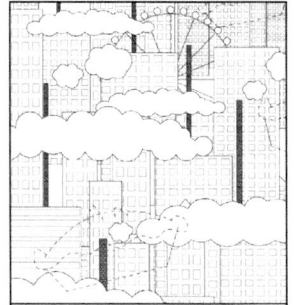

Pie Houses waking up

Children help prepare vegetables for Grandma's tasty and nutritious pies.

Let's wash…

Then chop

The pies are baking in the oven

Now that's a tasty looking pie!

The tall chimneys of each Pie House billow a 'comfort blanket' of social security over London's skyline…

The chimney drops down and rests upon their parallel road…

We need some chairs for the Pie Party!

Lets bring more chairs for Grandpa.

The party is almost ready!

Grandma and Grandpa serve their delicious and nutritious pies to the community.

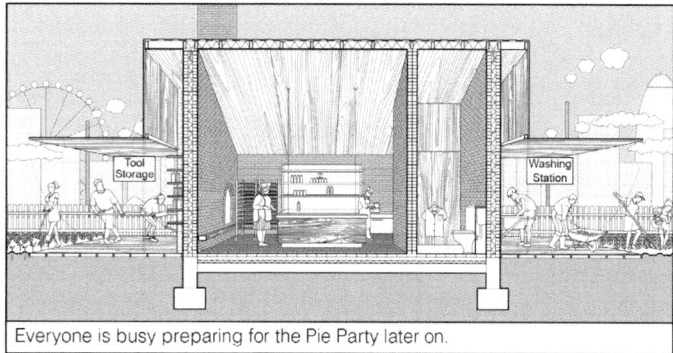
Everyone is busy preparing for the Pie Party later on.

235

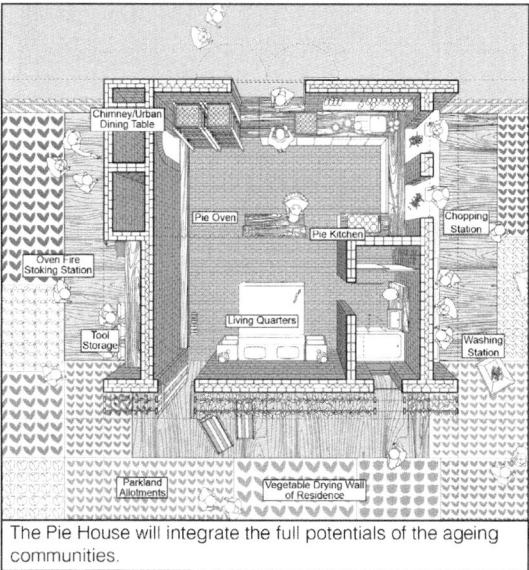
The Pie House will integrate the full potentials of the ageing communities.

Time to gather around and enjoy the pie!

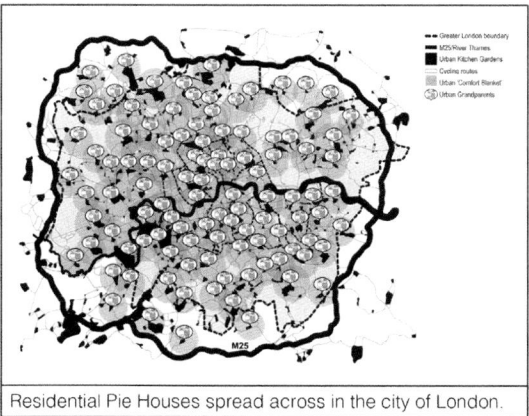
Residential Pie Houses spread across in the city of London.

The Pie House welcomes all and communities are rewarded with tasty pies!

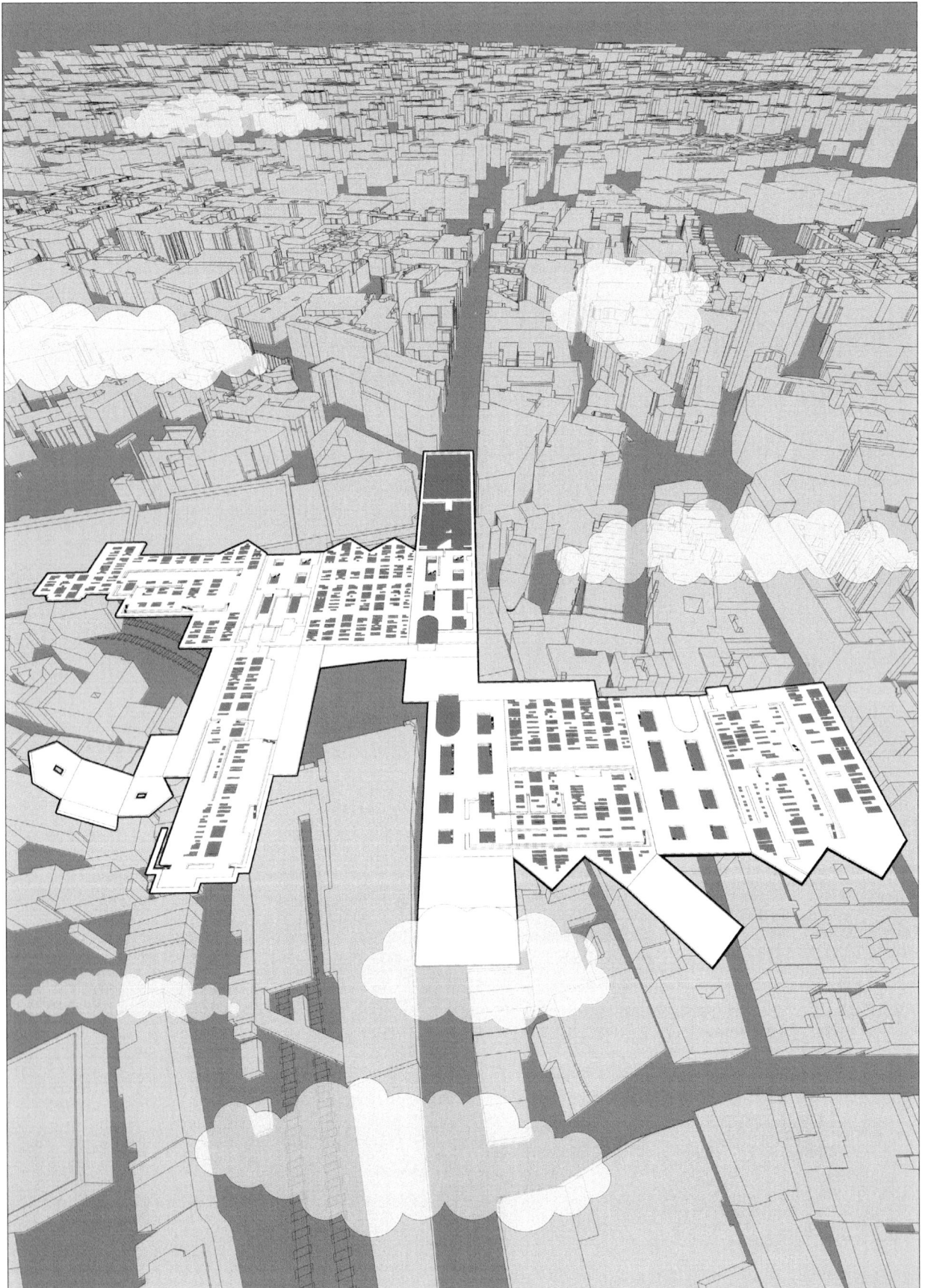

air pollution and heat stress, and chemical contents in water of the surrounding environments. The WHO Healthy Cities programme has recognised the benefits of urban agriculture and appealed to cities and their governments to incorporate food policies into urban plans.[8]

Toronto provides a good example of how urban planning can deliver such landscapes across the city and across different social groups. 'Toronto is known as "a city within a park", with 12 per cent of its surface devoted to green space. Diabetes has been reduced in areas with parks and other spaces conducive to physical activity', according to the Lancet Commission 'Shaping Cities for Health: The complexity of planning urban environments in the 21st century'.[9]

With gradual adaptation, the Green Pension Plan allows for greater flexibility and can take many different forms depending on climate, available technologies, and cultural preference. It commences with the assessment of spatial and ground suitability as well as engineering technologies necessary to host diverse urban agriculture recommendations ranging from hydroponics, irrigation systems, climate control and cultivar species. Workload, growing times and yield are measured to ascertain the appropriate physical and mental demands, tailored to individual communities and sites. Furthermore, gradual adaptation facilitates attitude changes through public involvement. Stakeholder consultations and pilot focus groups engage with local communities to promote positive experiences and the new notion of wealth.

Despite being a model for good physical and mental health, investments in landscape infrastructures and urban agriculture often need to be sunken, have low to medium returns, and require a long-term investment horizon. The global tightening of credit could discourage decision-makers and private investors from investing in sustainable programmes. The entrepreneurial ambition of the Green Pension Plan is to integrate with other national priorities and programmes in order to ensure support and funding by local and national authorities and their social, economic and environmental policies. According to the Committee on Climate Change, the UK's 2050 target to reduce emissions by at least 80 percent of 1990 levels will require reducing domestic emissions by at least three per cent per year.[10] More innovative measures alike the Green Pension Plan, will be able to supplement its existing progress.

While the Green Pension Plan establishes an emphasis to bridge generational demographics, it endeavours to feed other notions of citizenship and pluralistic participation forms to advocate a self-sustaining support network that contributes to wider community resilience. Ultimately, the Green Pension Plan represents an opportunity to help policymakers optimise the operation of a sustainable urban investment, and investors to understand the social economic relationship between a resilient landscape system and its society.

previous pages: The 'pie house' and its associated social and urban impacts.

facing page: The 'urban bandage' allotment landscape repairs urban scars caused by the networks of train-lines.

8. K Morgan, 'Feeding the City: The Challenge of Urban Food Planning', International Planning Studies, vol.14, no.4, 2010, pp.341–348

9. Y Rydin et al., 'Shaping Cities for Health: The complexity of planning urban environments in the 21st century', The Lancet Commissions, May 2012, p.24

10. Committee on Climate Change, 'How the UK Is Progressing' [https://www.theccc.org.uk/tackling-climate-change/reducing-carbon-emissions/how-the-uk-is-progressing], retrieved 2 June 2018

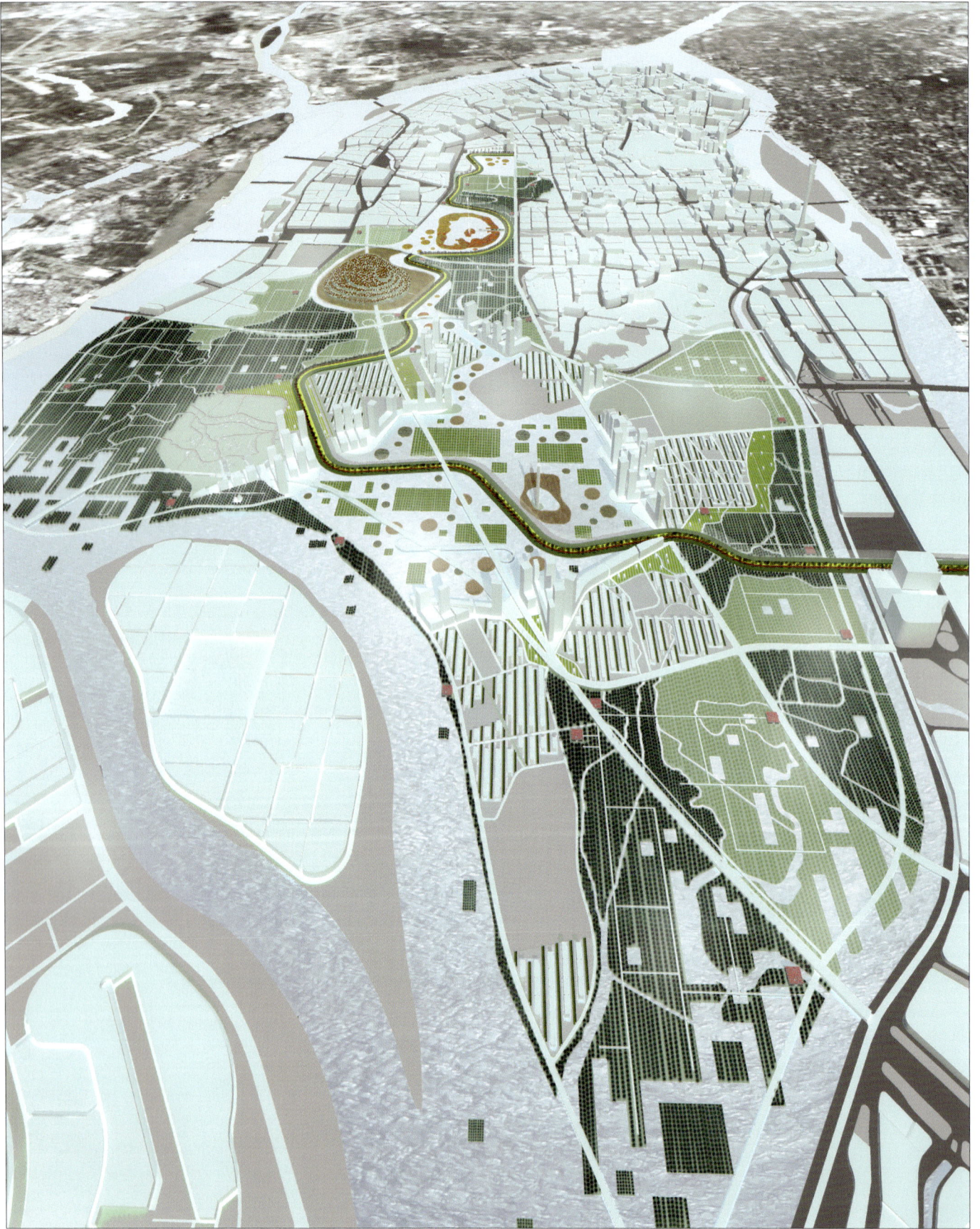

Wanmu Orchard Wetland China

'Now I see the secret of making the best persons; it is to grow in the open air and to eat and sleep with
the earth.'

        – Walt Whitman, 'Song of the Open Road', 1856

China's urbanisation has drastically rendered city-bound children fewer opportunities to get close to nature than their rural counterparts. In his book 'Last Child in the Woods', Richard Louv warned that the wellbeing of urban children is compromised by 'nature deficit disorder'.[1] Addressing this, Wanmu's resilient landscapes invite children to build an emotional connection to nature by igniting their curiosity and passion for learning in the same way Chinese NGOs and education groups, like Friends of Nature, Green Hope and the Antelope Bus, spark play and learning through bird-watching and gardening activities. Like these programmes, the Wanmu Orchard Wetland (WOW) project provides links between the urban public and the natural world. As Beijing Brooks Institute emphasised, 'nature is a treasure-house of knowledge, a palace of art, a spring of imagination and creativity. Children who know how to appreciate beauty will be happier, and creative children will be more successful.'[2]

One can only think about the story of the Monkey King, the trickster hero of the classical Chinese novel 'Journey to the West', to understand the mythic resonance behind the natural world as a setting for teaching childhood lessons. In the 16th century tale, the Monkey King is seduced by the peaches of the Heavenly Orchard that promise millennia of life for each forbidden bite. Disregarding the prohibition, the Monkey King abandoned his duty as the guardian of fruit to become their despoiler. Exchanging responsibility for thoughtless greed, the as yet unreformed hero makes the orchard the setting of a private feast. Imprisoned 500 years for theft and the rebellions his actions caused, the Monkey King was fed iron pellets for his hunger, and given molten copper for his thirst. The famous pilgrimage to the West to retrieve the sutras became the atonement for the trespass in the orchard.[3]

The legend is a cautionary tale inculcating the precepts of discipline and responsibility. Its real-world counterpart, the Wanmu Orchard Wetland (WOW), located south of Guangzhou's New City axial, is also a space that tells stories about the stewardship of nature. An open-air classroom, laboratory and playground, WOW cultivates childhood along with the bounty of nature's fragrant fruit-bearing trees and the variegated ecosystem that such plantings encourage. Adjacent to Guangzhou Biology Island and Guangzhou University City, the 25.96km2 orchard is a protected site and includes Wanmu Orchard, Ruibao Park, and Haizhu Lake Park. Within the Haizhu District of Guangzhou, the site is surrounded by the tributaries of the Pearl River. Adapting the orchard from its single function, which is to act as an

facing page: The WOW landscape promotes the region's ecology by nurturing the resilient imaginations of future generations.

following pages: The adaptive plans of WOW.

1. R Louv, 'Last Child in the Woods: Why children need nature, how it was taken from them, and how to get it back', Algonquin Books of Chapel Hill, 2005

2. X Liu, 'How Children in China's Urban Jungle are Reconnecting with Nature', Environment Network: Wildlife, The Guardian, 11 January 2012

3. C Wu, 'The Journey to the West', AC Yu (trans.), University of Chicago Press, 2012

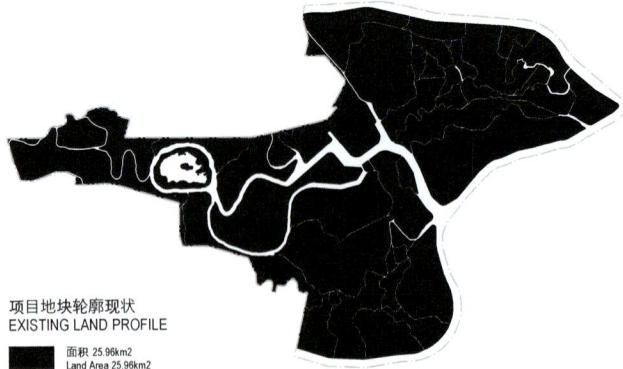

项目地块轮廓现状
EXISTING LAND PROFILE

▪ 面积 25.96km2
　Land Area 25.96km2

现有的建筑
EXISTING BUILDINGS

▪ 住宅
　Residential

▪ 教育中心
　Education

▪ 商 业
　Commercial

▪ 交通设施
　Transport Infrastructure

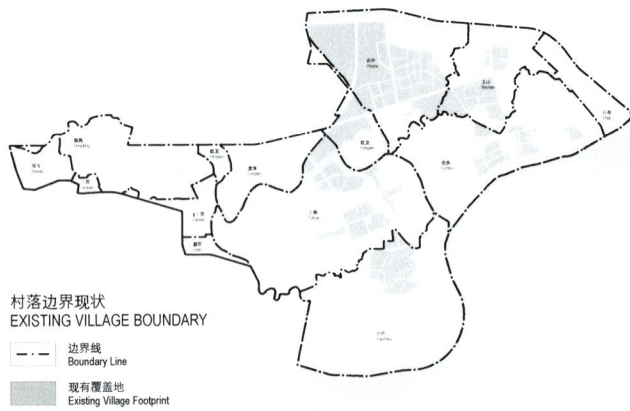

村落边界现状
EXISTING VILLAGE BOUNDARY

－‧－ 边界线
　　　Boundary Line

▪ 现有覆盖地
　Existing Village Footprint

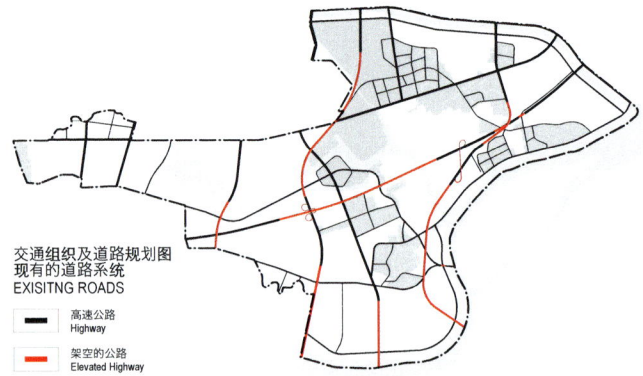

交通组织及道路规划图
现有的道路系统
EXISITNG ROADS

▪ 高速公路
　Highway

▬ 架空的公路
　Elevated Highway

项目地块情况
LANDPROFILE

▪ 面积
　LandArea17.32km2

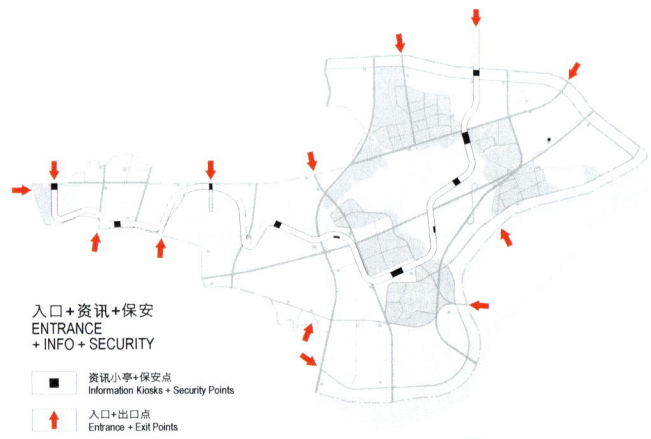

入口+资讯+保安
ENTRANCE
+ INFO + SECURITY

▪ 资讯小亭+保安点
　Information Kiosks + Security Points

↑ 入口+出口点
　Entrance + Exit Points

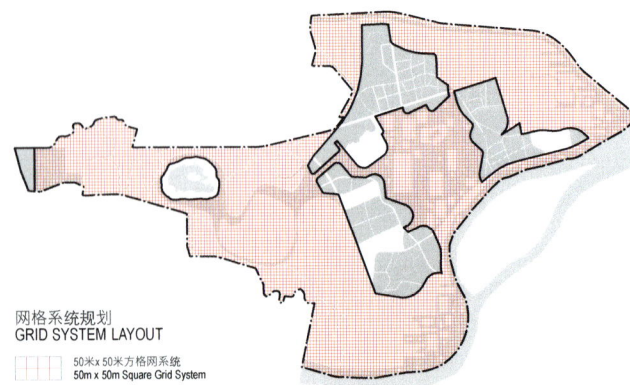

网格系统规划
GRID SYSTEM LAYOUT

▦ 50米x 50米方格网系统
　50m x 50m Square Grid System

交通组织及道路规划图
绿色林荫大道
GREEN BOULEVARD

∿ 100米宽的绿色林荫大道
　100m width Green Boulevard

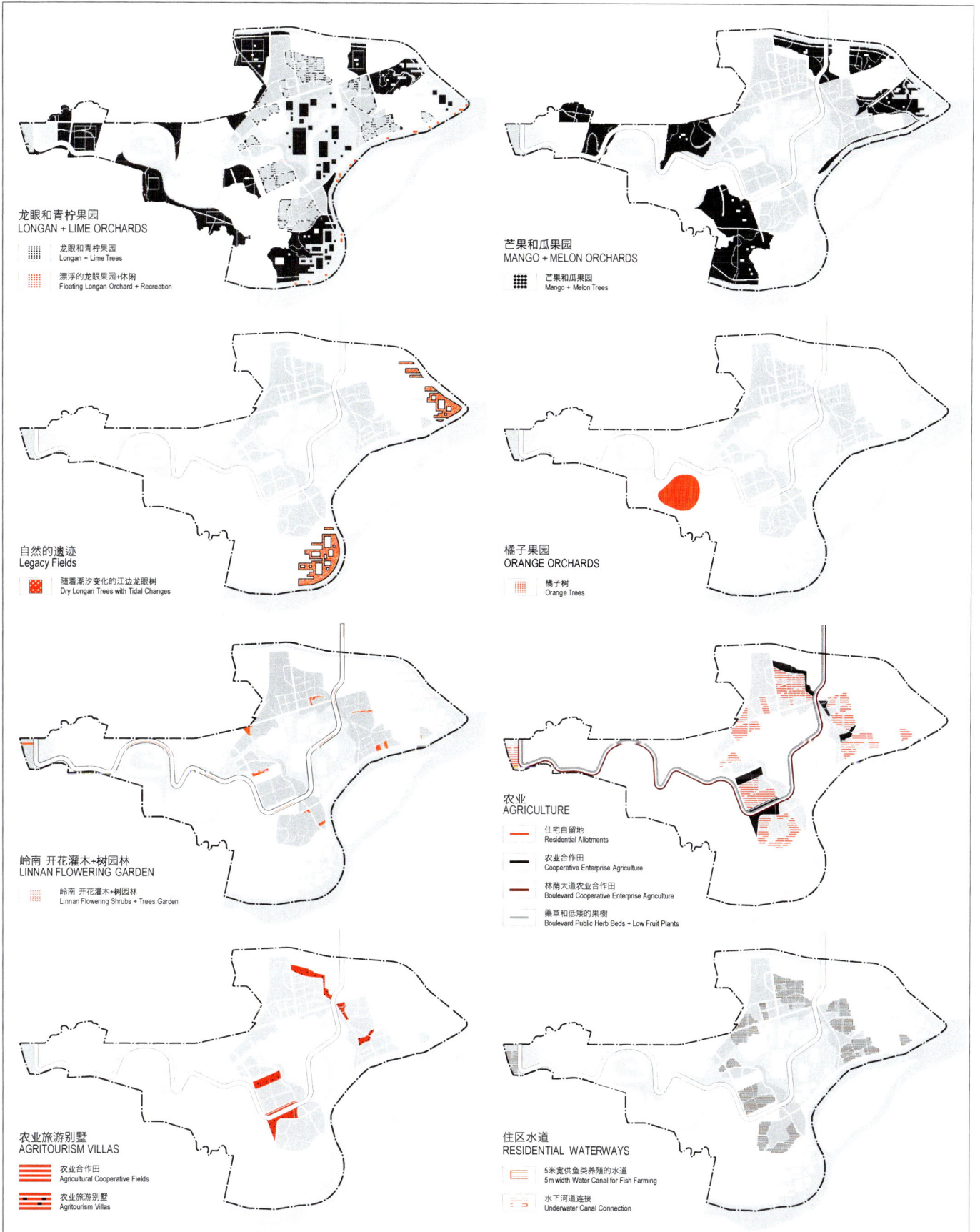

龙眼和青柠果园
LONGAN + LIME ORCHARDS

龙眼和青柠果园
Longan + Lime Trees

漂浮的龙眼果园+休闲
Floating Longan Orchard + Recreation

芒果和瓜果园
MANGO + MELON ORCHARDS

芒果和瓜果园
Mango + Melon Trees

自然的遗迹
Legacy Fields

随着潮汐变化的江边龙眼树
Dry Longan Trees with Tidal Changes

橘子果园
ORANGE ORCHARDS

橘子树
Orange Trees

岭南 开花灌木+树园林
LINNAN FLOWERING GARDEN

岭南 开花灌木+树园林
Linnan Flowering Shrubs + Trees Garden

农业
AGRICULTURE

住宅自留地
Residential Allotments

农业合作田
Cooperative Enterprise Agriculture

林荫大道农业合作田
Boulevard Cooperative Enterprise Agriculture

药草和低矮的果树
Boulevard Public Herb Beds + Low Fruit Plants

农业旅游别墅
AGRITOURISM VILLAS

农业合作田
Agricultural Cooperative Fields

农业旅游别墅
Agritourism Villas

住区水道
RESIDENTIAL WATERWAYS

5米宽供鱼类养殖的水道
5m width Water Canal for Fish Farming

水下河道连接
Underwater Canal Connection

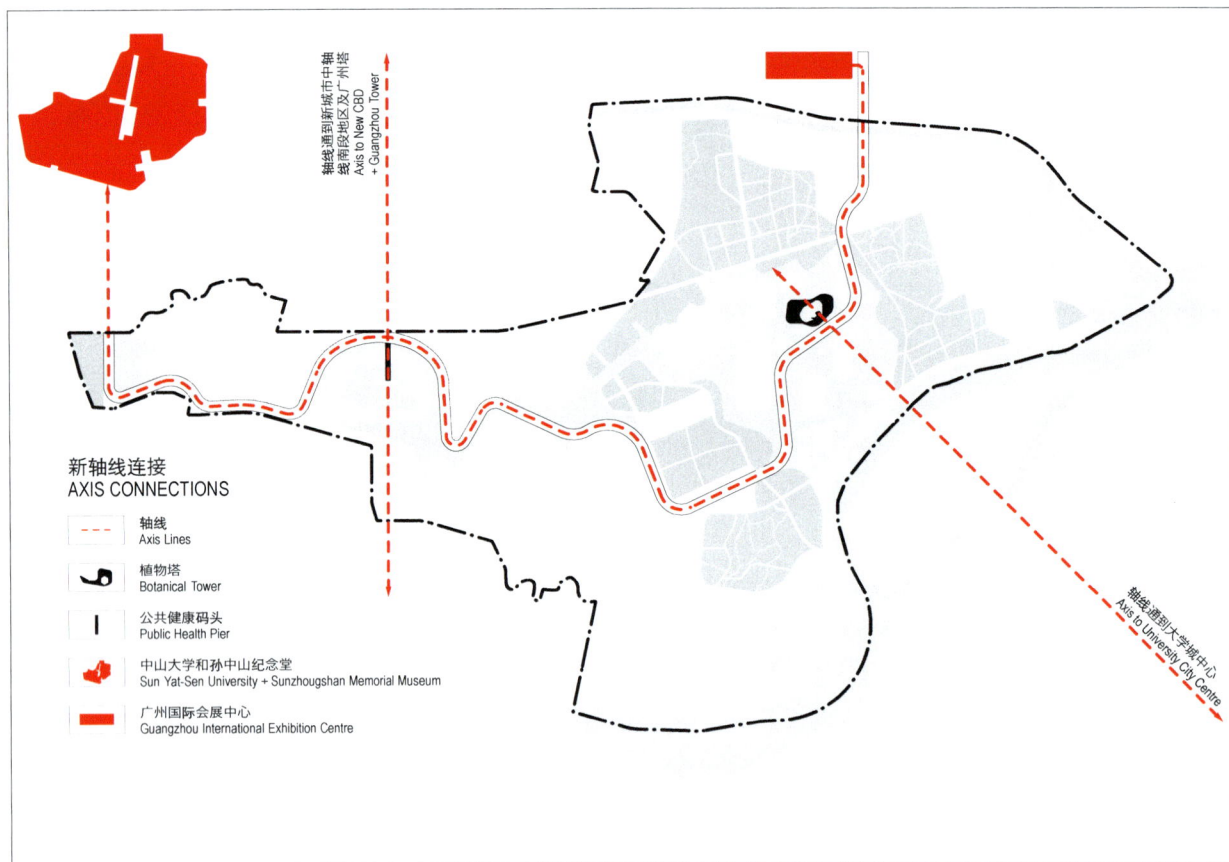

新轴线连接
AXIS CONNECTIONS

- - - - 轴线
Axis Lines

植物塔
Botanical Tower

公共健康码头
Public Health Pier

中山大学和孙中山纪念堂
Sun Yat-Sen University + Sunzhougshan Memorial Museum

广州国际会展中心
Guangzhou International Exhibition Centre

轴线通到新城市中轴
线南段地区及广州塔
Axis to New CBD
+ Guangzhou Tower

轴线通到大学城中心
Axis to University City Centre

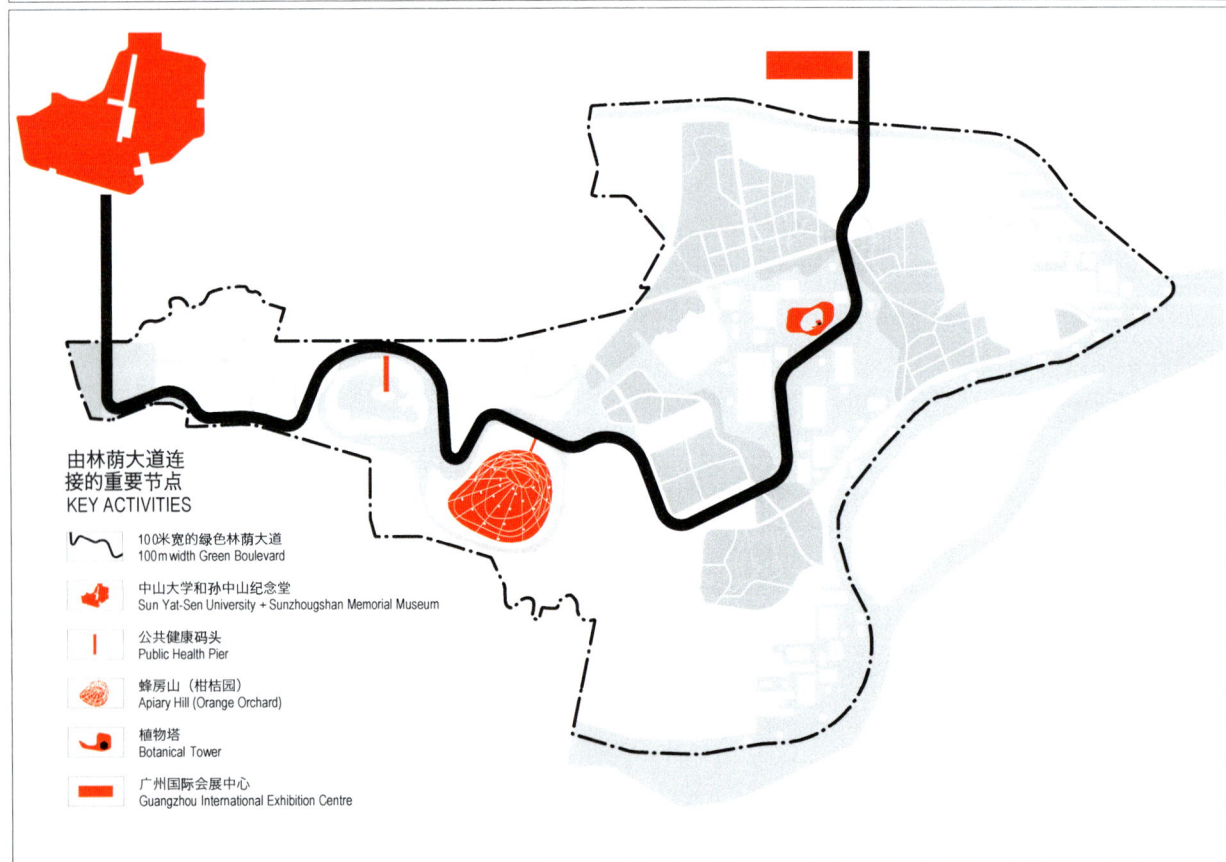

由林荫大道连
接的重要节点
KEY ACTIVITIES

100米宽的绿色林荫大道
100m width Green Boulevard

中山大学和孙中山纪念堂
Sun Yat-Sen University + Sunzhougshan Memorial Museum

公共健康码头
Public Health Pier

蜂房山（柑桔园）
Apiary Hill (Orange Orchard)

植物塔
Botanical Tower

广州国际会展中心
Guangzhou International Exhibition Centre

adaptiveness: Wanmu Orchard Wetland

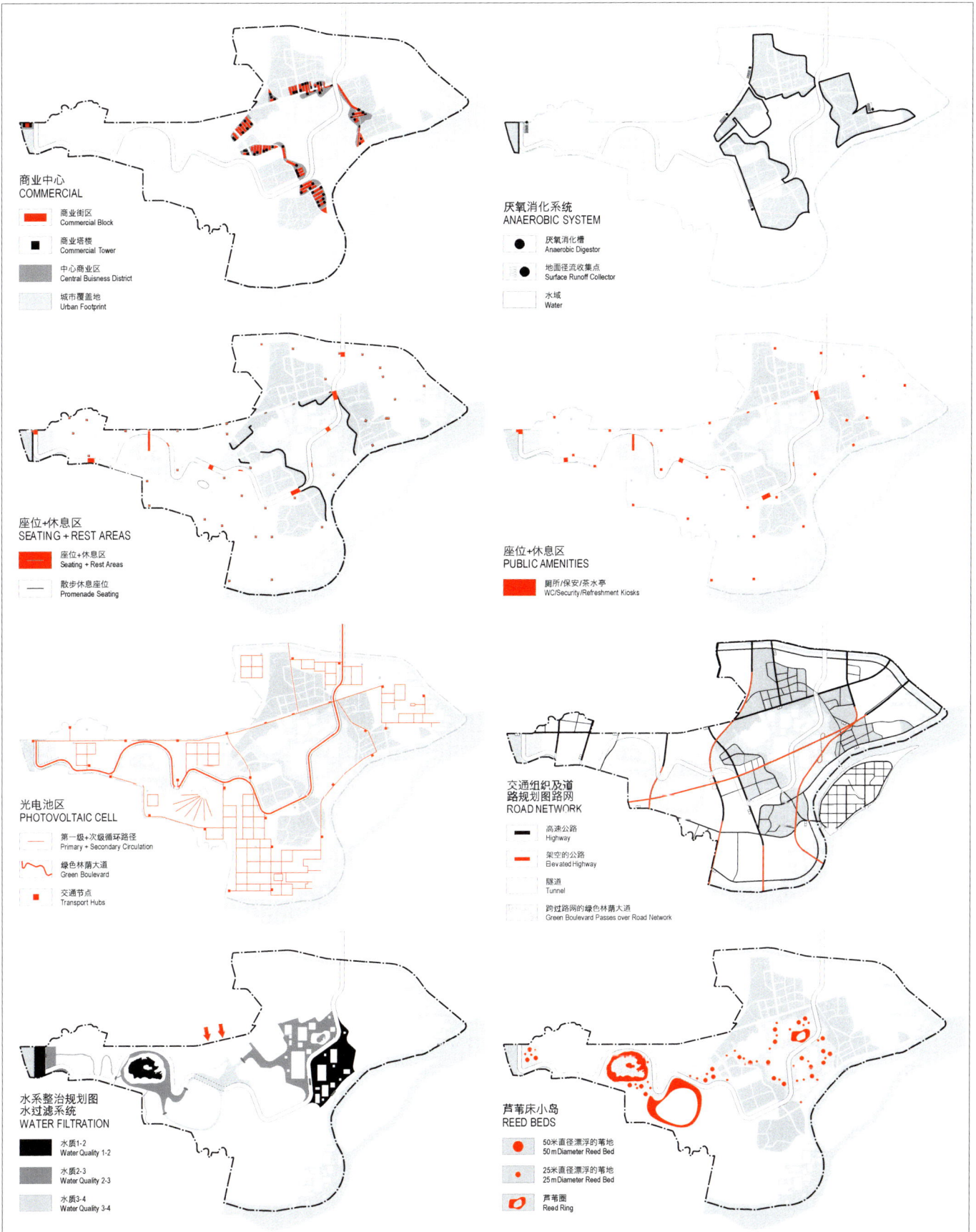

商业中心
COMMERCIAL

- 商业街区 Commercial Block
- 商业塔楼 Commercial Tower
- 中心商业区 Central Buisness District
- 城市覆盖地 Urban Footprint

厌氧消化系统
ANAEROBIC SYSTEM

- 厌氧消化槽 Anaerobic Digestor
- 地面径流收集点 Surface Runoff Collector
- 水域 Water

座位+休息区
SEATING + REST AREAS

- 座位+休息区 Seating + Rest Areas
- 散步休息座位 Promenade Seating

座位+休息区
PUBLIC AMENITIES

- 厕所/保安/茶水亭 WC/Security/Refreshment Kiosks

光电池区
PHOTOVOLTAIC CELL

- 第一级+次级循环路径 Primary + Secondary Circulation
- 绿色林荫大道 Green Boulevard
- 交通节点 Transport Hubs

交通组织及道路规划图路网
ROAD NETWORK

- 高速公路 Highway
- 架空的公路 Elevated Highway
- 隧道 Tunnel
- 跨过路网的绿色林荫大道 Green Boulevard Passes over Road Network

水系整治规划图 水过滤系统
WATER FILTRATION

- 水质1-2 Water Quality 1-2
- 水质2-3 Water Quality 2-3
- 水质3-4 Water Quality 3-4

芦苇床小岛
REED BEDS

- 50米直径漂浮的苇地 50 m Diameter Reed Bed
- 25米直径漂浮的苇地 25 m Diameter Reed Bed
- 芦苇圈 Reed Ring

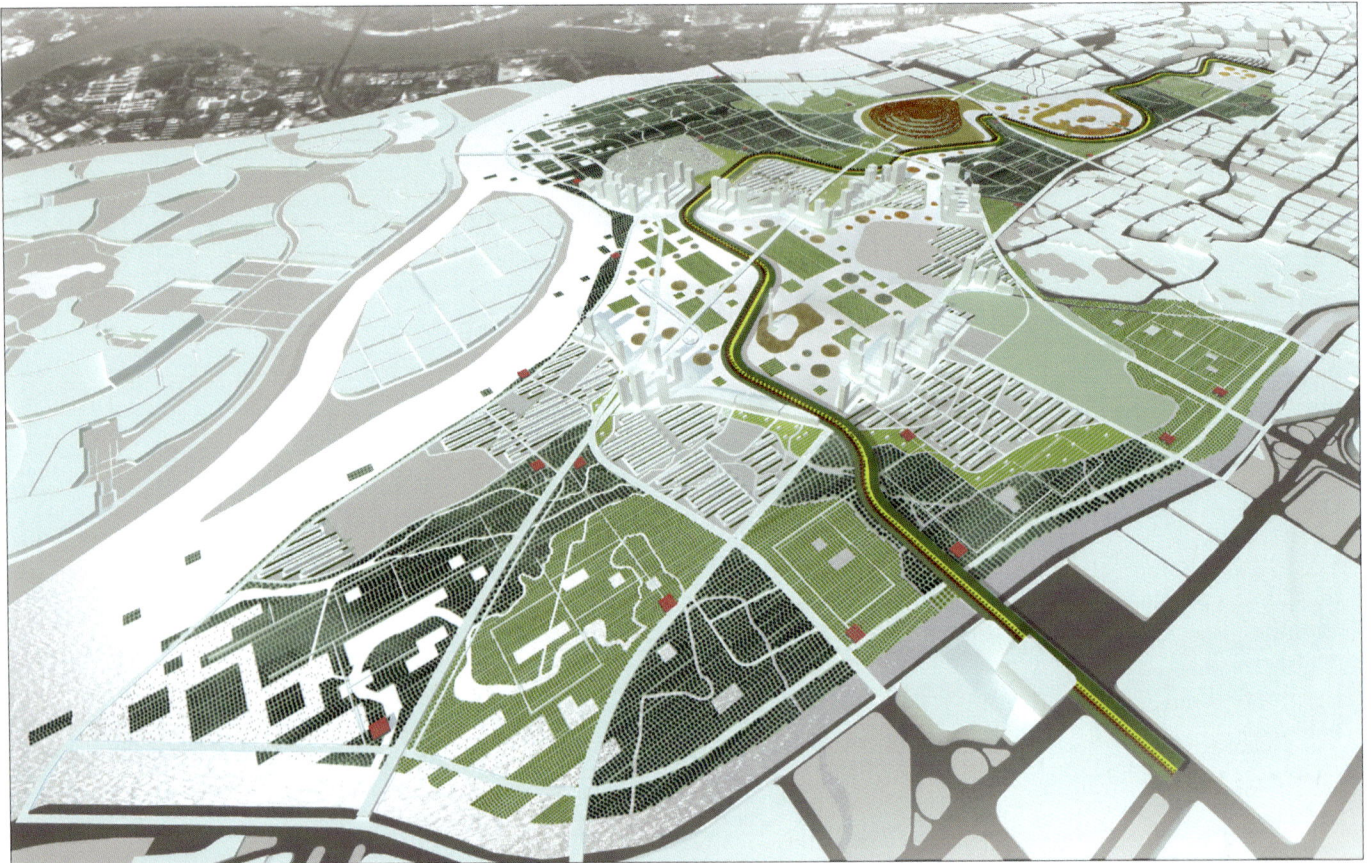
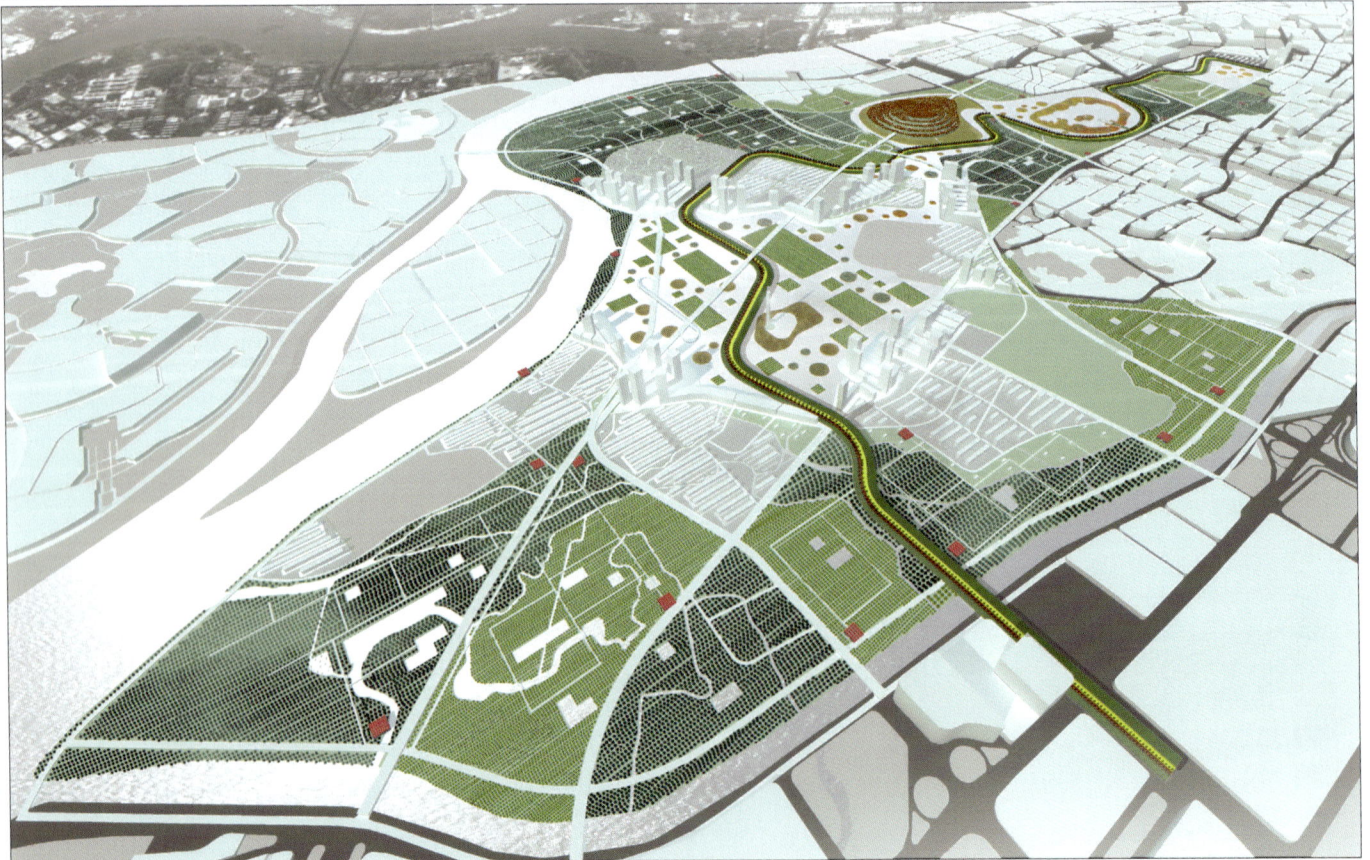

agent for food security, the new orchard is envisioned as a multi-programme national knowledge hub. Located in the middle of prime real estate in Guangzhou, the plan turns to nature education as a way to ensure the adaptation and survival of the orchard and wetlands, guarding against its likely erasure by unchecked urbanisation. From this educational platform, WOW achieves its other goals: to conserve and regenerate existing villages. Embodying the wisdom of the deities who reformed the Monkey King, WOW is a site that looks toward the future of the region's ecology by nurturing the resilient imaginations of future generations.

The programmes for WOW act as the green generative seed for the Haizhu district. Ecological, cultural and heritage preservation, and long-term socio-economic sustainability are goals that play a significant role within the context of fostering a relationship between the city inhabitants, especially the young ones, and the natural environment. WOW creates encounters that cultivate a spatial affinity with the 'vernacular' landscape of Guangzhou – children are introduced to the inimitable histories of adaptation that grew out of the practical needs of the community and the constraints of place and climate. In fact, limitations are the salient factors that imprint a space with its identity: 'regional identity has to do with where one stays, where one's roots are, and consequently with where long-standing social traditions can develop.'[4] Guangzhou's fruit orchards are a trove not only of produce but also of cultural wisdom and practices, and thus are of great importance for the city's heritage of resilience and green credentials.

WOW is fully aware of its ecological responsibilities. Another natural enclave in Guangzhou, the National Baiyun Mountain Scenic Area (BMSA), has experienced severe biotic pauperisation since the 1950s, with the loss of close to 500 floral and 50 faunal species.[5] To avoid the same fate, WOW serves to educate its users about the benefits of urban biodiversity, from improving soil, water and air quality to the preservation of floral and faunal species, in addition to a careful strategy to facilitate non-indigenous fruits cultivation, research and education. The best conservation motive actually lies in opportunities to learn and connect to nature, and to understand cultural and historical values underwritten by the diversity of the biota.

By retaining existing local practices for cultivating and farming fruits and vegetables, WOW transmits knowledge about local traditions of food production that accompany the bounty provided by the land. Local restaurants can showcase local fresh produce to visitors, while organic fruits and herbs can be sold fresh or packaged as preserves, cakes, sweets and drinks. Behind the circulation of these products, new sustainable technologies such as nutrient waste recycling operate as agricultural production and are only profitable if sustainably managed. Existing farming communities can become essential players as the park's gardening and organic agriculture workforce. Like a table laden with the harvest of a generous year that invites all to partake, the orchards of WOW should be conceived as a social and economic tool drawing the community together, increasing local employment and establishing WOW's branding as a local, regional and international tourist destination.

WOW also aims to disseminate environmental and wellbeing knowledge to the larger public, especially through the potential afforded by the growing ubiquity of agritourism. A recent study about how to assess

245

facing page top + bottom: Low and high tide configurations of the legacy floating orchards.

4. M Hough, 'Out of Place: Restoring identity to the regional landscape', Yale University Press, 1992, p.58

5. WY Chen & CY Jim, 'Resident Motivations and Willingness-to-Pay for Urban Biodiversity Conservation in Guangzhou (China)', Environmental Management, vol.45, no.5, 2010, pp.1052–1064

the economic value of landscapes asserts that while 'it is very difficult, if not impossible, to value the environment...it must be possible to value preferences for environmental goods, such as the agricultural landscape, and so produce measures directly comparable with the values of marketed goods'.[6] Clearly, landscapes are key to ecotourism and agritourism, significant economic generators for the local communities in many countries. A weekend jaunt or an ideal daytrip for the young and the old, WOW's set of activities are educational and informative while offering scenic beauty and relaxation to visitors and residents. Visitors stay in villa accommodations located on farms where they can assist with farming tasks, pick their own fruit and vegetables or simply relax and take in fresh country air. Recreational activities on the public Health Pier, fresh water fishing in the wetlands, eco-education in the Botanical Tower and honey production on the Apiary Hill are a few of the attractions that WOW can offer.

247

Biodiversity conservation is inextricably linked with the conservation of the imaginative and emotional values that attach us to the environment. Current research on conservation actually draws attention to the role human communities play in embodying and transmitting value for the environment through teaching cultural practices. Case studies documenting these in situ or in vivo conservation practices show that 'all these modes of countering homogeneity in modernity involve human sovereignty and resilience in marginal spaces where a sense of place can be elaborated within a milieu of memory'.[7] From orchard cultivators in the American south preserving seeds of heirloom apple trees to Quechua potato farmers fighting to reintroduce indigenous varietals, biodiversity's eco-warriors are the 'local custodians steeped in the cultural significance' of the flora and fauna they cultivate, and are ultimately inspired by the desire to transmit the 'idea of place'.[8] Thus, WOW's educational goals draw from the expertise within communities in existing villages.

facing page top: The Green Boulevard accommodates the key activities and event spaces, and is the main circulation.

facing page bottom: Haizhu Lake is an oasis for wildlife.

The respect for human diversity drives the careful conservation and regeneration of the area, where regeneration might actually be minimal to highlight the heritage values of places like Xiaozhou, for example. The strategy for regeneration imagines two key urban hubs: (1) Ruibao with the east of Haizhu district, and (2) Xiaozhou, Tuhua, Hongwei, Chisha, Beishan and Luntou as a new hub connected by a strip of commercial development around the main wetland of WOW. Collectively, the villages can prosper as the vision brings new economic development to the two hubs. A new residential typology, inspired by the Xiaozhou historical canal housing design, is developed for each village's regeneration. Vegetable allotments and fish farming channels are introduced between housing units to promote community wellbeing and local healthy eating. Little modification of villages occurs in Phase 1 with a few new housing units; subsequent construction of residential units is carefully phased with minimal disturbance to village communities. By strengthening communities, WOW contributes to the repository of knowledge that ultimately serves to preserve an ecological affinity with the land.

6. I Vanslembrouck & GV Huylenbroeck, 'Landscape Amenities: Economic assessment of agricultural landscapes', Springer, 2005, p.55

7+8. VD Nazarea et al. (eds.), 'Seeds of Resistance, Seeds of Hope: Place and agency in the conservation of biodiversity', University of Arizona Press, 2013, p.5

The conceptual motif for WOW is a necklace of eight pearls set in priceless emeralds – eight ecological, educational and sustainable living hubs within the green orchards. The concept continues the goal of preserving the area's historical and cultural identity through minimal modification of existing villages and orchards. Instead, innovative and sensitive urban spatial landscape moves accentuate the site and enhance the capacity of nature to teach and inspire, drawing similar values to those of the City of a

烤肉区
Barbeque Area
5m x 5m

潜水通气管租赁点
Snorkel Hire
5m x 5m

木铺板
Wooden Decking
10m x 10m

收成储存棚
Harvest Storage Shed
10m x 10m

艺术家工作室1
Artist Studio 1
15m x 5m

有座位的茶室
Teahouse with
Seating
5m x 5m

绿色球车停车
Green Buggy Parking
5m x 7.5m

Parking Paving
5m x 7.5m

小船租赁点
Kayak Hire
5m x 5m

沙滩椅及阳伞租赁点
Deckchair & Parasol Hire
5m x 5m

艺术家工作室2
Artist Studio 2
15m x 10m

食物摊贩
Food Stall
2.5m x 1.25m

Cycle Parking
5m x 5m

帐蓬租赁
Tent Rent
5m x 5m

鱼饵店
Bait Shop
2.5m x 2.5m

广场铺地
Plaza Paving
10m x 10m

工具房
Toolshed
10m x 10m

货车停车棚
Goods Trailer
Parking Shed
2.5m x 5m

咖啡厅
Cafe
5m x 20m

桑拿浴
Sauna
10m x 5m

水果加工点
Orchard Fruit Processing Point
40m x 12m

蜜蜂点
Honey point
10m x 10m

苗圃
Nursery
10m x 10m

售票处
Ticketdesk
5m x 5m

座位
Seating
5m x 5m

厕所
Toilets
5m x 5m

咖啡厅
Cafe
5m x 5m

淋浴
Showers
5m x 5m

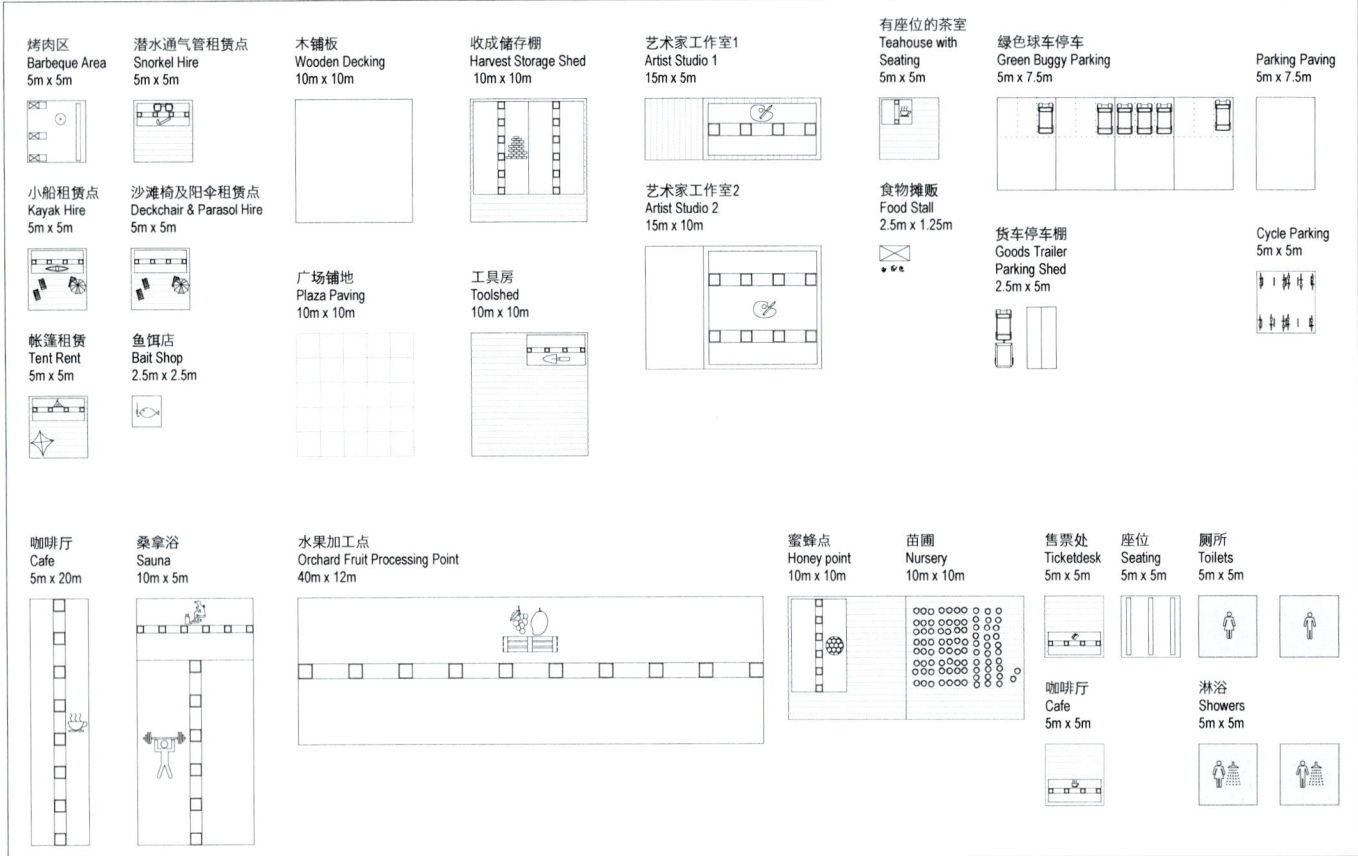

公共健康码头
PUBLIC HEALTH PIER
300m x 60m

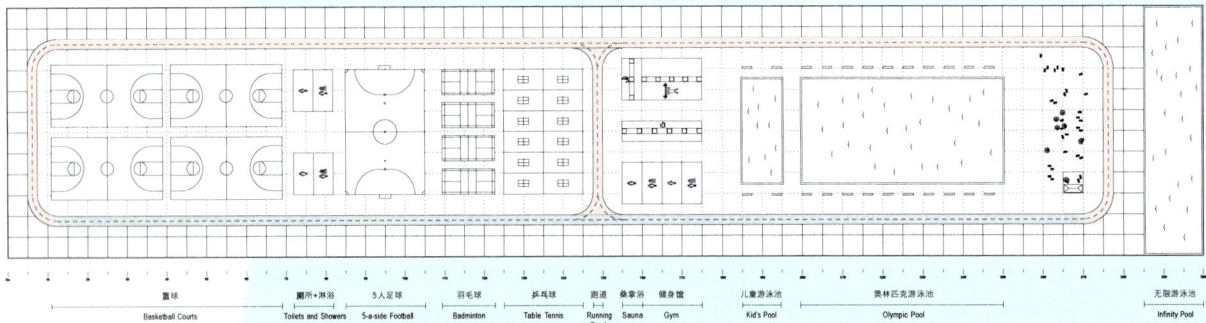

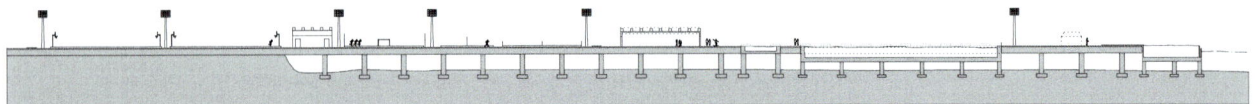

篮球
Basketball Courts

厕所+淋浴
Toilets and Showers

5人足球
5-a-side Football

羽毛球
Badminton

乒乓球
Table Tennis

逃道
Running
Track

桑拿浴
Sauna

健身馆
Gym

儿童游泳池
Kid's Pool

奥林匹克游泳池
Olympic Pool

无限游泳池
Infinity Pool

Thousand Lakes in Gaochun. WOW's hubs are organised by two fully interacting urban groundwork systems, each different but complimentary: (1) the 50 x 50m Ecological Grid, and (2) the Green Boulevard.

In the **Ecological Grid**, all hydrology, ecology and landscape dynamics of the site are supported. A kit-of-parts of activities is applied to the living grid to strategically organise and retrofit the existing orchards and new wetlands.

249

The fruit orchards are retrofitted with minimal disruption and little waste to enhance biodiversity and visual enjoyment of the productive landscape. As existing fruit trees were planted too close together, a solution can be to increase the space between each individual tree. Fruit shrubs and ground cover of different heights could be planted along the strips in-between the tree rows, creating different canopies and micro-habitats within a small area for local fauna to use as shelter or feeding ground. The resulting increase of habitat complexity could enhance the productivity of terrestrial habitat and improve the suitability of orchard habitat for the use of local fauna groups. The orchards provide dramatic seasonal colours from emerald green to white, yellow and pink blossoms emitting fragrances that vivify the landscape's sensorium.

Additionally, wetlands are employed to reinforce the hydrology and ecology dynamics of the site. The collection of wetlands on the site filters water from the Pearl River for agriculture and recreation. Rivers and canals link the new wetland lakes and ponds, while providing for fresh water fish and shrimp farming and recreational fishing and boating. The increased expanse of water on site encourages displacement cooling of the surrounding areas.

The site is predominantly a car-free zone. Transport hubs connect the pedestrian boardwalk and the bicycle and buggy hire system, to allow visitors to enjoy WOW at their own pace. Hire points are available from villages, car parks and transport hubs for the convenience of the users. The roofs of transport hubs are planted with Cercis chinensis, and are perfect to watch the sun set.

Additionally, a network of timber decking routes provides secondary circulation available for cycling, buggies and walking. The network is also used for buggies to transport fruit harvests from orchards to food production and retail units. Surrounding roads and existing highways are kept, weaving across the site, connecting villages, and public transport hubs.

The public transport system within WOW aims to integrate with existing citywide routing of metro, tram and water networks, making WOW and its activities easily accessible. For this proposal, a buggy and bicycle route connects Guangzhou East Train Station to Guangzhou Exhibition Centre, via the new Guangzhou CBD axis and WOW's green boulevard. The route is also proposed to act as the Guangzhou eco and cultural tourist route.

The **Green Boulevard**, on the other hand, accommodates the main activities, event spaces and main circulation. The new CBD axis physically joins the boulevard at the public Health Pier, while the Botanical

facing page top: A kit-of-parts of activities is applied to the Ecological Grid.

facing page bottom: Public Health Pier hosts sports activities.

following pages: The adaptiveness of the Ecological Grid offers an ecomonic strategy to host diverse functions.

具有次级循环路径的龙眼和青柠果园
LONGAN AND LIME ORCHARD WITH SECONDARY CIRCULATION PATH
50m x 50m

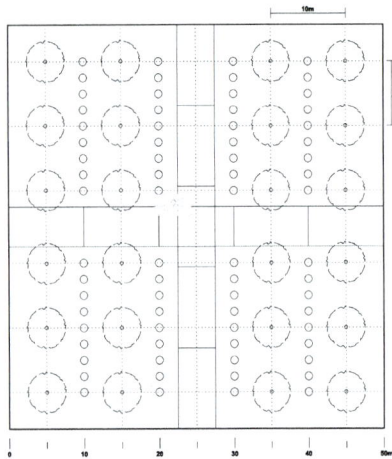

芒果和龙眼漂浮果园
MANGO AND LONGAN FLOATING ORCHARD
100m x 50m

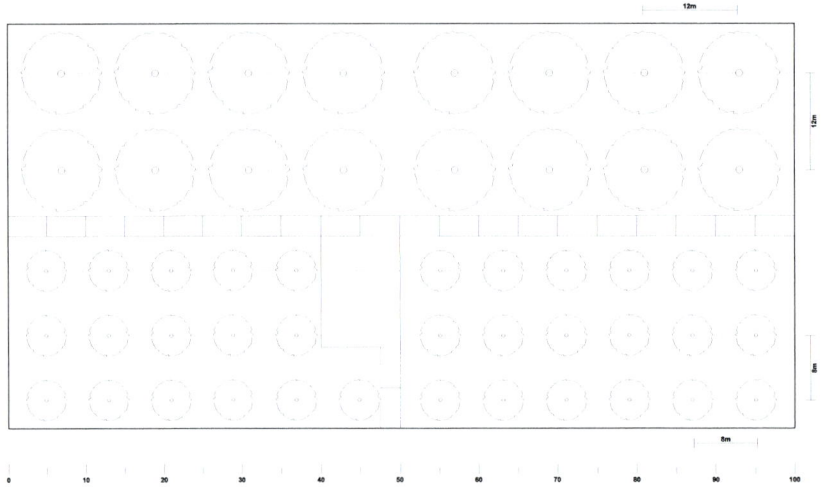

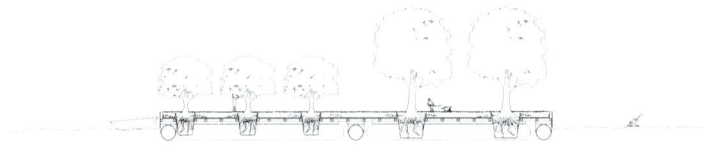

漂浮的果园
FLOATING ORCHARD

林荫大道 样式1
BOULEVARD TYPE 1
100m x 100m

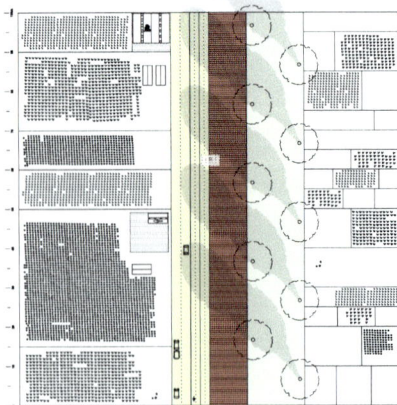

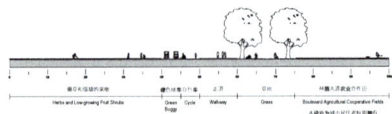

林荫大道 样式2 — 聚集区
BOULEVARD TYPE 2 - GATHERING AREAS
100m x 100m

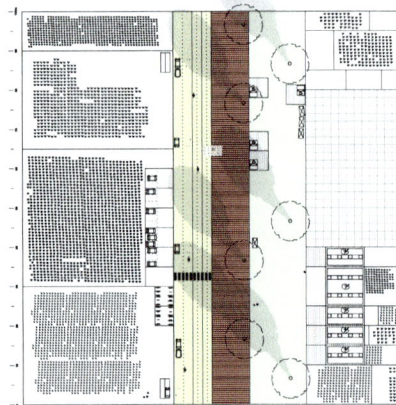

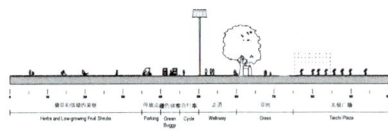

林荫大道 样式3 — 聚集区
BOULEVARD TYPE 3 - GATHERING AREAS
100m x 100m

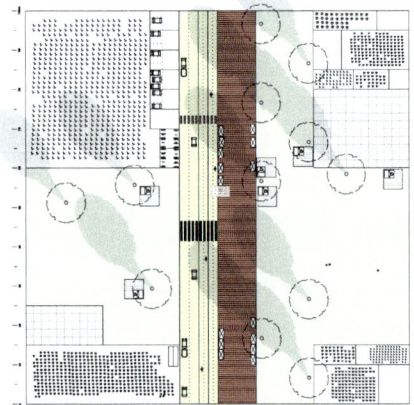

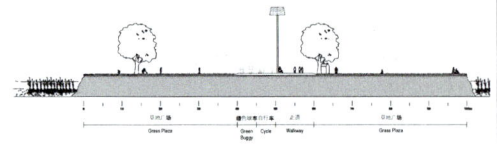

露营区
CAMPING SITE
50m x 50m

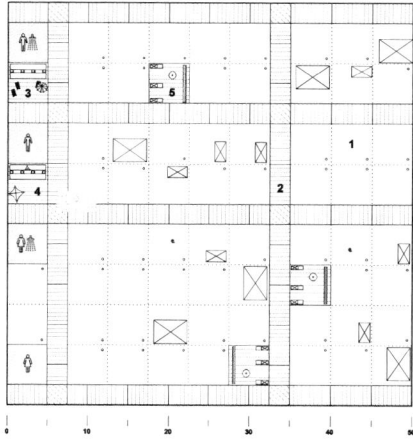

1 具有电力提供的单一露营区
Single Camping Area with Electric Point

2 人行道
Walkway

3 沙滩椅及阳伞租赁点
Deckchair and Parasol Hire

4 帐篷租赁
Tent rent

5 烤肉点
Barbeque spot

2人帐篷
2-man tent

4人帐篷
4-man tent

漂浮的海滩
FLOATING BEACH
50m x 50m

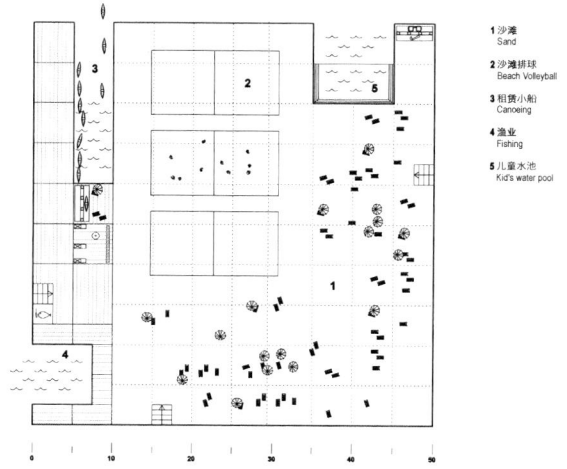

1 沙滩
Sand

2 沙滩排球
Beach Volleyball

3 租赁小船
Canoeing

4 渔业
Fishing

5 儿童水池
Kid's water pool

漂浮的沙滩可以放置在海珠湖，主要的苇池塘和珠江沿岸
Floating Beaches can be placed in Haizhu Lake, main reed pond and along the Pearl River

蜜蜂中心
BEE CENTRE
50m x 50m

具有循环路径的橙子开花果园
ORANGE BLOSSOM ORCHARD WITH CIRCULATION PATH
50m x 50m

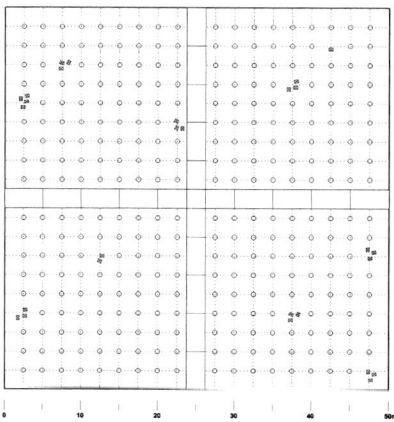

蜜蜂点
HONEY POINT
50m x 50m

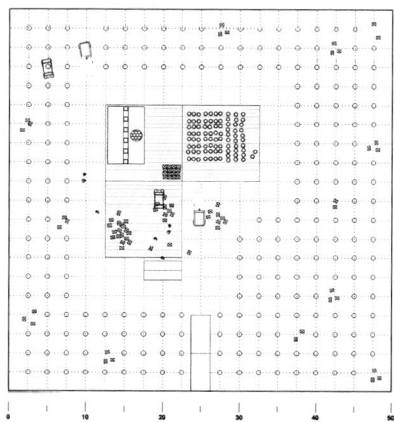

蜜蜂山顶
BEE HILL PEAK
50m x 50m

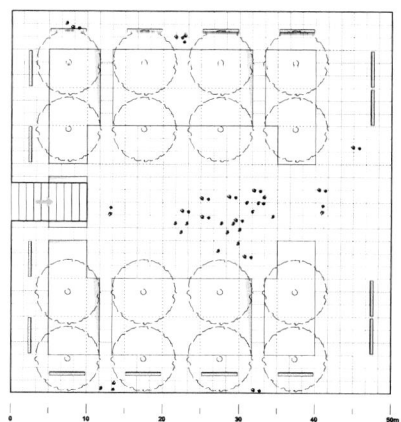

Tower is located where the axis from the University City intersects the boulevard. An urban link, this green circulation infrastructure connects the Health Pier, the Apiary 'Honey' Hill and the Botanical Tower, as it weaves across new wetlands and existing orchards and villages, to embrace 'knowledge' Sun Yat-sen University in the east and 'economy' Guangzhou Exhibition Centre in the west. The Apiary 'Honey' Hill and the Botanical Tower provides visitors with knowledge of sustainable food production and nature, and also the most spectacular views of WOW and Guangzhou.

Within its 100m width, the boulevard comprises herb/fruit shrubs, buggy/cycle lanes, pedestrian timber boardwalk, and short-term lease vegetable allotments for city dwellers. Lingnan flowering shrubs and diverse varieties of mango trees line the boulevard providing natural beauty and shade to pedestrians and cyclists. The uninterrupted endless stretches of manicured green lawn provide surfaces for picnics, sunbathing and games. The timber boardwalk acts as a catalyst for visitors to engage with local farmers and embraces the ideal opportunities of knowledge transfer on ecology, green living and organic food production.

Meanwhile, public toilets, tea kiosks, farmers' markets, photography opportunity locations, contemplation gardens and artists' studios provide a rhythm to the journey along the boulevard. The vendors and artists indirectly act as 'guardians' of the boulevard. Scattered through the fields, the sound gardens create moments of contemplation and quiet observation of some of the district's indigenous flora and fauna. Each activity platform has the flexibility to vary in size, furniture detailing and sound type depending on location and surrounding programmatic conditions.

The **Park Management** is modelled on sustainable energy and resource management practices. The opportunities to provide an exemplar open space in the middle of Guangzhou city include methods not only of providing sustainable energy and resource management for the park, but also of contributing to the city's energy grid. Integrated PV panels along the boardwalk provide shading for residents and visitors, as well as further increasing the park's electricity supply; for example, street lights and transport hubs. The park itself provides relatively high quantities of landscaping and agricultural organic waste. This biomass waste is carbon neutral and can be processed using anaerobic digestion to provide a biogas for energy production or could be processed to provide high quality compost. Biomass from wetland reeds can contribute directly to the park's biogas production for buggies. Inter-seasonal thermal storage is used extensively in both new and retrofit urban scale landscapes. The principal is that the system serves a collection of buildings (in this case not only the park buildings but also, more importantly, buildings in neighbouring villages). Energy supply for the Botanical Tower comes from this source.

The new water bodies work in harmony with the existing surface water network to provide the main water supply and storage for the whole development. In addition to the rain collected on the vast areas of wetlands, the water storage bodies include ponds and canals in the orchards, and surface runoff. During dry periods, the shortage can be mitigated by greywater and blackwater recycling. The site is also to be modified with areas of ground remodelling. The overall hydrology of the district is not altered but existing waterways and drainage patterns are channelled into an expanse of newly created wetland.

facing page top: WOW is a repository of knowledge that preserves an ecological affinity with the land.

facing page bottom: The legacy floating orchards at high tide.

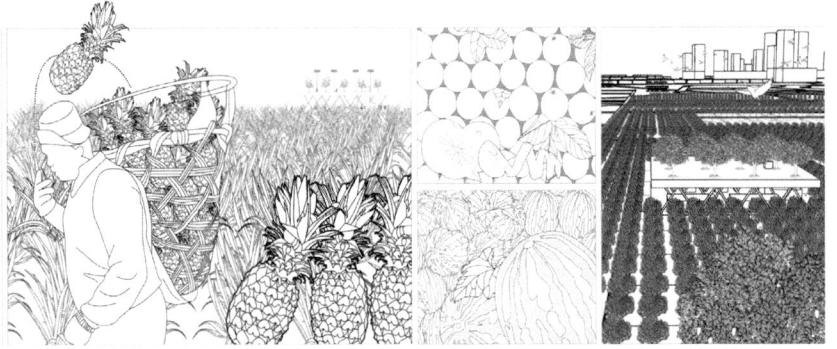

芒果园

龙眼园

漂浮果园

蜂房

柑桔园

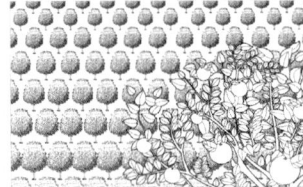

桔

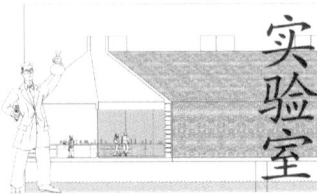

实验室

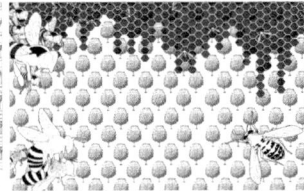

宴席

This amenity is managed as part of the flood control system of the Pearl River. A shallow two-metre deep wetland is created using grading machines to move earth to roadside locations where it is loaded onto trucks and delivered to a collection point. A conveyor system is then used to construct an artificial mound (Apiary Hill) with the excavated material. The new 'hill' has gently sloping sides, less than the angle of repose of the soil so that no special ground reinforcement is required. Wick drains to control the sub-surface movement of water are installed and the land is terraced with simple gabion walls built up from rocks collected across the site and from the excavations.

255

The success of the project depends on the careful management of water resources. Temporary channels and culverts are installed at the beginning of the work to maintain surface water flows and drainage. Excavation and earthmoving can then commence in several locations at once and material moved steadily over a period of months. Special precautions, spraying, enclosed excavators and haulage are used to minimise dusting. Areas of the wetlands may need sealing, and clay is imported to areas where a puddled layer is required. Once the final levels are achieved water is redistributed over a period of two years to ensure the establishment of the required growth pattern. The water levels are maintained by a series of weirs emptying into established watercourses or storm drains. The edges of the wetlands are generally naturally banked and stabilised by planting. Where levelled surfaces, pavements or grasslands abut the water, a shallow edging of reinforced concrete crib-walling makes the boundary. Sheet pile walls are provided where vertical surfaces are needed; moorings for boats and deeper embankments.

Finally, a large causeway crosses the wetland areas. The carriageway is carried on a low embankment with internal drainage and stabilised slopes. The vehicle restraint system (crash-barrier) incorporates low-level lighting to provide adequate carriageway lighting while avoiding light-pollution. Services are buried along each edge walkway. To the south of the site, a road swings out over the river and is ramped up to allow sufficient air-draft for boats to enter the lagoon behind. The bridge supporting the road is a reinforced concrete viaduct on piled piers in the riverbed. Low-level lighting is provided and services pass along the soffit. Spans of 25 metres allow the structure to be fully pre-fabricated and rapidly erected.

facing page top: The floating 50 x 50m Ecological Grid of the fruit orchard.

facing page bottom: The Apiary 'Honey' Hill provides knowledge of sustainable food production and nature.

Landscapes, by forces of globalisation, make striking brushstrokes of place, climate and resilience: from the wide particoloured swathes of vermilion and ocher of tulips in Keukenhof to the undulating ribbons of glistening Balinese rice terraces, from the wildly twisting branches of old olive trees etching the hillsides of Tuscany to the old growth vineyards of Bordeaux. Fritz Steele writes, 'we can shape settings so that they will have a strong spirit of place through their sharp identity, human vitality, rich symbolic messages that stimulate fantasies and memories, structure that shape people's experiences in certain patterns or sequences, and special opportunities'.[9] Looking beyond the visual register, what ties luxuriant sweeps of aesthetic pleasure to what Fritz Steele calls 'a spirit of a place' are actually the details of life that these landscapes evoke: the kinds of occupations that sustain them, the economies they generate, the customs that connect human beings who actually tend fields not only to the land itself, but also to the future of those groves in the hearts and minds of their children.

9. F Steele, 'The Sense of Place', CBI Pub Co, 1981, p.199

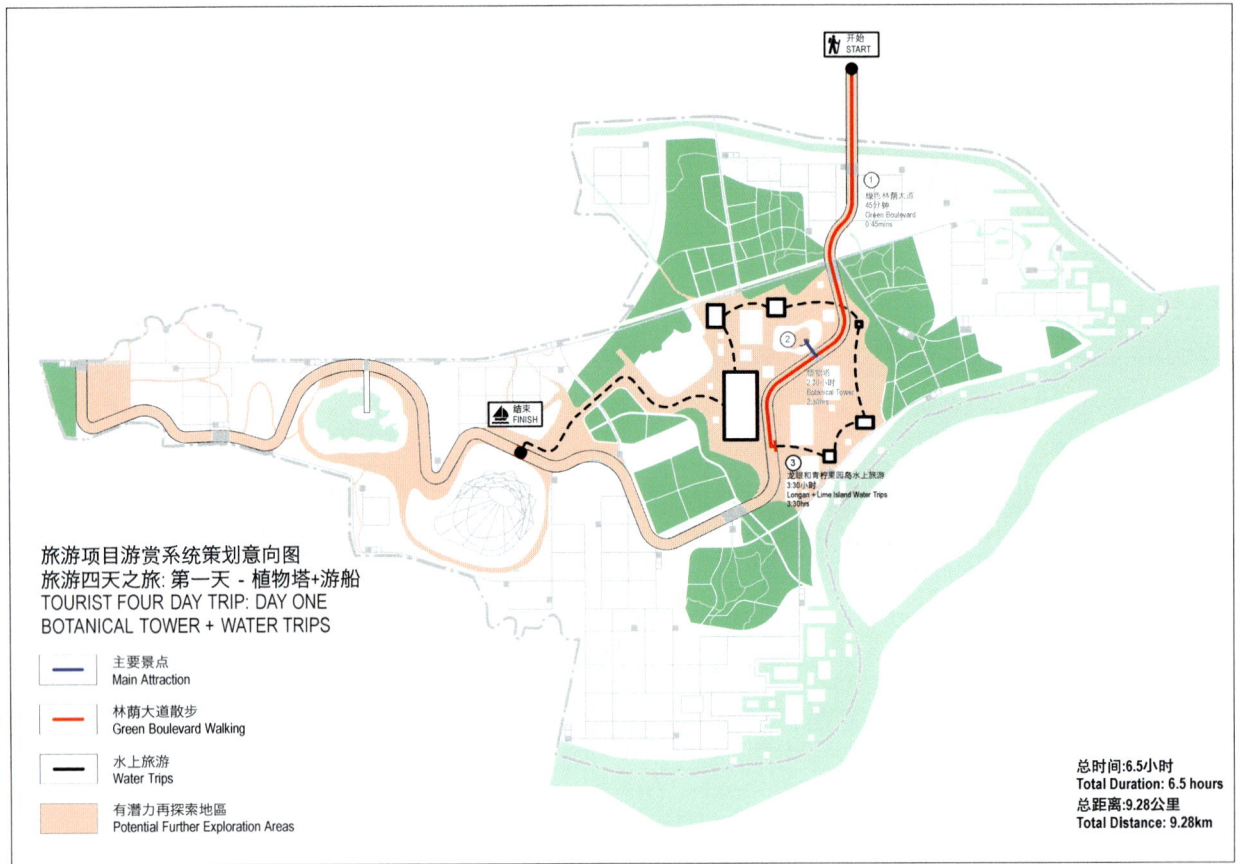

旅游项目游赏系统策划意向图
旅游四天之旅: 第一天 - 植物塔+游船
TOURIST FOUR DAY TRIP: DAY ONE
BOTANICAL TOWER + WATER TRIPS

主要景点
Main Attraction

林荫大道散步
Green Boulevard Walking

水上旅游
Water Trips

有潜力再探索地区
Potential Further Exploration Areas

开始
START

绿色林荫大点
45分钟
Green Boulevard
0:45mins

植物塔
2:30小时
Botanical Tower
2:30hrs

结束
FINISH

龙眼和青柠果园岛水上预游
3:30小时
Longan + Lime Island Water Trips
3:30hrs

总时间:6.5小时
Total Duration: 6.5 hours
总距离:9.28公里
Total Distance: 9.28km

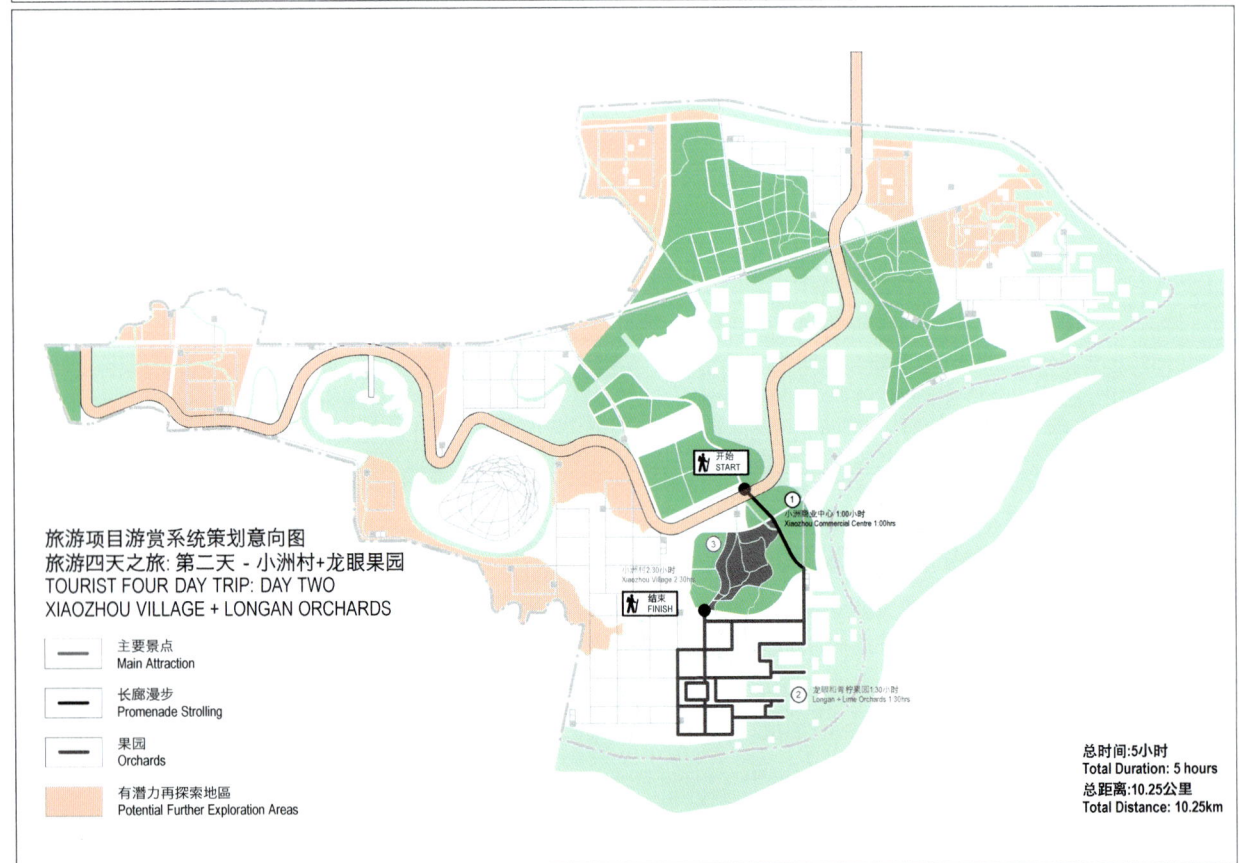

旅游项目游赏系统策划意向图
旅游四天之旅: 第二天 - 小洲村+龙眼果园
TOURIST FOUR DAY TRIP: DAY TWO
XIAOZHOU VILLAGE + LONGAN ORCHARDS

主要景点
Main Attraction

长廊漫步
Promenade Strolling

果园
Orchards

有潜力再探索地区
Potential Further Exploration Areas

开始
START

小洲商业中心 1:00小时
Xiaozhou Commercial Centre 1:00hrs

小洲村 2:30小时
Xiaozhou Village 2:30hrs

结束
FINISH

龙眼和青柠果园 1:30小时
Longan + Lime Orchards 1:30hrs

总时间:5小时
Total Duration: 5 hours
总距离:10.25公里
Total Distance: 10.25km

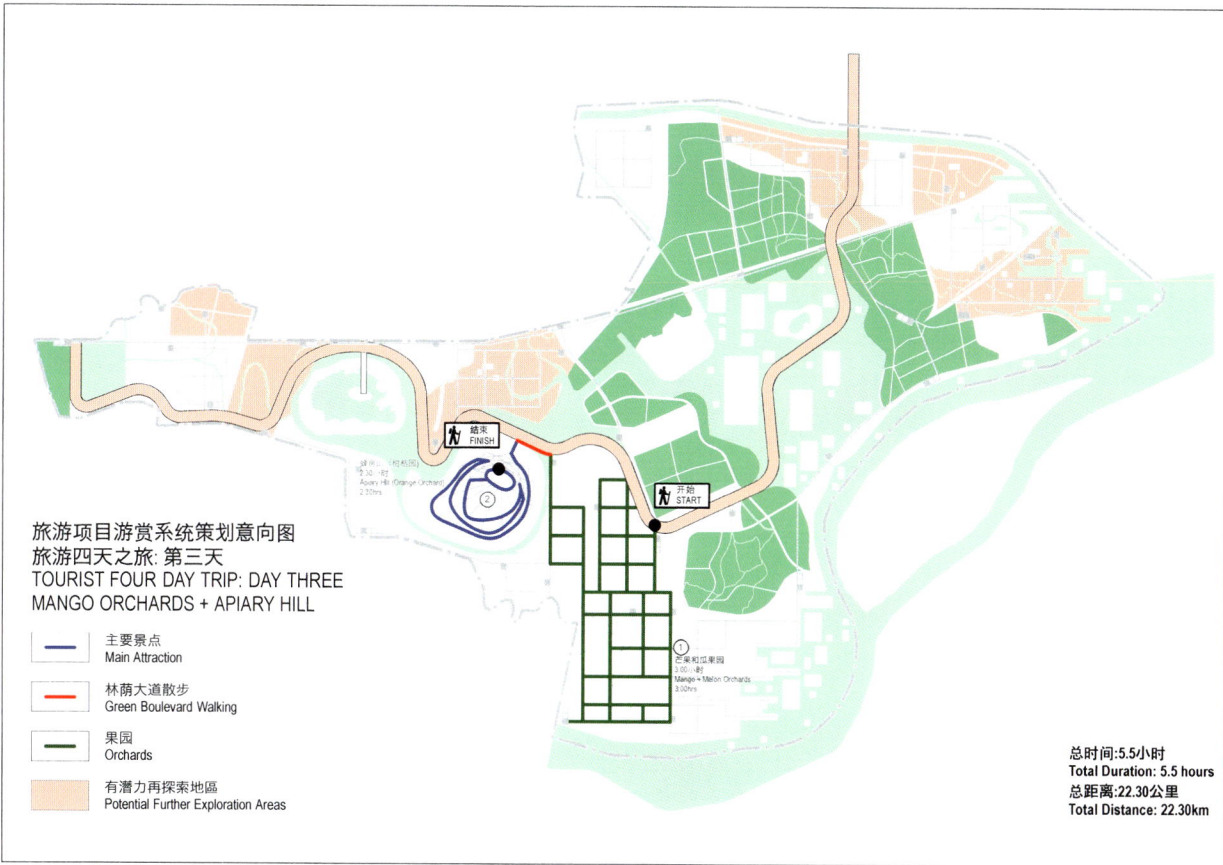

旅游项目游赏系统策划意向图
旅游四天之旅: 第三天
TOURIST FOUR DAY TRIP: DAY THREE
MANGO ORCHARDS + APIARY HILL

主要景点
Main Attraction

林荫大道散步
Green Boulevard Walking

果园
Orchards

有潜力再探索地區
Potential Further Exploration Areas

总时间:5.5小时
Total Duration: 5.5 hours
总距离:22.30公里
Total Distance: 22.30km

旅游项目游赏系统策划意向图
旅游四天之旅: 第四天
TOURIST FOUR DAY TRIP: DAY FOUR
GREEN BOULEVARD + PUBLIC HEALTH PIER

主要景点
Main Attraction

林荫大道散步
Green Boulevard Walking

长廊漫步
Promenade Strolling

果园
Orchards

有潜力再探索地區
Potential Further Exploration Areas

总时间:11小时
Total Duration: 11 hours
总距离:23.34公里
Total Distance: 23.34km

258

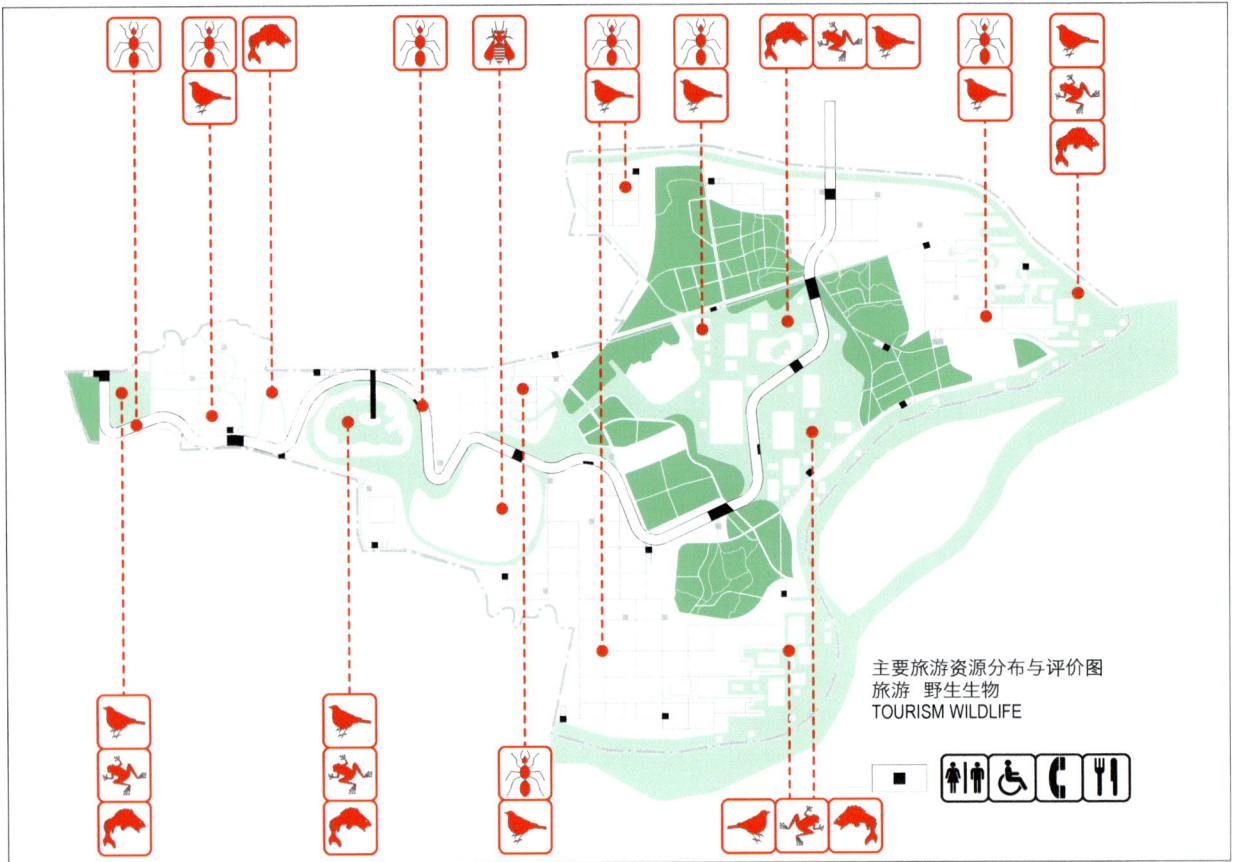

主要旅游资源分布与评价图
旅游 野生生物
TOURISM WILDLIFE

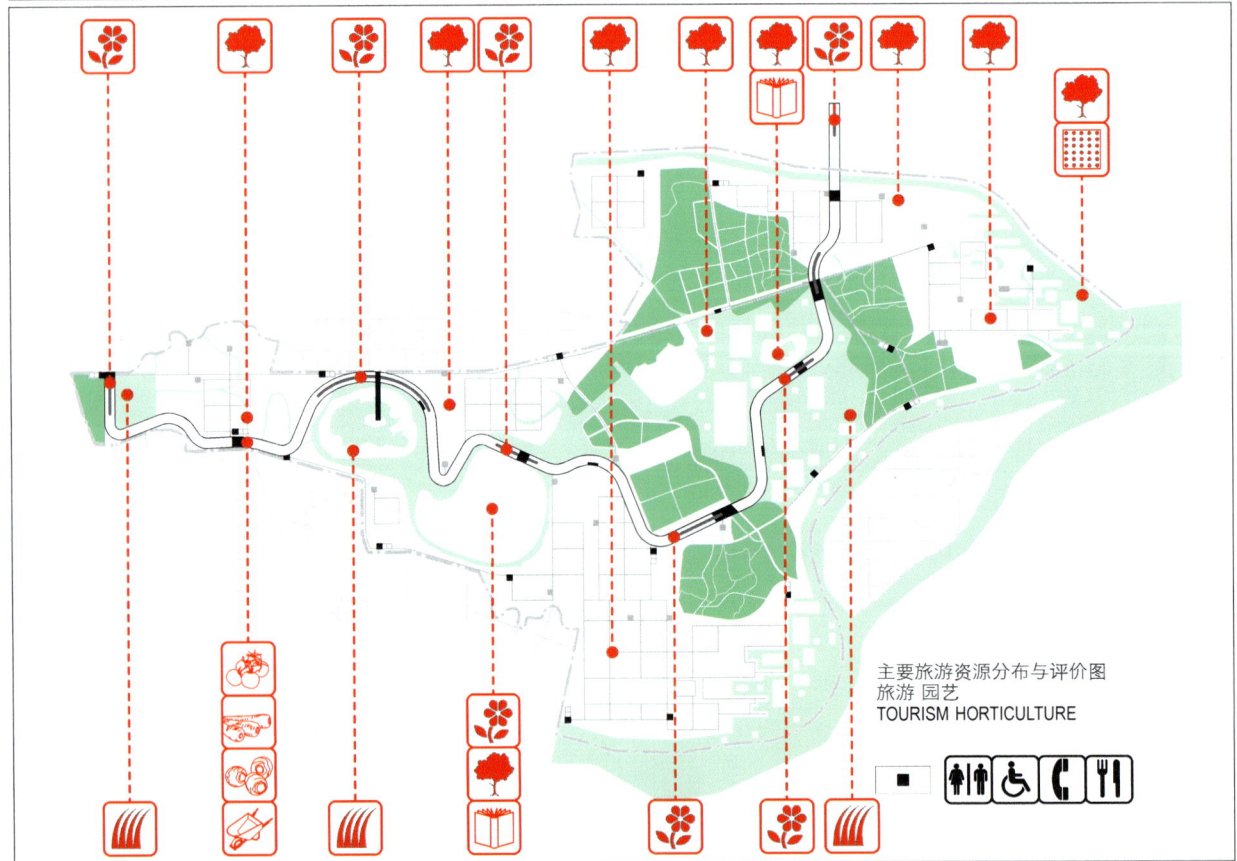

主要旅游资源分布与评价图
旅游 园艺
TOURISM HORTICULTURE

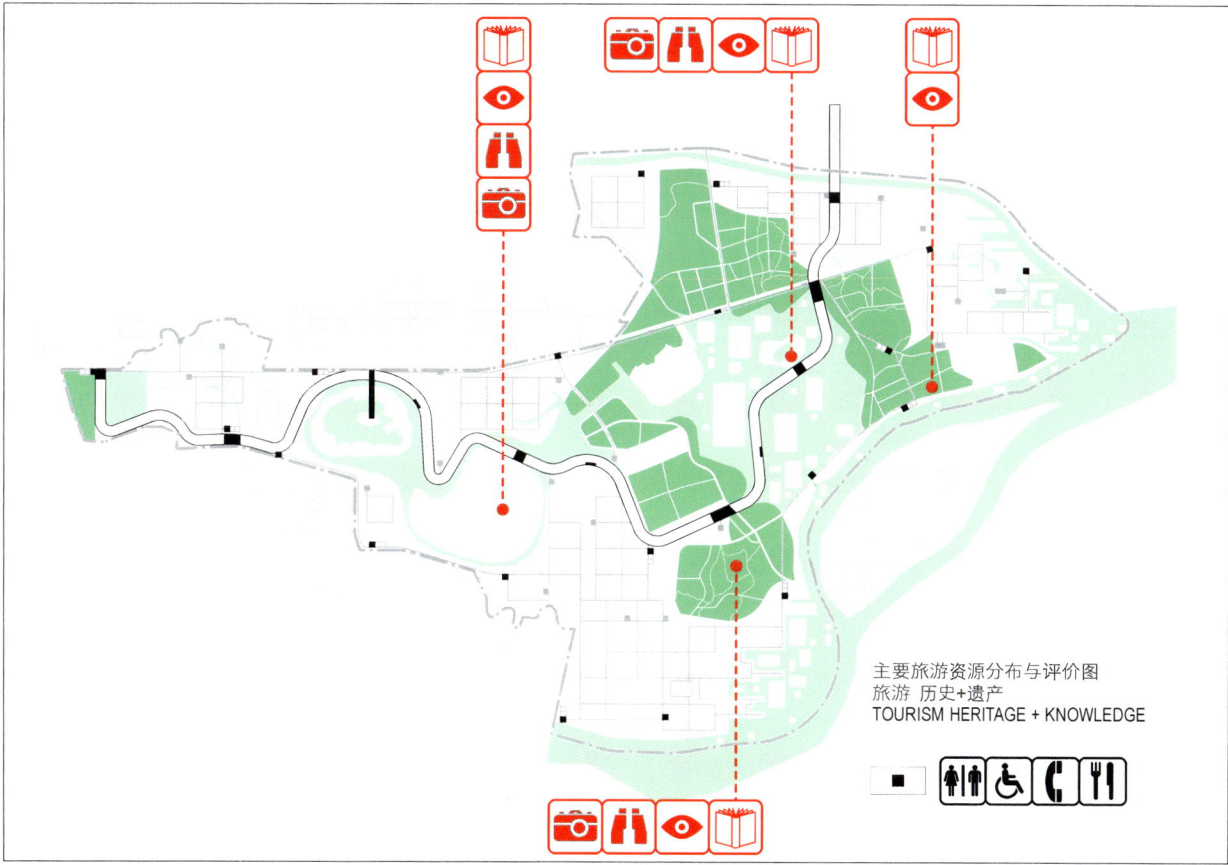

主要旅游资源分布与评价图
旅游　历史+遗产
TOURISM HERITAGE + KNOWLEDGE

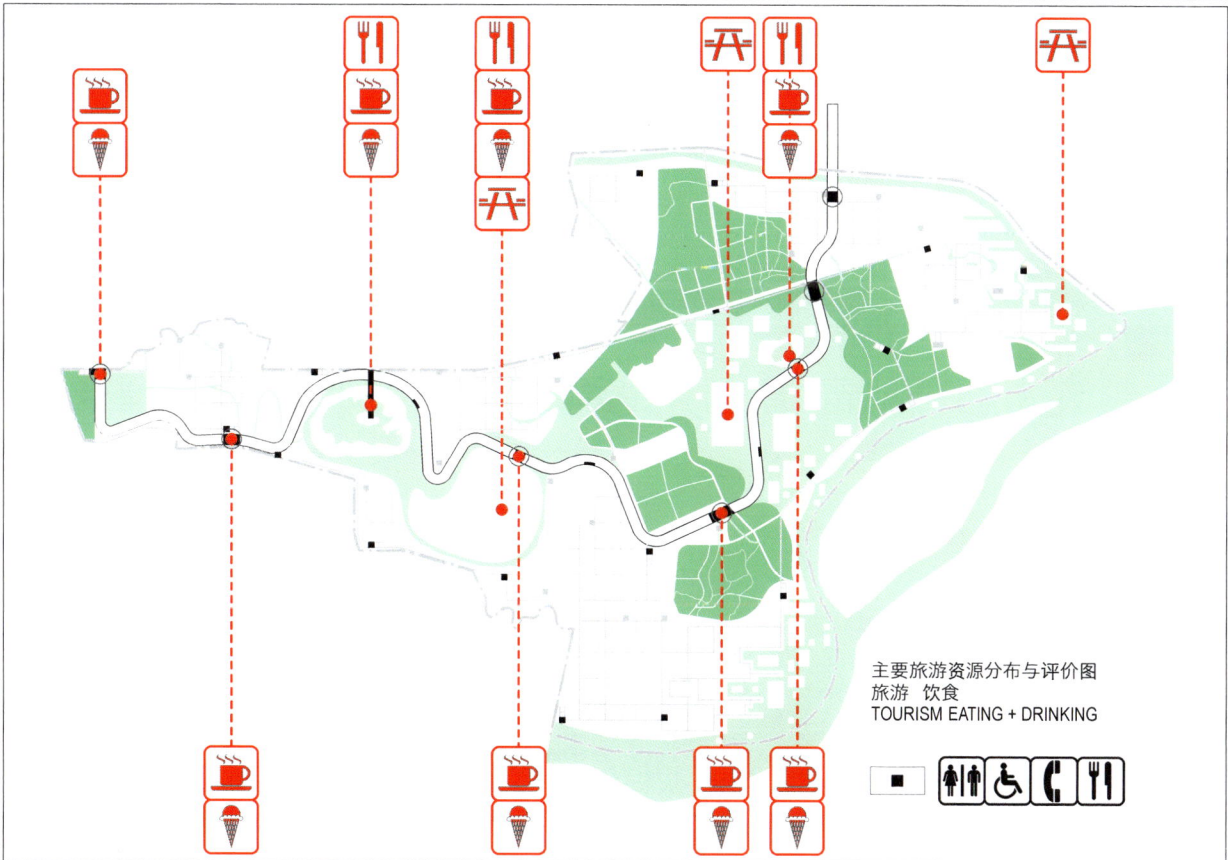

主要旅游资源分布与评价图
旅游　饮食
TOURISM EATING + DRINKING

生态功能植物
ECOLOGICAL FUNCTION

河岸/河流/河道
Embankment/Rivers/Canals

分散的池塘/湖
Isolated Ponds/Lakes

湿地
Wetlands

果园
Orchards

有机农场
Agriculture/Organic Farms

林荫大道
Boulevard

珠江湿地堤岸
Pearl River Swamp Embankment

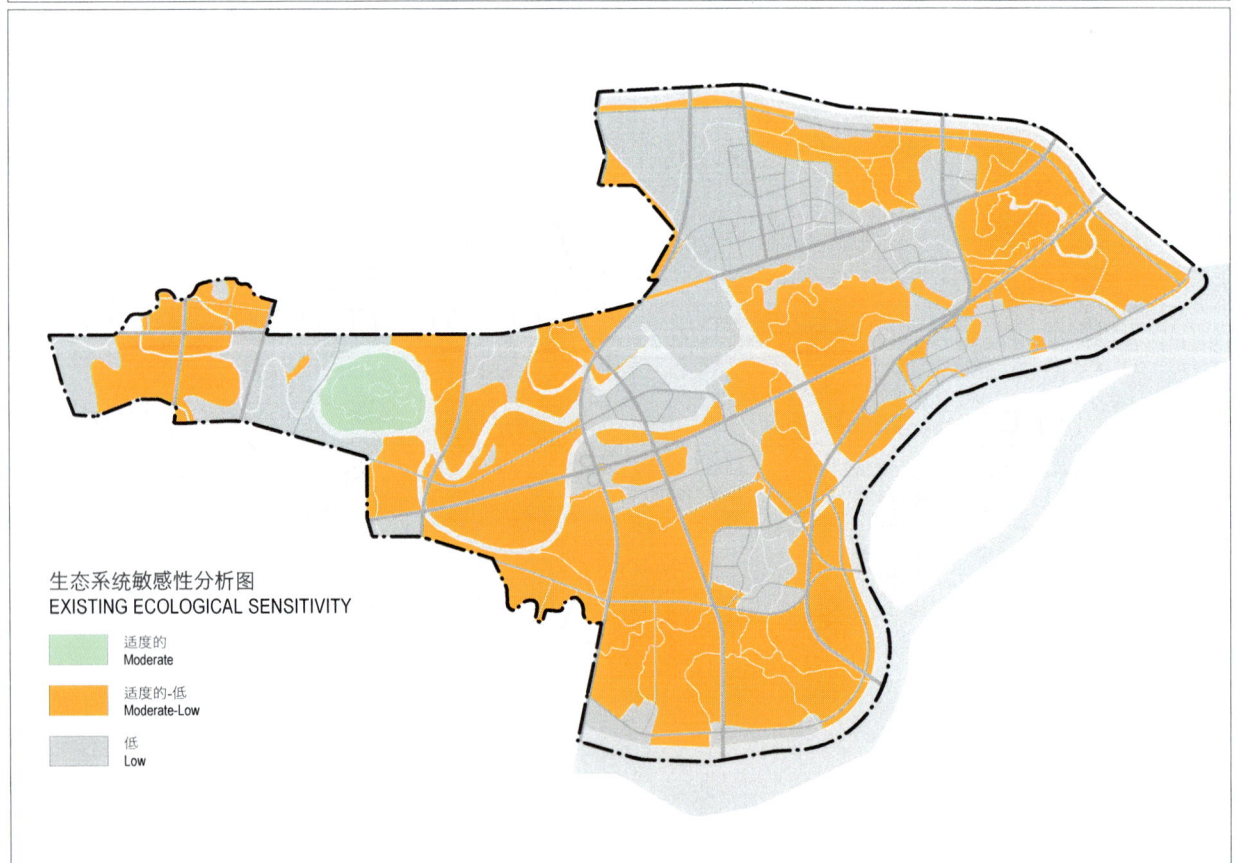

生态系统敏感性分析图
EXISTING ECOLOGICAL SENSITIVITY

适度的
Moderate

适度的-低
Moderate-Low

低
Low

Ecological Baseline Conditions

The ecological sustainability study covers ecological baseline, ecological sensitivity analysis of existing ecosystem, ecological function and design criteria of wetland and orchard habitats, and ecological assessment for ecological construction plan.

The project area of WOW is located within the Haizhu region of Guangzhou, China. The Haizhu region is surrounded by tributaries of the Pearl River. A complex network of watercourses intersects the project area. Extensive orchard farms and pockets of agricultural land are present in the form of 'Yingzhou Ecological Park', 'Fulin Farm', 'Longtan Fruit Tree Park', etc. The Haizhu Lake located at the western portion of the project area has recently undergone modification. The lake is interconnected with six tributaries from the surrounding area, providing capacity for water storage for effective water management for the surrounding area.

Orchard: This is the dominant habitat and land use in the project area. A large proportion of the project area is covered by orchard habitat. It is extensively planted with fruit trees including Dimocarpus longan (Longan), Musa paradisiaca (Banana), Averrhoa carambola (Carambola), Mangifera indica (Mango), and Clausena lansium (Wampi). As the orchards are run as ecological parks, farms, and orchard parks for commercial, educational, recreational and landscaping purposes, this habitat is heavily managed and frequently disturbed by humans. For the purpose of high production, the density of the fruit tree planting is very high. The spacing of the fruit trees on average is around three to four metres apart. Although orchard is usually not considered an important habitat for wildlife, their flowers are favourite food sources to some insects, particularly bees.

Agricultural Area: Pockets of agriculture production are scattered across the project area, where some are interspersed with orchard habitat and village houses. Crop plants are cultivated within the agricultural area. In order to have better pest control, it is not uncommon that local farmers apply a high dosage of insecticides onto the crop plants. The agricultural land is therefore rarely utilised by insects and other wildlife.

Watercourse: A network of watercourses intersects the project area, providing water resource for the irrigation of the extensive orchard and agricultural area. Most of the existing watercourses are heavily polluted by direct domestic discharge from adjacent villages. Problems such as anoxic condition, and blockage of the water flow due to accumulation of sediment and rubbish occasionally occur at some sections. Their ecological value is affected by the runoff contaminated by insecticide and pesticide residues from the nearby orchards and agricultural lands.

Pond + Lake: Pond habitat is present at numerous locations across the project area, where the largest one is Haizhu Lake (water surface area ~50ha) while others are much smaller in size. The ponds are generally used for water storage for irrigation in the orchard and agricultural areas, as well as fish farming. Haizhu Lake is being used by the locals for water management of the surrounding area, where it is interconnected with six water channels. The water level within the lake is affected by the tidal action

previous pages: Plans of activities and places of interests.

facing page top: Plan of ecological function – Flora.

facing page bottom: Plan of existing ecological sensitivity.

表 1：果园的生态评价 Table 1: Ecological Evaluation of Orchard

准则 Criteria	评价 Orchard
自然性 Naturalness	完全人造的栖息地 Wholly man-made habitat
规模 Size	大约1050公顷 Approximately 1,050 ha
多样性 Diversity	植物多样性低：由低植物多样性，单调及经常被扰乱的栖息地本质来推测动物多样性低 Low flora diversity; fauna diversity expected to be low due to low flora diversity and monotonous and frequently disturbed nature of habitat
稀有性 Rarity	普通 Common
再生性 Re-creatability	很容易再生 Readily re-creatable
分散性 Fragmentation	虽然分散，但每片面积都很大 Although fragmented, but each patch is large in size
生态联系 Ecological Linkage	与任何高生态价值的栖息地都没联系 Not linked to any habitats of high ecological value
潜在价值 Potential Value	中低 Moderate-low
整体评价 Overall Value	中低 Moderate-low

表 2：农业区的生态评价 Table 2: Ecological Evaluation of Agricultural Area

准则 Criteria	评价 Agricultural Area
自然性 Naturalness	完全人造的栖息地 Wholly man-made habitat
规模 Size	大约140公顷 Approximately 140 ha
多样性 Diversity	植物多样性低：由低植物多样性，单调及经常被扰乱的栖息地本质来推测动物多样性低 Low flora diversity; fauna diversity expected to be low due to low flora diversity and monotonous and frequently disturbed nature of habitat
稀有性 Rarity	普通 Common
再生性 Re-creatability	很容易再生 Readily re-creatable
分散性 Fragmentation	在整个项目基地中小块零星分布 Fragmented into small pockets throughout the project area
生态联系 Ecological Linkage	与任何高生态价值的栖息地都没联系 Not linked to any habitats of high ecological value
潜在价值 Potential Value	低 Low
整体评价 Overall Value	中低 Moderate-low

表 3：水道的生态评价 Table 3: Ecological Evaluation of Watercourse

准则 Criteria	评价 Watercourse
自然性 Naturalness	部分修改为防洪、居住区、果园和耕地的排水道 Partly modified as drainage channels for flood control, human settlement, orchard and cultivated areas
规模 Size	总长大约66千米 Approximately 66 km in total length
多样性 Diversity	由于居民区和农业、果园区的直接排放，预计动植物种的多样性低 Expected to have low flora and fauna diversity due to direct discharge from human settlement and agricultural/orchard areas
稀有性 Rarity	普通 Common
再生性 Re-creatability	难以再生 Difficult to be re-created
分散性 Fragmentation	不适用 N/A
生态联系 Ecological Linkage	结构上与珠江相连 Structurally linked to Pearl River
潜在价值 Potential Value	可通过改善水质和适当的河畔种植来提升价值 Can be enhanced through water quality improvement and proper riparian planting
整体评价 Overall Value	中低 Moderate-low

表 4：水塘的生态评价 Table 4: Ecological Evaluation of Pond

准则	评价
自然性 Naturalness	完全人造的栖息地，用来灌溉农田和养殖鱼类 Wholly man-made habitat for irrigation and/or fish farming
规模 Size	散布在基地中：0.5-0.8公顷 Scattered; ranges 0.5-0.8 ha
多样性 Diversity	植物多样性低：由于其面积小及位置相互孤立的特点来推测动物多样性也低 Low flora diversity; fauna diversity expected to be low due to small size and isolated distribution
稀有性 Rarity	非常常见 Very common
再生性 Re-creatability	很容易再生 Readily re-creatable
分散性 Fragmentation	高度分散，在整个基地中以小规模分布 Highly fragmented and present in small sizes across the project area
生态联系 Ecological Linkage	与任何高生态价值的栖息地都没联系 Not linked to any habitats of high ecological value
潜在价值 Potential Value	可通过种植适当的湿地植物被吸引野生动物，从而提升其生态价值 Can be enhanced by planting proper wetland plant species to attract wildlife
整体评价 Overall Value	中低 Moderate-low

at the water channels. The lake serves the function of water storage and provision during wet and dry conditions; providing water security for the surrounding area and extensive orchard and agricultural areas.

Wasteland: This is a modified habitat abandoned by humans. Due to the lack of disturbance and open nature, vegetation found in wasteland habitat is often dominated by invasive, fast growing and highly adaptive weeds, eliminating the establishment of other less adaptive flora species. As the wasteland habitat within the project area is in proximity to a watercourse, it is likely to face occasional flooding during heavy rainfall events. Wasteland is low in both ecological and landscape value.

Developed Area: The habitat consists of urbanised areas including village houses, shopping centres, parks, infrastructures and construction sites. Roadside plantation, ornamental plantings and landscaping areas are also included in this habitat.

Flora: The area of Haizhu region has been highly modified by human activities. Natural habitats with vegetation native to South China basically do not exist. Wild plants are listed as Category I State protected species, and recorded in China Plant Red Data Book. They are rare in the wild nowadays but still common in cultivation, for ornamental use or riverbank stabilisation and soil erosion control.

Fauna: Generally, the interconnected river estuary is a suitable habitat for aquatic fauna such as fishes and amphibians. However, owing to the urban development, heavily polluted river water and over-exploitation of natural resources, the estuary area is now not considered as a natural habitat for wildlife. Those fish fauna that currently live in the pond/lake area are exotic species or of captive origin, which are of lower ecological value.

In regards to terrestrial habitat, the extensive orchard in the area eliminates the potential of colonisation of native fauna species. Those fauna species currently inhabiting the area are highly adaptive to the anthropogenic environment and therefore not of high ecological value. Common bird species can be found in this area, but their propagation is limited by lack of suitable natural habitat.

Migratory Bird Flight Path: The project area is not identified as an optimal habitat for migratory birds, owing to the urban development in the surrounding areas. Therefore, the potential for migratory bird flight path is very low. Owing to its proximity to riverine habitat, wetland-dependent species such as kingfisher can be found in the area, but in limited numbers. However, it is expected that the wetland can be transformed into a more attractive wetland area to waterbirds through habitat enhancement.

Ecological Function

This section explores potential opportunities for improving the ecological function of existing habitats or creating new types of habitat/land use which are considered beneficial to the local ecology. Based on existing land use in Haizhu district, different ecological function zones/habitats are suggested. The aim of these proposals is to enhance the ecological, aesthetical and social value of each identified functional zone.

(1) Objective: As the site is located in an urbanised area, the Haizhu district is utilised by local people for various purposes. Nevertheless, the development within it usually lacks any form of principle planning. The principle objective of the ecological function is to provide recommendations for environmental enhancement with an aim to form an integrated land use district, which eventually lead to an environmentally sustainably development:

- To improve the environmental condition of the district and connectivity of the existing and proposed environmental enhancements.
- To take current ecological conditions (both flora and fauna) into ecological consideration as part of the function zoning.
- To restore the wetland ecosystem function and enhance the ecological value.
- To enhance the potential for an ecotourism development.
- To consider Sustainable Urban Drainage (SUDs) as a key principle of the associated urban areas.
- To enhance the integrity of the wetland system and orchard habitat, and plan for a sustainable future.

(2) Site Context: As a precondition for the review of zoning formulation, the major environment constraints that the Haizhu district currently encounters are identified. It forms a background basis for the zoning and land use proposal.
Constraints:

- Scattered distribution of wetland – lack of ecological linkage.
- Poor water quality – heavily polluted water sources (anoxic condition, unpleasant colour and odour).
- Low ecological value of wetland.
- Heavy use of chemicals during crop production for pest control leads to pollution of nearby water bodies and rivers.

(3) Identification of Potential Ecological Land Use/Habitat: In order to meet the objective listed above, habitat enhancement measures are recommended below which comprise environmentally friendly river channel, integrated wetland, orchard, organic farms and planting alongside boulevard.

Environmentally Friendly River Channel

River channelisation plays a key role in urban development in waterfront areas as it alleviates the flooding risk and makes more land available for residential, industrial or agricultural development. Traditional engineering design of concrete paving alongside the channel is neither ecologically nor aesthetically desirable, since it replaces greenery with visually intrusive concrete structure. Aside from direct widening and deepening of channelisation works, environmental consideration can be incorporated in the engineering design to enhance the ecological and aesthetic value of the channel. Examples include the use of grassed cellular concrete paving and geo-fabric reinforced grass lining for river embankments and bed; creating steps, curves and aquatic planting ponds/bays in the channel; creating marshland and ecological reed beds alongside the channel, etc.

Design Objective + Potential Ecological Function

- Alleviate flooding of river channel by adopting an engineering design that has incorporated practicable ecological considerations and features.

- Enhance the environmental value of river bank zone through installing various habitat and landscape features.

Ecological Design Considerations

Flood Storage Area: The concept of flood storage is to intercept the runoff at the upstream area and temporarily store in a flood storage pond/area. When the water level in the downstream river recedes, water at the flood storage pond/area will be re-diverted into the river. This will substantially reduce the volume of runoff discharged into the downstream river during heavy rainstorms and flooding in low-lying/downstream areas will remain undisturbed and with constant water flow at all times. Flood storage can alleviate flood risk and preserve the riverine habitats. The flood storage area provides a potential roosting and foraging habitat for wetland-dependent fauna species. Its ecological function can be further enhanced by design of permanently wet lagoons for birds.

River Bank Measures: There is a wide range of soft revetment materials and bio-engineering alternatives, ranging from the softer natural vegetative treatment (e.g. grass) to the harder gabion wall. More examples such as brush mattresses and coconut fibre can be considered. In addition to the choice of soft materials for bank revetment, the retention and creation of river margin are also essential. Margins, which are the damp areas between the normal water level and the terrestrial habitat, are very important habitats for wildlife as many specialised plants and associated animals occur only in this zone. These damp areas are usually considered as part of the riparian zone and are important for wildlife because of the special refuge habitats provided during both the flooding and non-flooding periods. They also serve as interface connecting aquatic and terrestrial habitats and are particularly important for amphibians, which rely on both types of habitats to complete their life cycles.

Creation of Margin Habitats: The margin should be varied in height and profile as well as width so that the diversity of the habitat can be ensured. Appropriate vegetation variability at the re-profiled margins can also provide more benefits to the wildlife using the water edge. The increased riparian zone is also suitable for colonisation of wetland associated insect species.

Indents in the Banks: Articulations in the land trap water which provides fish with spawning locations away from main flows and small ponds to create areas of still water. Creating riffles in the riverbed aerates the water and enhances the riverine habitat. A weir and lock structure can be used to provide fish access so species can move past human-made blockages in the river system.

265

following page: The diverse habitats in WOW enhance the productivity of local fauna groups.

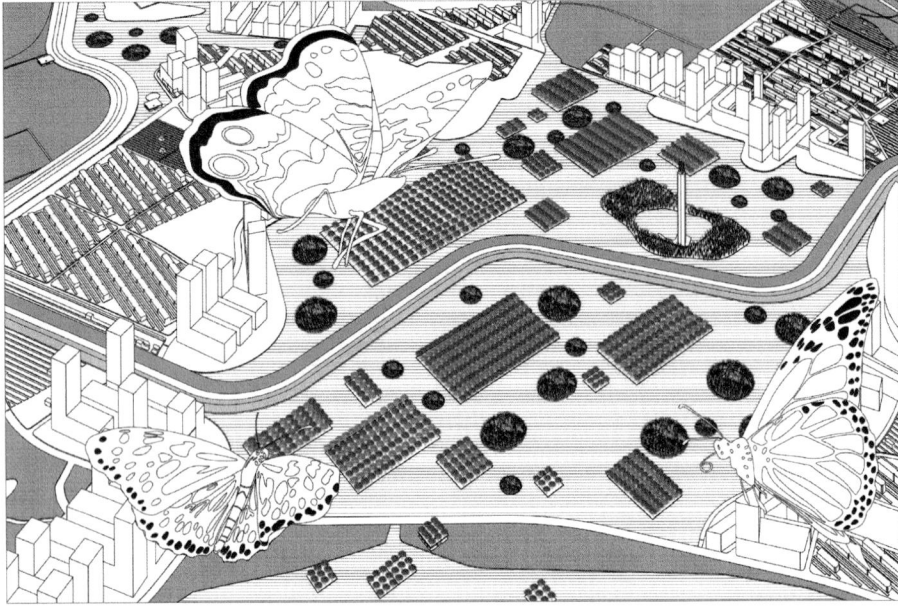
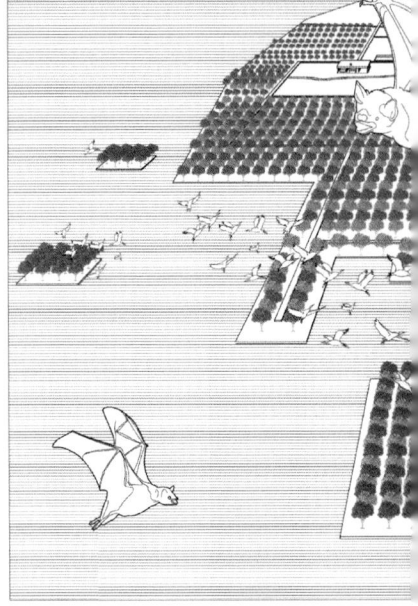

湿地景观

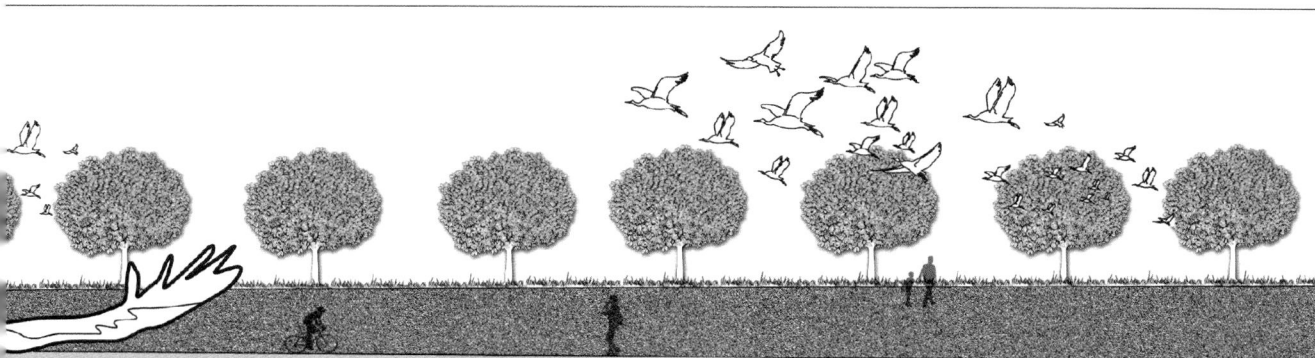

物种名称 Species Name	中文名称	Name	动物相 Fauna Group
Stenopsychidae	石蛾	Caddisfly	水生昆虫 Aquatic Insect
Ephemera spilosa	斑点蜉	-	水生昆虫 Aquatic Insect
Ptilomera tigrina	水黾	-	水生昆虫 Aquatic Insect
Diplonychus rusticum	负子蝽	-	水生昆虫 Aquatic Insect
Cybister	真龙虱	-	水生昆虫 Aquatic Insect
Ctenopharyngodon idellus	鲩鱼	Grass carp	淡水鱼 Freshwater Fish
Mugil cephalus	乌头鱼	Grey Mullet	淡水鱼 Freshwater Fish
Parasilurus asotus	鲶鱼	Chinese Catfish	淡水鱼 Freshwater Fish
Channa maculata	生鱼	Spotted Snakehead	淡水鱼 Freshwater Fish
Plecoglossus altivelis altivelis	香鱼	Ayu fish	淡水鱼 Freshwater Fish
Hypophthalmichthys molitrix	鲢鱼	Silver carp	淡水鱼 Freshwater Fish
Carassius carassius	鲫鱼	Crucian Carp	淡水鱼 Freshwater Fish
Metapenaeus ensis	刀额新对虾	Greasyback shrimp	淡水鱼 Freshwater Shrimp
Corbicula fluminea	淡水沙蚌	-	淡水贝类 Clam
Eriocheir japonicus	日本绒毛蟹	Mitten Crab	淡水蟹 Freshwater Crab
Semisulcospira libertinaw	大河螺	Large Stream Snail	淡水螺 Snail
Brotia hainanensis	海南蜷螺	-	淡水螺 Snail

表1b 水道堤岸植物、灌木、鲜花景观 Table 1b: Amphibious river plants/shrubs/flowers

物种名称 Species Name	中文名称	植物相 Flora Group
Commelina diffusa	节节草	药草 Herb
Alternanthera philoxeroides	空心苋	药草 Herb
Leersia hexandra	李氏禾	药草 Herb
Rotala rotundifolia	圆叶节节菜	药草 Herb
Polygonum hydropiper	水蓼	药草 Herb
Acorus gramineus	金钱蒲	药草 Herb
Eleocharis dulcis	荸荠	药草 Herb
Ludwigia octovalvis	毛草龙	灌木 Shrub

表1a 水道堤岸 Table 1a: Rivers/Canals Embankments

物种名称 Species Name	中文名称	Name	动物相 Fauna Group
Bufo melanostictus	黑眶蟾蜍	Asian Common Toad	两栖类 Amphibian
Rana guentheri	沼蛙	Gunther's Frog	两栖类 Amphibian
Fejervarya limnocharis	泽蛙	Paddy Frog	两栖类 Amphibian
Hoplobatrachus chinensis	虎纹蛙	Chinese Bullfrog	两栖类 Amphibian
Ardea cinerea	苍鹭	Grey Heron	鸟类 Bird
Egretta alba	大白鹭	Great Egret	鸟类 Bird
Egretta intermedia	中白鹭	Intermediate Egret	鸟类 Bird
Egretta garzetta	小白鹭	Little Egret	鸟类 Bird
Ardeola bacchus	池鹭	Chinese Pond Heron	鸟类 Bird
Nycticorax nycticorax	夜鹭	Black-crowned Night Heron	鸟类 Bird
Amaurornis phoenicurus	白胸苦恶鸟	White-breasted Waterhen	鸟类 Bird
Motacilla cinerea	灰鹡鸰	Grey Wagtail	鸟类 Bird
Motacilla alba	白鹡鸰	White Wagtail	鸟类 Bird
Phylloscopus fuscatus	褐柳莺	Dusky Warbler	鸟类 Bird
Agricnemis femina oryzae	杯斑小蟌	Orange-tailed Midget	昆虫 Insect
Paracercion melanotum	黑背尾蟌	Eastern Lilysquatter	昆虫 Insect
Ischnura senegalensis	褐斑异痣蟌	Common Bluetail	昆虫 Insect
Crocothemis servilia servilia	红蜻	Crimson Darter	昆虫 Insect
Hydrobasileus croceus	臀斑楔翅蜻	Amber-winged Glider	昆虫 Insect
Orthetrum glaucum	黑尾灰蜻	Common Blue Skimmer	昆虫 Insect
Orthetrum luzonicum	吕宋灰蜻	Marsh Skimmer	昆虫 Insect
Orthetrum pruinosum neglectum	赤褐灰蜻	Common Red Skimmer	昆虫 Insect
Orthetrum sabina sabina	狭腹灰蜻	Green Skimmer	昆虫 Insect
Pantala flavescens	黄蜻	Wandering Glider	昆虫 Insect
Pseudothemis zonata	玉带蜻	Pied Skimmer	昆虫 Insect
Rhodothemis rufa	红胭蜻	Ruby Darter	昆虫 Insect
Rhyothemis triangularis	三角丽翅蜻	Sapphire Flutterer	昆虫 Insect
Zygonyx iris insignis	彩虹蜻	Emerald Cascader	昆虫 Insect
Papilionidae	凤蝶	Papilionidae	昆虫 Insect
Pieridae	粉蝶	Pieridae	昆虫 Insect

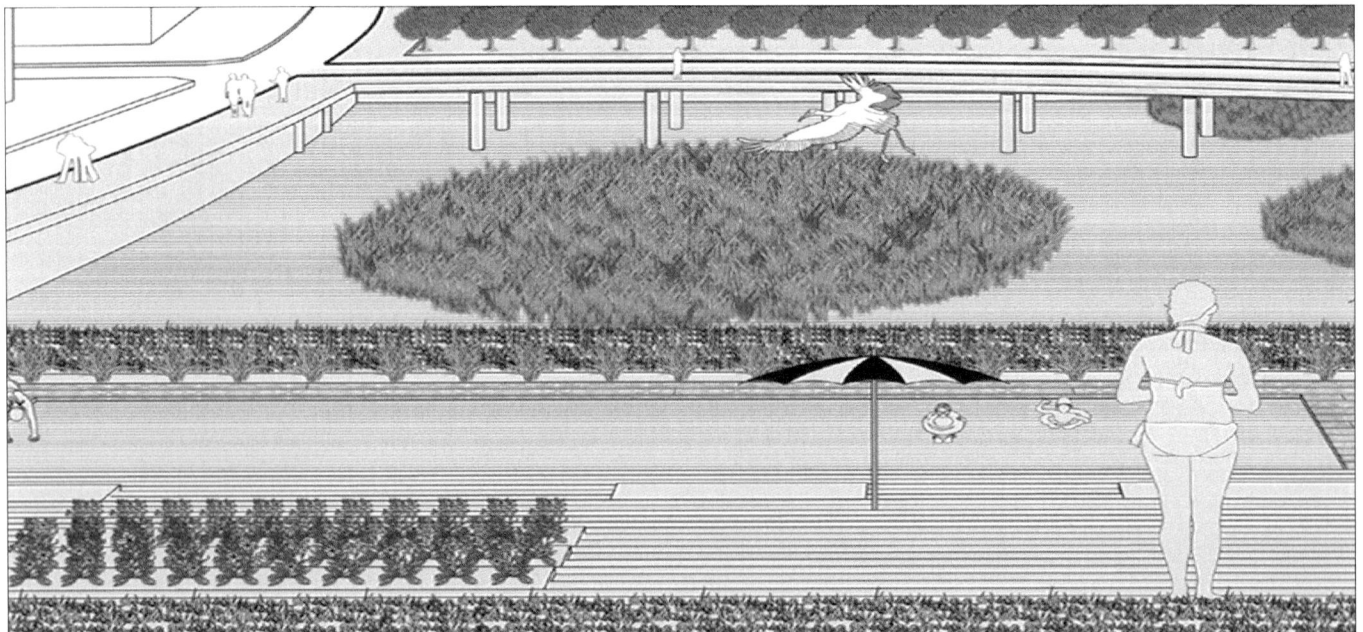

Integrated Wetland Habitat

Wetland is a transitional habitat between terrestrial and aquatic communities that can exist in various forms depending on the topography, hydrology, soil, vegetation and management regime. The wetland habitat is valuable in a number of ways, including flood storage and fishery activities, and as an area of visual beauty and a sanctuary for wildlife. Wetland is a naturally occurring habitat but can also be artificially created and modified for ecological enhancement. In a highly urbanised area, wetland can be created to serve both a recreational and an ecological purpose.

269

Design Objective + Potential Ecological Function

- Increase the area of wetland, with some areas allocated for biodiversity and other areas to serve as flood storage.
- Improve the aesthetical value of the area.
- Provide more diverse wetland habitat and increase the biodiversity for both flora and fauna.

Ecological Design Considerations

Fishery Pond: The pond is a wetland habitat composed of static water installed with an aeration system. It serves as an aesthetic feature in many urban parks whilst it can also provide an oasis for wildlife. Pond currently exists in the Haizhu area in the form of Haizhu Lake and other forms of scattered pond. With careful planning, the ecological and landscape value of the ponds can be improved.

Utilising the ponds for fishery activities can preserve the wetland function, and on the other hand add economical value to the area. Fishery pond is primarily used for fish cultivation where the fish stock can be harvested as food or used for leisure-fishing. Ecological considerations for the design of the fishery pond are recommended below:

- Increase of wetland area by installation of more ponds or connecting existing wetland to form a larger pond area or a series of interlinked waterbodies. It is desirable to connect the pond to the canal system so that it can be used for flood storage.
- Increase the water depth of the ponds (at the centre of the ponds) to increase the water-carrying capacity, which in turn improves the flood storage capacity and the fishery productivity. Also deep pools help keep water temperatures low during summer months.
- Stocking of fish or shrimp population can help initiate an aquatic community at the initial stage. However, exotic species must be avoided to ensure compliance with international agreements on bio-security.

Isolated Ponds: The ponds can be embellished to improve their aesthetic and ecological value. Apart from aesthetic decoration, ecological considerations can be taken into account during the pond design so that the ecological function of the wetland can also be enhanced.

- Shape of the ponds can be irregular to resemble a natural system and maximise the shore area. It increases the transition zone, which is an optimal habitat for wetland associated flora and fauna.
- Design for a shallow fringe of wetland surrounding a pond, and avoid creating a steep-slope sided pond. It provides a transitional zone for aquatic wildlife and waterbird use.

following page top: The Health Pier connects the Green Boulevard to the serene Haizhu Lake.

following page bottom: The Apiary 'Honey' Hill offers the perfect location to enjoy the setting sun.

- Designing for interspersion of vegetated area and open water can enhance aesthetic value whilst increasing the micro-habitat diversity. Partially submerged wetland plants such as lotus can be planted in open water area to enhance the aesthetic value.
- Creation of small islands within the ponds as nesting habitats for birds; open water around the islands needs to be deep to ensure predators cannot access the island.
- Partially submerged wetland plants also provide an optimal environment for most aquatic invertebrates and juvenile fish communities. Floating vegetation which is suitable for local waterbirds to use, such as Euryale ferox or Jacana, can be taken into consideration. Vegetation which is attractive to aquatic invertebrates will aid in sustaining an abundance of fish species dependent on those invertebrates. However, very dense vegetation should be avoided.
- Cultivation of Ginger Lily (Hedychium coronarium) embellishes the isolated ponds. It also has ecological value such as attraction of butterfly species Grass Demon (Udaspes folus).
- Logs or large rocks can be provided in the shallow water area for wildlife use, such as perches for birds and sunning spots for turtles. Trees and shrubs along the shore can provide perches for kingfishers and other species of birds that feed upon fish.
- Nest boxes, such as bat boxes or artificial nests for swallows, can be provided at pond site for cavity nesting wildlife.

表2a 独立的河塘 Table 2a: Isolated Ponds and Lakes

物种名称 Species Name	中文名称	Name	动物种类 Fauna Group
Tringa ochropus	白腰草鹬	Green Sandpiper	鸟类 Bird
Tringa glareola	林鹬	Wood Sandpiper	鸟类 Bird
Chlidonias hybrida	须浮鸥	Whiskered Tern	鸟类 Bird
Chlidonias leucopterus	白翅浮鸥	White-winged Tern	鸟类 Bird
Apus nipalensis	小白腰雨燕	House Swift	鸟类 Bird
Alcedo atthis	普通翠鸟	Common Kingfisher	鸟类 Bird
Halcyon smyrnensis	白胸翡翠	White-throated Kingfisher	鸟类 Bird
Motacilla cinerea	灰鹡鸰	Grey Wagtail	鸟类 Bird
Motacilla alba	白鹡鸰	White Wagtail	鸟类 Bird
Anthus hodgsoni	树鹨	Olive-backed Pipit	鸟类 Bird
Acrocephalus orientalis	东方大苇莺	Oriental Reed Warbler	鸟类 Bird
Phylloscopus fuscatus	褐柳莺	Dusky Warbler	鸟类 Bird
Spodiopsar sericeus	丝光椋鸟	Red-billed Starling	鸟类 Bird
Lamprigera sp.	扁萤	Firefly	昆虫 Insect
Pyrocoelia sp.	窗萤	Firefly	昆虫 Insect
Luciola sp.	黑端熠萤	Firefly	昆虫 Insect
Diaphanes citrinus	橙萤	Firefly	昆虫 Insect
Sinictinogomphus clavatus	大团扇春蜓	Golden Flangetail	昆虫 Insect
Epophthalmia elegans	闪蓝丽大蜻	Regal Pond Cruiser	昆虫 Insect
Neurothemis fulvia	网脉蜻	Russet Percher	昆虫 Insect
Pantala flavescens	黄蜻	Wandering Glider	昆虫 Insect
Rhodothemis rufa	红胭蜻	Ruby Darter	昆虫 Insect
Rhyothemis triangularis	三角丽翅蜻	Sapphire Flutterer	昆虫 Insect
Tramea virginia	华斜痣蜻	Saddlebag Glider	昆虫 Insect
Trithemis festiva	晓褐蜻	Indigo Dropwing	昆虫 Insect
Zygonyx iris insignis	彩虹蜻	Emerald Cascader	昆虫 Insect
Ctenopharyngodon idellus	鲩鱼	Grass carp	淡水鱼 Freshwater Fish
Mugil cephalus	乌头鱼	Grey Mullet	淡水鱼 Freshwater Fish
Clarias fuscus	塘虱鱼	Walking catfish	淡水鱼 Freshwater Fish
Parasilurus asotus	鲶鱼	Chinese Catfish	淡水鱼 Freshwater Fish
Carassius carassius	鲫鱼	Crucian Carp	淡水鱼 Freshwater Fish
Misgurnus anguillicaudatus	泥鳅	Eel-like Loach	淡水鱼 Freshwater Fish
Metapenaeus ensis	刀额新对虾	Greasyback shrimp	淡水虾 Shrimp

物种名称 Species Name	中文名称	Name	动物种类 Fauna Group
Pipistrellus abramus	东亚家蝠	Japanese Pipistrelle	哺乳类 Mammal
Occidozyga lima	尖舌浮蛙	Floating Frog	两栖类 Amphibian
Rana livida	大绿蛙	Green Cascade Frog	两栖类 Amphibian
Hoplobatrachus chinensis	虎纹蛙	Chinese Bullfrog	两栖类 Amphibian
Ardea cinerea	苍鹭	Grey Heron	鸟类 Bird
Egretta garzetta	小白鹭	Little Egret	鸟类 Bird
Ardeola bacchus	池鹭	Chinese Pond Heron	鸟类 Bird
Nycticorax nycticorax	夜鹭	Night Heron	鸟类 Bird
Amaurornis phoenicurus	白胸苦恶鸟	White-breasted Waterhen	鸟类 Bird
Gallicrex cinerea	董鸡	Watercock	鸟类 Bird
Gallinula chloropus	黑水鸡	Common Moorhen	鸟类 Bird
Hydrophasianus chirurgus	水雉	Pheasant-tailed Jacana	鸟类 Bird
Rostratula benghalensis	彩鹬	Greater Painted-Snipe	鸟类 Bird
Glareola maldivarum	普通燕鸻	Oriental Pratincole	鸟类 Bird
Vanellus vanellus	凤头麦鸡	Northern Lapwing	鸟类 Bird
Charadrius dubius	金眶鸻	Little Ringed Plover	鸟类 Bird

表2b 独立的河塘滤水植物相 Table 2b: Isolated Ponds + Lakes Filtration Flora

物种名称 Species Name	中文名称	植物相 Flora Group
Ludwigia adscendens	水龙	药草 Herb
Sagittaria sagittifolia	慈菇	药草 Herb
Ipomoea aquatica	蕹菜	药草 Herb
Pontederia cordata	梭鱼草	药草 Herb
Lythrum salicaria Linn	千屈菜	药草 Herb
Phragmites australis	芦苇	灌木 Shrub
Cyperus malaccensis	茳芏	灌木 Shrub
Pistia stratiotes	大薸	灌木 Shrub

物种名称 Species Name	中文名称	Name	动物种类 Fauna Group
Prinia flaviventris	灰头鹪莺	Yellow-bellied Prinia	鸟类 Bird
Spodiopsar sericeus	丝光椋鸟	Red-billed Starling	鸟类 Bird
Cisticola juncidis	棕扇尾莺	Zitting Cisticola	鸟类 Bird
Emberiza tristrami	白眉鹀	Tristram's Bunting	鸟类 Bird
Emberiza pusilla	小鹀	Little Bunting	鸟类 Bird
Emberiza rustica	田鹀	Rustic Bunting	鸟类 Bird
Lonchura striata	白腰文鸟	White-rumped Munia	鸟类 Bird
Lonchura punctulata	斑文鸟	Scaly-breasted Munia	鸟类 Bird
Photuris lucicrescens	萤火虫	Firefly	昆虫 Insect
Ischnura senegalensis	褐斑异痣蟌	Common Bluetail	昆虫 Insect
Anax guttatus	斑伟蜓	Pale-spotted Emperor	昆虫 Insect
Anax immaculifrons	黄伟蜓	Fiery Emperor	昆虫 Insect
Ictinogomphus pertinax	霸王叶春蜓	Common Flangetail	昆虫 Insect
Sinictinogomphus clavatus	大团扇春蜓	Golden Flangetail	昆虫 Insect
Epophthalmia elegans	闪蓝丽大蜻	Regal Pond Cruiser	昆虫 Insect
Acisoma panorpoides	锥腹蜻	Asian Pintail	昆虫 Insect
Neurothemis fulvia	网脉蜻	Russet Percher	昆虫 Insect
Rhodothemis rufa	红胭蜻	Ruby Darter	昆虫 Insect
Rhyothemis triangularis	三角丽翅蜻	Sapphire Flutterer	昆虫 Insect
Tramea virginia	华斜痣蜻	Saddlebag Glider	昆虫 Insect
Trithemis festiva	晓褐蜻	Indigo Dropwing	昆虫 Insect
Zygonyx iris insignis	彩虹蜻	Emerald Cascader	昆虫 Insect
Ypthima baldus baldus	矍眼蝶	Common Five-ring	昆虫 Insect
Zizeeria maha serica	酢浆灰蝶	Pale Grass Blue	昆虫 Insect
Ampittia dioscorides etura	黄斑弄蝶	Bush Hopper	昆虫 Insect
Udaspes folus	姜弄蝶	Grass Demon	昆虫 Insect
Clarias fuscus	塘虱鱼	Walking catfish	淡水鱼 Freshwater Fish
Channa maculata	生鱼	Spotted Snakehead	淡水鱼 Freshwater Fish
Carassius carassius	鲫鱼	Crucian Carp	淡水鱼 Freshwater Fish
Misgurnus anguillicaudatus	泥鳅	Eel-like Loach	淡水鱼 Freshwater Fish
Macropodus opercularis	叉尾斗鱼	Paradise Fish	淡水鱼 Freshwater Fish
Monopterus albus	黄鳝	Swampy Eel	淡水鱼 Freshwater Fish
Sinotaia quadrata	小田螺	-	淡水螺 Freshwater Snail

表3a 湿地 Table 3a: Wetlands

物种名称 Species Name	中文名称	Name	动物种类 Fauna Group
Pipistrellus abramus	东亚家蝠	Japanese Pipistrelle	哺乳类 Mammal
Bufo melanostictus	黑眶蟾蜍	Asian Common Toad	两栖类 Amphibian
Occidozyga lima	尖舌浮蛙	Floating Frog	两栖类 Amphibian
Rana livida	大绿蛙	Green Cascade Frog	两栖类 Amphibian
Rana kuhlii	大头蛙	Big-headed Frog	两栖类 Amphibian
Ardea cinerea	苍鹭	Grey Heron	鸟类 Bird
Egretta garzetta	小白鹭	Little Egret	鸟类 Bird
Ardeola bacchus	池鹭	Chinese Pond Heron	鸟类 Bird
Butorides striata	绿鹭	Striated Heron	鸟类 Bird
Nycticorax nycticorax	夜鹭	Night Heron	鸟类 Bird
Ixobrychus sinensis	黄斑苇鳽	Yellow Bittern	鸟类 Bird
Ixobrychus eurhythmus	紫背苇鳽	Von Schrenck's Bittern	鸟类 Bird
Amaurornis phoenicurus	白胸苦恶鸟	White-breasted Waterhen	鸟类 Bird
Gallicrex cinerea	董鸡	Watercock	鸟类 Bird
Gallinula chloropus	黑水鸡	Common Moorhen	鸟类 Bird
Hydrophasianus chirurgus	水雉	Pheasant-tailed Jacana	鸟类 Bird
Rostratula benghalensis	彩鹬	Greater Painted-Snipe	鸟类 Bird
Glareola maldivarum	普通燕鸻	Oriental Pratincole	鸟类 Bird
Vanellus vanellus	凤头麦鸡	Northern Lapwing	鸟类 Bird
Charadrius dubius	金眶鸻	Little Ringed Plover	鸟类 Bird
Vanellus cinereus	灰头麦鸡	Grey-headed Lapwing	鸟类 Bird
Tringa ochropus	白腰草鹬	Green Sandpiper	鸟类 Bird
Tringa glareola	林鹬	Wood Sandpiper	鸟类 Bird
Chlidonias hybrida	须浮鸥	Whiskered Tern	鸟类 Bird
Chlidonias leucopterus	白翅浮鸥	White-winged Tern	鸟类 Bird
Apus nipalensis	小白腰雨燕	House Swift	鸟类 Bird
Alcedo atthis	普通翠鸟	Common Kingfisher	鸟类 Bird
Halcyon smyrnensis	白胸翡翠	White-throated Kingfisher	鸟类 Bird
Motacilla cinerea	灰鹡鸰	Grey Wagtail	鸟类 Bird
Motacilla alba	白鹡鸰	White Wagtail	鸟类 Bird
Anthus hodgsoni	树鹨	Olive-backed Pipit	鸟类 Bird
Acrocephalus orientalis	东方大苇莺	Oriental Reed Warbler	鸟类 Bird

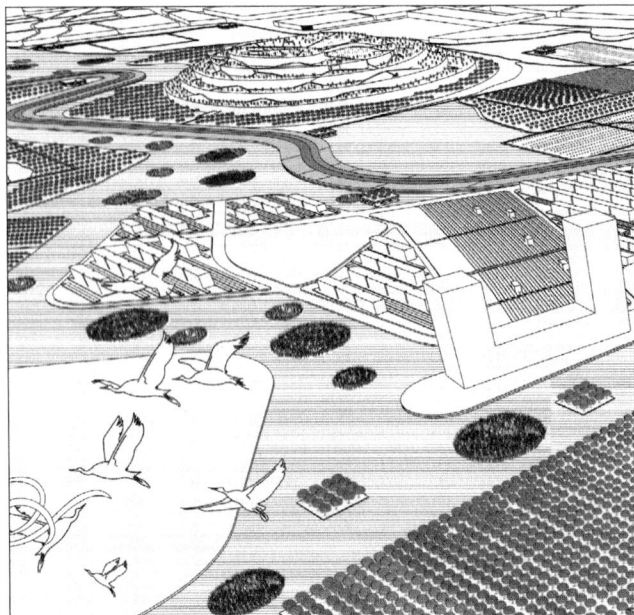

表3b 湿地滤水植物相 Table 3b: Wetlands Water Filtration Flora

物种名称 Species Name	中文名称	Name	植物相 Flora Group
Phragmites australis	芦苇	Reed	灌木 Shrub
Typha orientalis Presl	香蒲	Cattail	灌木 Shrub
Zizania aquatica	茭白	Water Bamboo	灌木 Shrub
Canna indica	水生美人蕉	Canna	灌木 Shrub
Acorus calamus Linn	菖蒲	Calamus	灌木 Shrub
Scirpus validus Vahl	水葱	Soft stem bulrush	灌木 Shrub
Pontederia cordata	梭鱼草	Pickerelweed	灌木 Shrub
Cyperus papyrus	纸莎草	Paper reed	灌木 Shrub
Iris tectorum Maxim	鸢尾	Wall Iris	灌木 Shrub
Iris ensata Thunb	花菖蒲	Japanese Iris	灌木 Shrub
Vetiveria zizanioides	香根草	Vetiver	灌木 Shrub
Colocasia esculenta	芋	Taro	灌木 Shrub
Camellia japoica	茶花	Japanese Camellia	灌木 Shrub
Eleocharis tuberosa	荸荠	Chinese water chestnut	药草 Herb
Lythrum salicaria Linn	千屈菜	Purple lythrum	药草 Herb
Ipomoea aquatica	蕹菜	Water Spinach	药草 Herb
Radix Aucklandiae	木贼	Horsetail	药草 Herb
Sagittaria sagittifolia	慈姑	Sagittaria	药草 Herb
Acorus gramineu	石菖蒲	Japanese sweet flag	药草 Herb
Juncus effusus	灯心草	Soft Rush	药草 Herb

Wetland offers a constructed large-scale, low-tech system to treat polluted river water and stormwater runoff of Haizhu region. Rapid economic development of the Pearl River Delta Economic Zone (PRDEZ) has impacted the water quality and the natural environment of the area. The main cause of river pollution is untreated urban and rural sewage, followed by industrial and agricultural wastewater. The water of the Pearl River suffers from high levels of ammonium, phosphorus and organic compounds.

Wetland reed bed filtration makes use of reeds (and other emergent water plants) to transfer large amounts of oxygen down into the root system (rhizosphere). This oxygen providing process allows natural bacteria and microbes to multiply, significantly increasing their resident capacity to break down the polluted water. These bacteria break down pollutants, releasing minerals, which the reeds take up and use to grow. The larger the reed, the better the oxygenating capacity, the more bacteria resulting in more water being cleaned: a natural and simple cycle to clean water. Phragmites australis is an ideal form of vegetation for this purpose, while the wetland can be embellished through provision of flowering wetland plants. Mature reeds are harvested for bio-energy production using anaerobic digesters.

In WOW, a floating reed bed system is used instead of the traditional gravel-pit systems. This system delivers an effective treatment of pollutants by creating anaerobic and aerobic pockets of water, providing a micro-treatment zone within the floating reed bed zone. The reed beds were spaced apart to keep the water viable for aquatic wildlife and biota. The floating reed beds also provided safe havens amongst the wetlands for nesting birds. It is favourable for colonisation of wetland associated insect species, notably dragonfly and damselfly. The nocturnal and photophobic firefly is also another potential target. Floating biofiltration does not require extra or specialist maintenance above a conventional wetland, nor is it more expensive.

The floating reed bed system requires a 30 per cent coverage ratio of floating reed beds to water surface, at a dry matter biomass of 7500g/m3. Once matured, this wetland ecology should be capable of filtering water in four cycles annually, each being able to rise by two classes of the water quality index as described in 'Environmental Quality Standard for Surface Water GB 3838-2002', which is the standard that China is working towards. Water from the Pearl River is brought into the wetland lakes at increments of half a metre depth per year to a maximum of two metres to allow the reeds, as well as other flora and fauna, to mature. Reeds will be matured at the lake perimeter before being transplated to floating pontoons. The pontoon-floated reed beds should be attached by a chain to a concrete anchor to ensure the reed beds are kept strategically positioned around the lake whereby the chain is long enough to allow for sufficient rise and fall in water in seasons of high rainfall.

River water will be systematically filtered through a series of wetland lakes within the site. Water will be cleaned progressively, passing through a maximum of three lakes, at which point it can be classed as very clean (water quality standard I). Water can be siphoned off from each lake for agricultural, recreational, fish farming, and other ecological uses. A series of locks at natural boundaries will allow the flow of water between different environments – the locks will separate cleaner waters from contamination whilst allowing boats from the naval yard to access the river.

Orchard

The dominant habitat type within the project area, the orchards take up about 40 percent of the total area and play an important role in recreation, landscaping and income generation for the Haizhu region. Despite the large scale, the ecological value of the current orchard habitat is not optimum due to the low flora diversity, monotonous nature and dense planting pattern. Only a small group of fauna species is expected to benefit from this habitat type. Ecological factors should be incorporated in the detailed design with the aim to enhance the flora and fauna biodiversity of the area. Examples include diversifying fruit tree species, change of operation practice and management, e.g. organic farming, creation of different canopies, increasing vegetation ground coverage.

Design Objective + Potential Ecological Function
- Diversify flora species and subsequently increase the attractiveness to fauna species.
- Increase the productivity of terrestrial habitat.
- Increase the value for ecotourism and enhance leisure enjoyment.
- Increase educational value.

Design Criteria
- Increase flora species diversity.
- Provide sufficient spacing between fruit trees to allow planting and tourist visits.
- Provide artificial shelters for fauna groups (e.g. bats and honeybees).
- Install access network, e.g. footpath, cycle route, for visitors' use.
- Incorporate leisure facilities and open space within the orchard to provide opportunity for ecotourism.

Ecological Design Considerations

Increase Spacing between Fruit Trees: By increasing the space between individual trees, vegetation of different heights could be planted along strips in-between the tree rows, creating different canopies and micro-habitats for local fauna to use as shelter or feeding ground. Canopies of different heights can be created by planting species of different forms, i.e. tall shrub, short shrub, herb and climber/ground cover. The resulted increase of habitat complexity could enhance the productivity of the terrestrial habitat and improve the suitability of the orchard habitat for the use of local fauna groups. Moreover, the additional vegetation cover would enhance the landscape value.

Provision of Artificial Shelter: Shelters for fauna could be provided with artificial structures specifically designed for their use. Bug boxes could be placed in areas with the previously proposed additional planting to attract insects, e.g. honeybees; and bat boxes could be provided to attract bats. Honeybee rearing could be considered for pollination and honey production.

Access Network: Access network, for example footpaths and cycle routes, shall be provided to ensure a positive and educational experience. With the well designed access network, other possibilities for leisure activities could be explored, e.g. fruit picking, fruit tree ownership and a landscaped picnic area. In addition, sufficient areas should be designated as 'no go' areas for use as wildlife reserves.

facing page: Artifice of nature painted on columns under the highway, akin to William Morris's famous wallpaper.

表4b 果园经济作物 Table 4b: Orchards Economic Crops

物种名称 Species Name	中文名称	Name	植物相 Flora Group
Dimocarpus longan lour	龙眼	Longan	树木 Tree
Mangifera indica 'Nan Klang Wan'	象牙芒	Mango	树木 Tree
Mangifera indica 'Deshehari'	鸡蛋芒	Mango	树木 Tree
Mangifera indica 'Tsar-Swain'	柴檨	Mango	树木 Tree
Mangifera indica 'Pung-Swain'	香檨	Mango	树木 Tree
Mangifera indica 'Da-Hung-Man'	大黄芒	Mango	树木 Tree
Mangifera indica 'Va-Swain'	肉檨	Mango	树木 Tree
Mangifera indica 'Carabao'	吕宋芒	Mango	树木 Tree
Mangifera indica 'Jidan'	鸡蛋芒	Mango	树木 Tree
Carica papaya	番木瓜	Papaya	树木 Tree
Averrhoa carambola	杨桃	Star-fruit	树木 Tree
Clausena lansium	黄皮	Wampee	树木 Tree
Ananas comosus	菠萝	Pineapple	灌木 Shrub
Citrullus lanatus	西瓜	Watermelon	灌木 Shrub
Citrus lime	柠檬	Lime	灌木 Shrub
Citrus sinensis	砂糖桔	Orange	灌木 Shrub

表5a 有机农业 Table 5a: Agriculture/Organic Farming

物种名称 Species Name	中文名称	Name	动物种类 Fauna Group
Bufo melanostictus	黑眶蟾蜍	Asian Common Toad	两栖类 Amphibian
Microhyla pulchra	花姬蛙	Marbled Pigmy Frog	两栖类 Amphibian
Rana guentheri	沼蛙	Gunther's Frog	两栖类 Amphibian
Anthus richardi	田鹨	Richard's Pitpit	鸟类 Bird
Ardeola bacchus	池鹭	Chinese Pond Heron	鸟类 Bird
Bubulcus ibis	牛背鹭	Cattle Egret	鸟类 Bird
Streptopelia chinensis	珠颈斑鸠	Spotted Dove	鸟类 Bird
Motacilla flava	黄鹡鸰	Yellow Wagtail	鸟类 Bird
Pieris canidia canidia	东方菜粉蝶	Indian Cabbage White	昆虫 Insect
Graphium agamemnon agamemnon	统帅青凤蝶	Tailed Jay	昆虫 Insect
Eurema hecabe hecabe	宽边黄粉蝶	Common Grass Yellow	昆虫 Insect
Hypolimnas bolina kezia	幻紫斑蛱蝶	Great Egg-fly	昆虫 Insect
Chondracris rosea	大棉蝗	Grasshopper	昆虫 Insect
Acrida turrita	短角蚱蜢	Grasshopper	昆虫 Insect

表5b 常见的当地蔬菜 Table 5b: Common Local Vegetables

物种名称 Species Name	中文名称	季节 Season
Brassica parachinensis	菜心	整年 Whole year
Brassica alboglabra	芥兰	夏季 Summer
Brassica chinensis	白菜	整年 Whole year
Lycium chinensis	枸杞	春、夏季 Spring, Summer
Vigna unguiculata	豆角	夏季 Summer
Zingiber officinale	姜	整年 Whole year
Lagenaria siceraria	葫芦	夏、秋季 Summer, autumn
Solanum melongena	茄子	整年 Whole year, but mainly summer
Zea mays	玉蜀黍	整年 Whole year
Sauropus spatulifolius	龙脷叶	整年 Whole year
Raphanus sativus	萝卜	冬、秋季 Winter, autumn
Benincasa hispida	节瓜	春、夏、秋季 Spring, Summer, Autumn
Ipomoea aquatica	通菜	夏季 Summer
Colocasia esculenta	芋	夏季 Summer

表4a 果园 Table 4a: Orchards

物种名称 Species Name	中文名称	Name	动物种类 Fauna Group
Cynopterus sphinx	短吻果蝠	Short-nosed Fruit Bat	哺乳类 Mammal
Callosciurus erythraeus	赤腹松鼠	Pallas's Squirrel	哺乳类 Mammal
Mustela kathiah	黄腹鼬	Yellow-bellied Weasel	哺乳类 Mammal
Herpestes javanicus	红颊獴	Small Asian Mongoose	哺乳类 Mammal
Phylloscopus inornatus	黄眉柳莺	Yellow-browed Warbler	鸟类 Bird
Zosterops japonica	暗绿绣眼鸟	Japanese White Eye	鸟类 Bird
Pycnonotus jocosus	红耳鹎	Red-whiskered Bulbul	鸟类 Bird
Pycnonotus sinensis	白头鹎	Chinese Bulbul	鸟类 Bird
Alectoris chukar	石鸡	Chukar	鸟类 Bird
Bambusicola thoracica	灰胸竹鸡	Chinese Bamboo Partridge	鸟类 Bird
Phasianus colchicus	雉鸡	Common Pheasant	鸟类 Bird
Deudorix epijarbas	玳灰蝶	Cornelian	昆虫 Insect
Euthalia aconthea	矛翠蛱蝶	Baron	昆虫 Insect
Euthalia phemius seitzi	尖翅翠蛱蝶	White-edged Blue Baron	昆虫 Insect
Papilio demoleus demoleus	达摩凤蝶	Lime Butterfly	昆虫 Insect
Papilio helenus helenus	玉斑凤蝶	Red Helen	昆虫 Insect
Papilio memnon agenor	美凤蝶	Great Mormon	昆虫 Insect
Papilio paris paris	巴黎翠凤蝶	Paris Peacock	昆虫 Insect
Papilio polytes polytes	玉带凤蝶	Common Mormon	昆虫 Insect
Papilio protenor protenor	蓝凤蝶	Spangle	昆虫 Insect
Tenodera sinensis	华大刀螳	Grasshopper	昆虫 Insect
Elimaea punctifera	长角蚱蜢	Grasshopper	昆虫 Insect
Chondracris rosea	大棉蝗	Grasshopper	昆虫 Insect
Acrida turrita	短角蚱蜢	Grasshopper	昆虫 Insect

Organic Agriculture

Agricultural areas currently occupy about six percent of land and are dispersed across the project area, interspersed with orchard habitat. A high level of human management is applied on this habitat due to its commercial purpose, resulting in low utilisation from fauna species. This habitat has the potential to be converted into a more environmentally friendly resource, through adoption of organic and environmentally friendly farming practice, promotion of an organic farm certification system to create brand identity and planting of insect repelling species for natural pest control. The yield of the organic agriculture helps provide a sustainable supply of safe food sources.

277

Design Objective + Potential Ecological Function
- Ecological enhancement for wildlife's use.
- Increase public involvement.
- Secure food safety.

Design Criteria
- Allocation of field margins for the planting of specific flora species for wildlife's use.
- Incorporate agricultural areas within residential areas.
- Provide an adequate water source for irrigation of farmlands.

Ecological Design Considerations
Organic Farming: Farmers could adopt organic farming practice and the government could promote awareness on this environmentally friendly practice. Facilities that would aid this practice, such as a composting area, should be provided. Practice that could minimise chemical input should be considered, for example crop rotation and land abandonment to allow soil recovery and nutrient recycling (hence minimising the use of fertiliser), and application of fertilisers produced from the on-site composting.

Good Management Practice: Installation of a well designed ditch system with appropriate topographical alteration (e.g. water canal within residential allotments) could minimise irrigation effort and risk of flooding/inundation, allowing a more manageable water supply for crops. Application of chemical fertilisers and pesticides/insecticides (if unavoidable) should not be excessive, as the leachate would enter the watercourse and result in water pollution and subsequent ecological impacts. The improvement in management would help to secure harvest, bringing positive impacts to income generation for farmland owners and securing food safety. In addition, field margins could be planted with wildlife-attracting plants to provide a niche. As the proposed water canal will be integrated with the residential development, this arrangement could allow fish farming and be considered to be attractive to some forms of wildlife such as dragonflies.

Public Engagement: Farming lessons could be organised to assist in the adoption of organic farming practice. Home-scale organic waste composting techniques could be demonstrated to the communities to promote the importance of waste minimisation, introducing the three Rs: Reduce, Reuse and Recycle. Organic farming can also be considered in residential allotments.

following page: The uninterrupted endless stretches of flora along the Green Boulevard act as a catalyst to embrace knowledge on ecology, green living and organic food production.

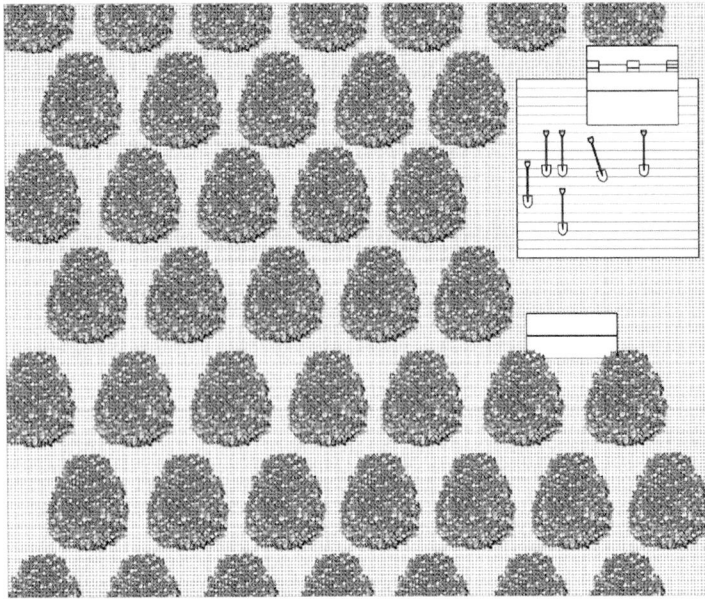
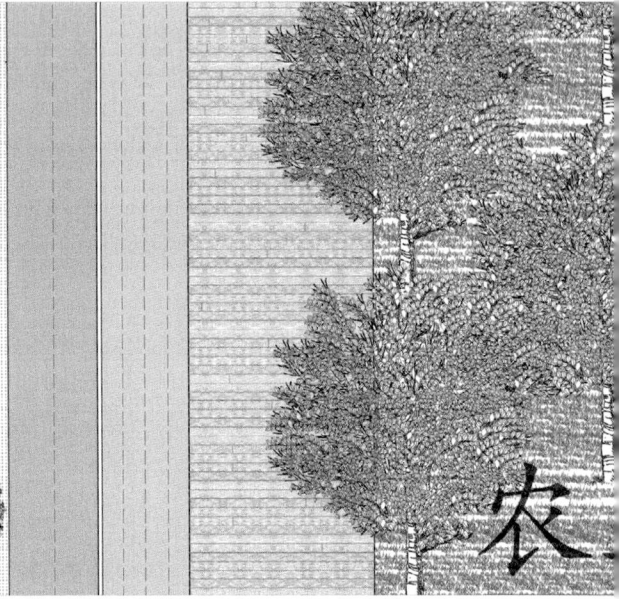

农

278

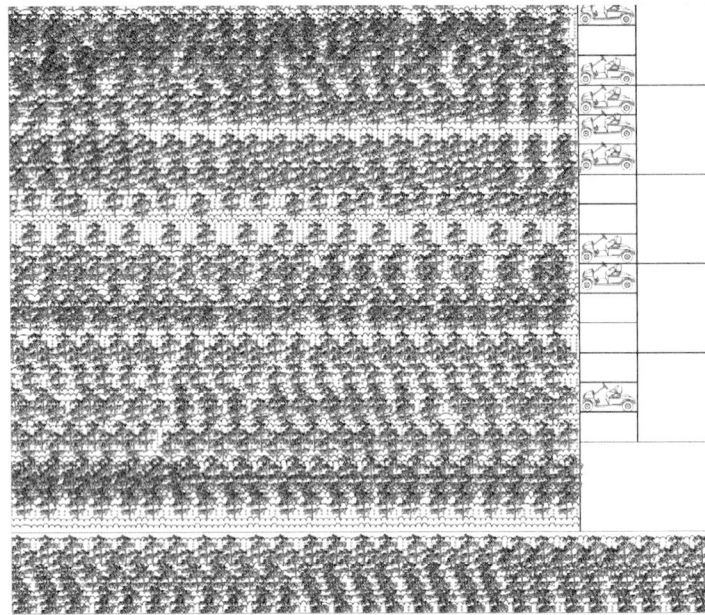
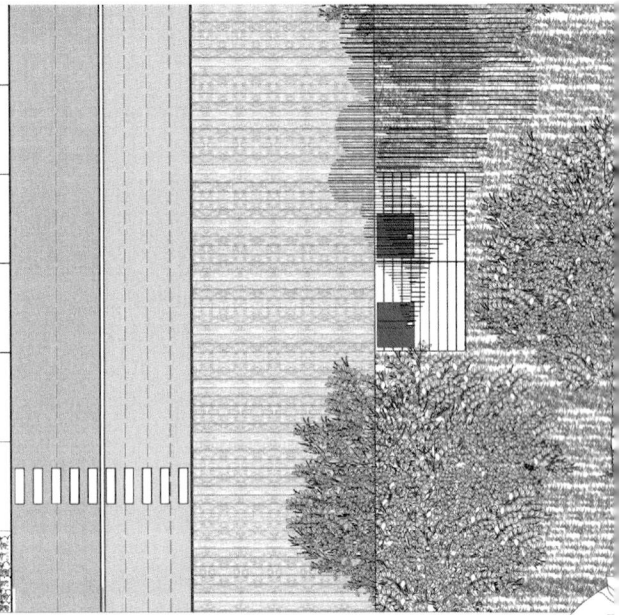

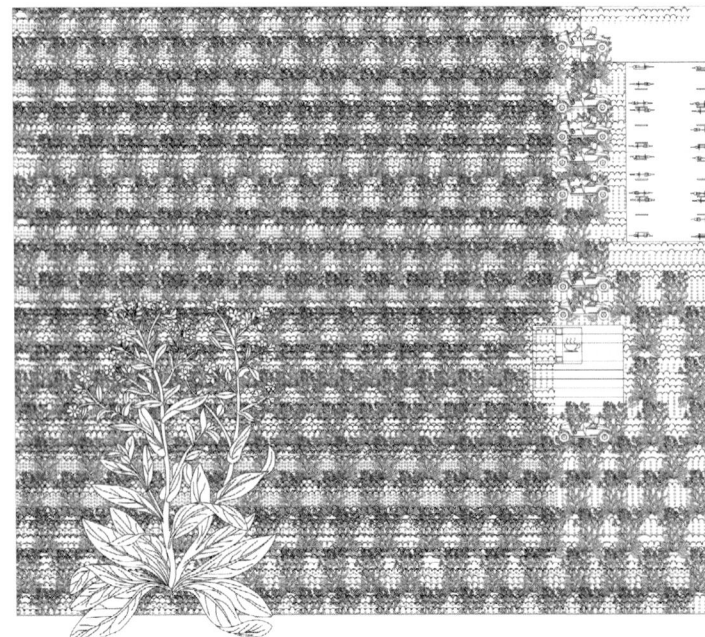
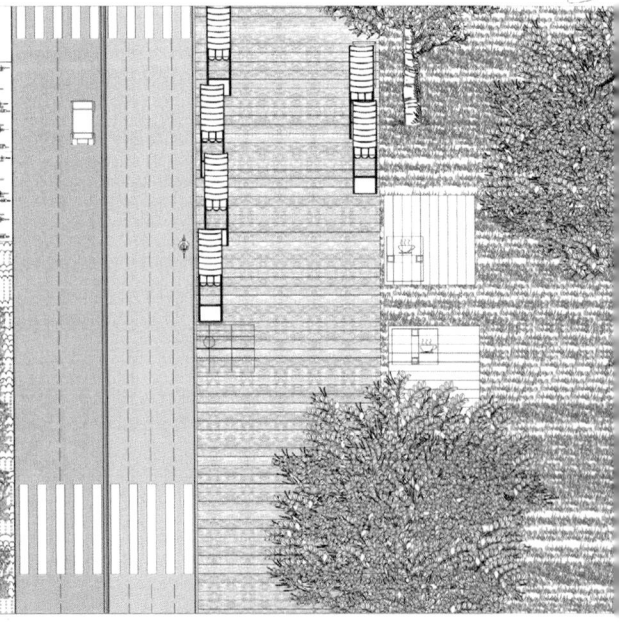

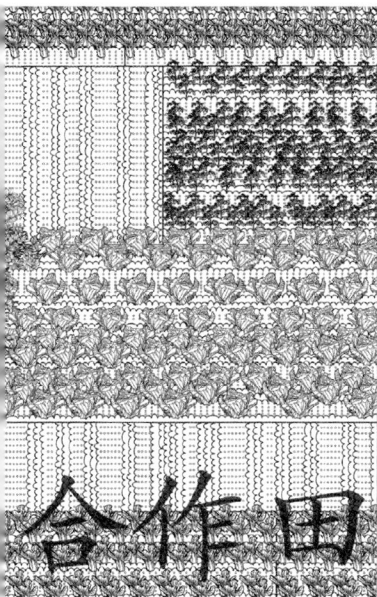

农场教育

合作田

太极

凉茶

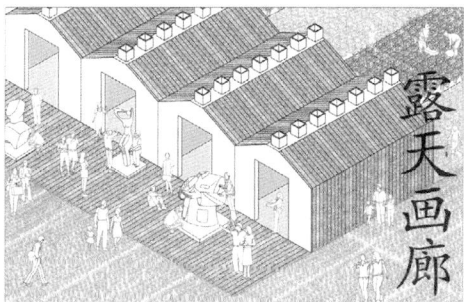

露天画廊

渔业

物摊贩

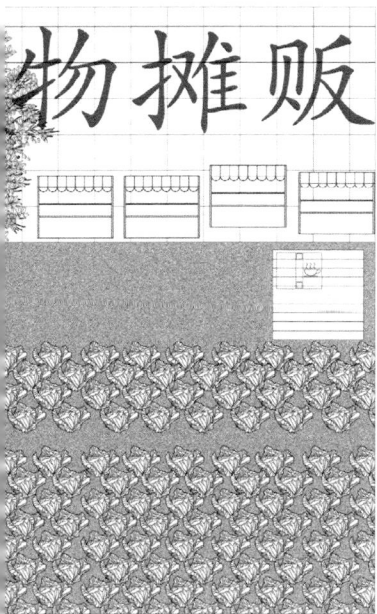

新鲜水果

表6b 岭南开花灌木+药草 Table 6b: Lingnan Flowering Shrubs + Herbs

物种名称 Species Name	中文名称	植物相 Flora Group
Michelia champaca	岭南黄兰	灌木 Shrub
Gordonia axillaris	大头茶	灌木 Shrub
Loropetalum chinense	纸末花	灌木 Shrub
Rhododendron sp.	杜鹃花	灌木 Shrub
Rhodomyrtus tomentosa	桃金娘	灌木 Shrub
Melastoma candidum	野牡丹	灌木 Shrub
Rhaphiolepis indica	舂花	灌木 Shrub
Desmodium styracifolium	金錢草	药草 Herb
Lophatherum gracile	淡竹葉	药草 Herb
Polygonum chinense	火炭母	药草 Herb
Helicteres angustifolia	山芝麻	药草 Herb
Elephantopus scaber	地膽草	药草 Herb
Phragmites australis	蘆葦	药草 Herb
Imperata koenigii	絲茅	药草 Herb

表6b 草 Table 6b: Grass/Lawn

物种名称 Species Name	中文名称	植物相 Flora Group
Axonopus compressus	地毯草	草 Grass
Cynodon dactylon	狗牙根	草 Grass
Zoysia matrella	沟叶结缕草	草 Grass
Eremochloa ophiuroides	假俭草	草 Grass
Chrysopogon aciculatus	竹节草	草 Grass

表8 候鸟类 Table 8: Migratory Birds

物种名称 Species Name	中文名称	Name	动物种类 Fauna Group
Circus spilonotus	白腹鹞	Eastern Marsh Harrier	候鸟类 Migratory Bird
Buteo Buteo	普通鵟	Common Buzzard	候鸟类 Migratory Bird
Botaurus stellaris	大麻鳽	Great Bittern	候鸟类 Migratory Bird
Ixobrychus cinnamomeus	栗苇鳽	Cinnamon Bittern	候鸟类 Migratory Bird
Ixobrychus eurhythmus	紫背苇鳽	Schrenck's Bittern	候鸟类 Migratory Bird
Gallicrex cinerea	董鸡	Watercock	候鸟类 Migratory Bird
Hydrophasianus chirurgus	水雉	Pheasant-tailed Jacana	候鸟类 Migratory Bird
Rostratula benghalensis	彩鹬	Greater Painted-Snipe	候鸟类 Migratory Bird
Himantopus himantopus	黑翅长脚鹬	Black-winged Stilt	候鸟类 Migratory Bird
Glareola maldivarum	普通燕鸻	Oriental Pratincole	候鸟类 Migratory Bird
Vanellus vanellus	凤头麦鸡	Northern Lapwing	候鸟类 Migratory Bird
Vanellus cinereus	灰头麦鸡	Grey-headed Lapwing	候鸟类 Migratory Bird
Chroicocephalus ridibundus	红嘴鸥	Black-headed Gull	候鸟类 Migratory Bird
Chlidonias hybrida	须浮鸥	Whiskered Tern	候鸟类 Migratory Bird
Chlidonias leucopterus	白翅浮鸥	White-winged Tern	候鸟类 Migratory Bird
Motacilla cinerea	灰鹡鸰	Grey Wagtail	候鸟类 Migratory Bird

表6a 林荫大道 Table 6a: Boulevard

物种名称 Species Name	中文名称	Name	动物种类 Fauna Group
Callosciurus erythraeus	赤腹松鼠	Pallas's Squirrel	哺乳类 Mammal
Calotes versicolor	变色树蜥	Changeable Lizard	两栖类 Amphibian
Bufo melanostictus	黑眶蟾蜍	Asian Common Toad	两栖类 Amphibian
Microhyla pulchra	花姬蛙	Marbled Pigmy Frog	两栖类 Amphibian
Rana guentheri	沼蛙	Gunther's Frog	两栖类 Amphibian
Anthus richardi	田鹨	Richard's Pitpit	鸟类 Bird
Ardeola bacchus	池鹭	Chinese Pond Heron	鸟类 Bird
Bubulcus ibis	牛背鹭	Cattle Egret	鸟类 Bird
Streptopelia chinensis	珠颈斑鸠	Spotted Dove	鸟类 Bird
Motacilla flava	黄鹡鸰	Yellow Wagtail	鸟类 Bird
Garrulax perspicillatus	黑脸噪鹛	Masked Laughing Thrush	鸟类 Bird
Zosterops japonicus	相思	Japanese White Eye	鸟类 Bird
Pycnonotus jocosus	红耳鹎	Red-whiskered Bulbul	鸟类 Bird
Pycnonotus sinensis	白头鹎	Chinese Bulbul	鸟类 Bird
Euthalia phemius	尖翅翠蛱蝶	White-edged Blue Baron	昆虫 Insect
Papilio demoleus	达摩凤蝶	Lime Butterfly	昆虫 Insect
Papilio helenus	玉斑凤蝶	Red Helen	昆虫 Insect
Papilio memnon	美凤蝶	Great Mormon	昆虫 Insect
Papilio paris	巴黎翠凤蝶	Paris Peacock	昆虫 Insect
Papilio polytes	玉带凤蝶	Common Mormon	昆虫 Insect
Pieris canidia canidia	东方菜粉蝶	Indian Cabbage White	昆虫 Insect
Graphium agamemnon agamemnon	统帅青凤蝶	Tailed Jay	昆虫 Insect
Eurema hecabe hecabe	宽边黄粉蝶	Common Grass Yellow	昆虫 Insect
Hypolimnas bolina kezia	幻紫斑蛱蝶	Great Egg-fly	昆虫 Insect
Pantala flavescens	黄蜻	Wandering Glider	昆虫 Insect
Zyxomma petiolatum	细腹绿眼蜻	Dingy Dusk-darter	昆虫 Insect
Neurothemis fulvia	网脉蜻	Russet Percher	昆虫 Insect
Tenodera sinensis	华大刀螂	Grasshopper	昆虫 Insect
Elimaea punctifera	长角蚱蜢	Grasshopper	昆虫 Insect
Chondracris rosea	大棉蝗	Grasshopper	昆虫 Insect
Acrida turrita	短角蚱蜢	Grasshopper	昆虫 Insect

表7 珠江湿地堤岸 Table 7: Pearl River Swampy Mud Embankment

物种名称 Species Name	中文名称	Name	动物种类 Fauna Group
Actitis hypoleucos	矶鹬	Common Sandpiper	鹬鸟类 Shore Birds
Charadrius dubius	金眶行鸻	Little Ringed Plover	鹬鸟类 Shore Birds
Gallinago gallinago	扇尾沙锥	Common Snipe	鹬鸟类 Shore Birds
Gallinula chloropus	黑水鸡	Common Moorhen	鹬鸟类 Shore Birds
Motacilla alba	白鹡鸰	White Wagtail	鹬鸟类 Shore Birds
Periophthalmus	弹涂鱼	Mudskipper	淡水鱼 Freshwater Fish
Onychargia atrocyana	毛面同痣蟌	Marsh Dancer	昆虫 Insect
Ischnura asiatica	東亞異痣蟌	Asian Bluetail	昆虫 Insect
Orthetrum chrysis	華麗灰蜻	Red-faced Skimmer	昆虫 Insect

Boulevard

A boulevard with trees and flowers is key for the proposed WOW. The alignment of the boulevard runs across the park. The boulevard is proposed not for vehicular use but environmentally friendly promenade and cycling. In order to enhance the greening function of the boulevard, planting strips along the two sides of the boulevard are recommended.

It can act as a main exhibition passage that allows visitors to observe different wildlife and flora as well as enjoying the natural beauty of the habitats of the park closely and safely. The boulevard also serves as a recreational area for cycling, jogging and leisure reading under the trees planted along it. Artist studios are placed at strategic positions along the length of the boulevard.

Design Objective + Potential Ecological Function
- To serve as a recreational area for cycling, jogging and leisure reading.
- To plant species native to and representative of the South China geographical region, which could enhance the ecological function.
- To provide an educational function by erection of signboards illustrating the plants species present along the planting area of the boulevard.

Ecological Design Considerations

A variety of native plant species that relate to human living is proposed along the boulevard to educate the visitors to appreciate the useful and, in some instances, finite resources provided by nature. On the boulevard, vegetable allotments and diverse varieties of mango trees are planted side by side with Lingnan flowering shrubs; low growing fruit plants, e.g. pineapple, are cultivated along with Lingnan herbal teas.

Drinking traditional herbal teas has been popular in the Lingnan region for hundreds of years. Herbal tea is a kind of traditional healthy drink made of native Chinese medicinal herbs in different formulas to cure or relieve diseases due to the heat and humid climate of the Lingnan region. Certain formulas of traditional herbal teas have been granted the status of national cultural heritage product in China.

Planting the ingredient plants of herbal teas in the planting areas along the boulevard allows visitors to understand and learn about the natural resources that have a long history of utilisation by local people. Herbal tea shops at kiosks for resting or snacks could also be considered along the long pedestrian boulevard. Herbal teas sold in the shops would be made of local produce planted within WOW. As these are native species to the region, they can not only act as excellent education materials which inform visitors of, and promote to them the benefits of cooling tea culture, but also serve an ecological function.

In addition, these plants have an amenity value. For instance: green foliage of Schefflera heptaphylla and Ilex asprella; colourful fruits of Ilex rotunda and Sarcandra glabra; attractive flowers of Cratoxylum cochinchinense and Lonicera japonica. Apart from the ecological, landscaping and medicinal functions of the plants, some species bear edible fruits, such as Morus alba and Phyllanthus emblica.

following page: The romance and resilience of WOW are gleaned from its imaginative landscape.

Romance + Resilience:
Landscapes of the imagination

'For now, just let me stand here and look at my trees.'

– Rabia of the Farza District in the Shomailu Valley, Afghanistan, 2004

Gazing at these 8000 poplars is Rabia's devotional act to assuage nothing less than a pilgrimage to Mecca, where she will offer prayers for her country and for peace. These groves were part of the 2004 initiative between Global Partnership for Afghanistan and the farmers and elders of the region, which sought to bring work and hope to the community by planting thousands of trees. Victimised by the scorched earth policies of the Soviets and Taliban, the plains' stark deforestation etched almost permanent scars into the landscape that once testified to cycles of human neglect and drought.[1]

Communities in Hiroshima preserve a collective knowledge about their environmental counterparts, the survivor trees of the atomic bombing. They marked each tree with a sign containing information about species and distance from ground zero, creating a map tagging their locations to direct visitors in the Peace Museum. What is remarkable, however, about this map is that it also exists in the minds of school children who can, when asked, point the visitor to the nearest arboreal monument to the catastrophe. The care for the survivor trees was initiated by a group of elementary students in 1974, who organised to save the hackberry tree in their schoolyard.[2] Surviving human technology at its most violently weaponised, the trees memorialise the troubled relationship we have with the work of our hands. Their very presence as living beings enduring calamity serves as a reminder of our precarity in the face of the danger we ultimately underwrite.

Resilience scholarship has begun to compile case studies like those above to recognise the resilience movement also as a metaphor, elaborated not only in the ecological processes of the natural world, but also in the interventions of the built environment, and the imagination and practices of human cultures. Urban resilience becomes a metaphor not only for the capacity to withstand or adapt to adversity, but also for the collective commitment to learn and achieve a stronger, more diverse and integrated revitalisation as codified by the Rockefeller Foundation's '100 Resilience Cities' initiative. For a Smartcity, the cultivation of ecological landscapes as a practice for urban resilience goes beyond merely thinking about the land in instrumental ways. Instead, the case studies in this book are explorations of the innumerable ways nature nurtures corresponding responses in the resilient imagination, and from here, collective action that rehabilitates wounded communities and ecologies.

1. S Thompson, '8,000 Trees: A Refuge from Ruins', in 'Greening in the Red Zone: Disaster, resilience and community greening', KG Tidball & ME Krasny (eds.), Springer, 2014, pp.125–128

2. J McBride et al., 'Restoration of the Urban Forests of Tokyo and Hiroshima Following World War II', in 'Greening in the Red Zone: Disaster, resilience and community greening', KG Tidball & ME Krasny (eds.), Springer, 2014, p.231

3. 'Oxford Pocket Dictionary', Oxford University Press, USA, 2009

4. CG Carus, 'Nine Letters on Landscape Painting: Written in the Years 1815-1824; with a Letter from Goethe by Way of Introduction', Getty Research Institute for the History of Art & the Humanities, 2003, p.7

5+6. T May, 'Prime Minister's Speech on the Environment: 11 January 2018', 2018 [www.gov.uk/government/speeches/prime-ministers-speech-on-the-environment-11-january-2017], retrieved 10 August 2018

7+8. N Nadkarni, 'Between Earth and Sky: Our Intimate Connections to Trees', University of California Press, 2009, p.11

9. A Barau, 'Restoring Indigenous Trees for Scaling Up City Resilience: The role of African millennials', The Nature of Cities, 24 September 2017 [www.thenatureofcities.com/2017/09/24/restoring-indigenous-trees-scaling-city-resilience-role-african-millennials], retrieved 10 August 2018

How might we further understand resilience as a tool to cultivate romantic imagination of the landscape itself? Romance 'appeals strongly to the imagination; an air, feeling, or sense of wonder'[3] and is the aspirational emotion associated with tales of heroes and knights, as well as its relationship to the European historical and cultural movement that is Romanticism. Arising in the late 18th century, Romanticism challenged the Western Enlightenment values of order and hierarchal systems that maintained it. Nature's creativity and undisciplined forces became key sources for imagining, understanding and celebrating human subjectivity, for approaching the divine and experiencing revelation. In that time of political and ecological tumult in industrialising Europe, Romantics '[aimed] at nothing less than the all-embracing ensoulment of nature'.[4]

The romance of resilience, like all stories, has a protagonist. Nature and her champions have eloquently and persistently assigned that role to the tree. Communities blighted by climatic and political catastrophe turn to trees to regenerate both their habitat and themselves, while recurring national policies centre around the care and recovery of the arboreal resources. As part of the UK's '25 Years Environmental Plan' in 2018, Prime Minister Theresa May announced that 'nothing is more emblematic of that natural environment than our trees'.[5] Alongside a commitment to plant millions more trees in urban and rural locations, the plan 'support[s] increased protections for existing trees and forests, while at the same time creating new habitats for wildlife'.[6]

Communities all over the world have turned to trees in their darkest hour for resilience. The tree recurs in multiple cosmologies as the axis mundi or 'the central pivot of the cosmos... the imaginary line linking heaven and earth'. Recall, for example, the trees of the Garden of Eden, or the Bodhi tree whose branches sheltered the Gautama Buddha when he achieved enlightenment.[7] Recall too that to the tree, we owe our opposable thumbs and our binocular vision, our love of wide sweeping vistas and horizons lined in green. Trees inspire us because they are our evolutionary and imaginary homes, our cradle and our castle.[8]

As our planet's ecological defenders, trees have enabled cities to attune themselves to a paradoxical situation where ever more subtle and dramatic shifts escape and overwhelm our lives. They stand at the vanguard of climate change efforts in cities, cooling cities down and absorbing pollution from the air. They have been a lasting symbol of diversity, their long life-span linking generations while their manifold speciation highlights the preciousness of individual identities, preserved yet interconnected. In Kano, for example, where the names of places are likewise names of indigenous tree species, young millennials have come into their own as contributors to their city's regeneration by restoring namesake local trees to their neighbourhood. Once ignorant about what these species even looked like, workers bearing seedlings of acacia or cottonwood have now reversed this toponymic amnesia, returning trees as sources of site-specific local industries to the city while bringing ecological purity to the urban environment.[9] The poetics of these gestures represent the attachment of people to their lands, returning living referents to the names of their indigenous landscape.

Like the enigmatic beloved of many tales, trees elicit our fascination because they confound and perplex expectations, bringing awareness of the mystery and potentialities of the natural world. In Africa, Nigerian workers returning from abroad, having hastily planted crops without clearing the land, discovered that trees had actually become their surprising allies in producing grain yields exceeding those of neighbouring cleared farmlands. The results heralded the start of agroforestry, a movement to restore farmlands through planting trees.[10] A similar intractable landscape is eastern Europe's dried up Aral Sea. The strange vistas of stranded fishing boats on the bone-dry sea bed form the backdrop to the stories of elderly Uzbek fishermen who lived to see their entire livelihood destroyed by Russia's cotton agriculture and the ecological disaster wrought by it. The lake waters were simultaneously drained for irrigation and turned into a receptacle for polluting chemicals. Winds whipped dust clouds from what remained, decimating the fishermen's living and bringing disease. Yet, the community clung tenaciously on, refusing to abandon the wasted sea for better prospects. Nothing but an enduring passion for their land can explain this decision that is poignantly clear in the present restoration of the Aral Sea through the planting of millions of saxaul trees. Undaunted by the fact that the forest will take hundreds of years to grow, the community returns in rain and shine to the sea, working for a healed landscape they will never see.[11]

In the romance of resilience, nurturing groves and restoring forests apprehends the temporality of the earth itself. When the Yangtze River flooded devastatingly in 1998 due to severe deforestation, China began one of the most ambitious ecological investments made by any nation to date, investing billions to plant trees across the country.[12] However, the anxiety for afforestation led the government to monetise the restoration of the landscape without considering the ramifications of monocultural tree plantations that would paradoxically speed up the cutting down of old growth forests.[13] What this effort misses is that resilience is also a deeply imaginative enterprise made possible by the poetic example of trees themselves. In the silent miracle of their endurance and generosity, they continually speak about the past and the future. We turn time and time again to the eloquent language of their expansive canopies and the stately columns of their trunks to re-express what we understand in our own creations, hoping that our own actions for the earth will be like a great tree itself, spanning generations beyond our own.

It is for this reason, novelists reaching for expression in a cosmos darkened by catastrophe have turned to trees. After witnessing the 'massacre of London's trees' during the Blitz, George Orwell penned an essay giving examples of less than moral individuals who nonetheless redeemed themselves by the simple act of planting trees. He writes, 'It might not be a bad idea, every time you commit an antisocial act, to make a note of it in your diary, and then, at the appropriate season, push an acorn into the ground. And, if even one in twenty of them came to maturity, you might do quite a lot of harm in your lifetime, and still... end up as a public benefactor after all.'[14]

In Kassel, the artist Joseph Beuys planted thousands of slow-growing oaks in his famous '7000 Eichen', a socio-political installation that endures in the city today, consisting of trees marked by rough-hewn basalt steles that testify to this permanent happening. He turned trees into beautiful inconveniences, planting oaks, birches and elms precisely at Kassel's city centre where pavements were pervasive, infrastructure essential and the citizenry willful. At this seemingly intractable nexus, however, Beuys's trees made his point: citizens could not be passive in their engagement with the environment, and each other.[15]

When Fukushima was reeling from the double cataclysm of a tsunami and a nuclear reactor meltdown, the Pritzker Prize architect Tadao Ando's impulse was to plant before he could build. His 'Chinkon no Mori', or 'Forest for the Repose of People's Souls', recalled an earlier participation in the Hyogo Green

10. J Vidal, 'A Eureka Moment for the Planet: We're finally planting trees again', The Guardian, 13 February 2018

11. R Qobil & P Harris, 'Restoring Life to the Aral Sea's Dead Zone', BBC News, 1 June 2018 [www.bbc.com/news/business-44159122], retrieved 12 August 2018

12. Greenpeace East Asia, 'China's Remaining Forests', Greenpeace International, 2012 [www.greenpeace.org/eastasia/campaigns/forests/problems/china-remaining-forests], retrieved 15 August 2018

13. Princeton University, Woodrow Wilson School of Public and International Affairs, 'Survival and Restoration of China's Native Forests Imperiled by Proliferating Tree Plantations', ScienceDaily, 2 May 2018 [www.sciencedaily.com releases/2018/05/180502153256.htm], retrieved 12 August 2018

14. G Orwell, 'A Good Word for the Vicar of Bray', in 'Fifty Essays' by G Orwell, Oxford City Press, 2010, pp.398–401

15. S Körner & F Bellin-Harder, 'The 7000 Eichen of Joseph Beuys – Experiences after Twenty-Five Years', Journal of Landscape Architecture, vol.4, no.2, 2009, pp.6–19

16. Arcspace, 'Tadao Ando: Regeneration - surroundings and architecture', Arcspace.com, 24 August 2014 [https://arcspace.com/exhibition/tadao-ando-regeneration-surroundings-and-architecture], retrieved 18 August 2018

17+18. WS Merwin, 'The Preservation of the Land and Home', The Merwin Conservancy, 3 March 2018 [https://merwinconservancy.org/land-and-home], retrieved 15 August 2018

19+20. C Haberman, 'The Lasting Legacy of a Fighter for the Amazon', The New York Times, 27 November 2016.

21+22. J Gettleman, 'Wangari Maathai, Nobel Peace Prize Laureate, Dies at 71', The New York Times, 26 September. 2011

Network, where Ando participated in a tree-planting effort after the Kobe earthquake, bringing white-blossoming trees to the interstices of the city: roads, home gardens, the forgotten spaces in-between buildings. As Ando remarked, 'While I create buildings, I dream the day will come when children gather and read books under huge magnolia trees. [This is why I] sometimes feel planting trees is my most important task.'[16]

In the Pe'ahi Valley on the north shore of Maui, Hawaii, US poet laureate William Stanley Merwin and his wife Paola planted 2740 palm trees. One of the world's most important botanical collections, the Merwin Palm Forest can be read also as Merwin's final ecological poem, a living analogue for the linguistic affinity for nature so prevalent in his writings. On surfaces of paper and on the surfaces of the earth, the poet composed lyrical expressions of resilience. The 19 acres of wasteland possessed a specific beauty that Merwin describes: 'I loved the wind-swept ridge, empty of the sounds of machines, just as it was, with its tawny, dry grass waving in the wind of late summer.'[17] Compelled by this beauty, Merwin and his wife brought 480 taxonomic species, more than 125 unique genera, and nearly 900 different horticultural varieties to the project, described as 'Hawaii's version of Walden Pond'.[18]

Trees are living monuments to our heroism forgotten or undervalued. When Brazilian ecological activist Chico Mendes was shot in front of his family for protecting the Amazon's rubber trees from rapacious business interests, he joined over 200 environmental defenders killed for 'simply... fighting to protect the environment'.[19] As Mendes had said, 'At first, I thought I was fighting to save rubber trees, then I thought I was fighting to save the Amazon rainforest. Now I realise I am fighting for humanity.'[20]

In this ecological story, women who took risks to defend the environment brought trees into the very heart of the struggle for power. Wangari Mathai, the 2004 Nobel Peace Prize Laureate, employed nearly 900 000 women to plant more than 30 million trees for her Green Belt Movement in Kenya, an action deemed 'subversive' by the government.[21] When she organised resistance against the building of a skyscraper in the centre of Nairobi's only public park, she was beaten unconscious by police soon after at a protest. Targeted by authorities, Mathai was nonetheless 'a force of nature', and was likened to Africa's acacia trees, 'strong in character and able to survive the harshest of conditions'.[22]

Just as the paintings of Romantic landscapes trained a revelatory disposition towards the natural, cultivating an outlook both whimsical and critical, imaginative and participatory, so too does the project of creating resilient landscapes. The work of our hands records our romance with trees and the ways our eyes have been ever-guided towards horizons, in landscapes that enlist and implicate the human observer with nature. Both environmental and cultural practices encode the poetics of resilience, which aims to project an idealised world which the human is both estranged from and organically connected to. Such an attunement is the deeper implication of a resilient imagination by a Smartcity, whose ambitions and scope include the nourishment not only of bodies, but also of minds, hearts and communities.

Thirty million trees in Kenya, the rubber trees of the Amazon. Eight thousand poplar in Afghanistan, seven thousand oaks in Kassel, the fragrant magnolia blossoms in Kobe, the inimitable branches of the acacia trees in Kano. The treasured survivor trees of Hiroshima, the tenacious saxaul trees of the Aral Sea. Trees are not only protagonists and poets, but are provocations to inspire us to think more expansively about how nature underwrites our resilience as a species.

The Grand Paris of Niger: Landscape of hope

The Grand Paris of Niger is imagined not merely as a landscape for restorative agriculture, but as a metaphor of hope, empowerment and independence. The holistic water management approach adapts the previously arid homeland into a planted shelterbelt grid of sustainability. The cultivated oasis dissolves decades of dependency on foreign aid and international politics, metaphorically forgiving its colonial past.

This seed of humanity is sown on the northern border of the Sahel region in Africa – Agadez in Niger. Sahel is the eco-climatic and biogeographic zone of transition between the Sahara to the north and the Sudanian Savannah to the south, stretching from the Atlantic Ocean to the Red Sea. Sahel has a fragile ecosystem, and is highly vulnerable to the many complex and interconnected challenges linked to climate fluctuations in the region. Land, impaired by weather extremes and multi-secular pastoralism, consistently loses its productivity. Securing access to water is a crucial prerequisite for pastoral mobility and is a common cause for farmer–herder conflicts. An estimated 500 million people live on land undergoing desertification – the most extreme form of land degradation – with the Sahelian belt most affected. Water scarcity has resulted in low agricultural productivity and poor diet, and has led to catastrophic malnutrition of millions. The hungry and weak population is caught in a 'poverty loop',[1] with few opportunities to study and work, and becomes even poorer.

At the same time, the region has experienced the most significant population growth: from 31 million in 1950 to more than 100 million today, and is forecast to reach more than 300 million in 2050.[2] The combination of resource scarcity and population growth has triggered one of the most massive migrations in history. Annually, over 100 000 migrants pass through Agadez to cross the scorching dusty Sahara Desert and swim across the Mediterranean Sea with only one simple longing – to reach Europe. Most of them, unfortunately, end up in the Libyan slave trade market, where one person could be sold for as little as US$800.[3] This human misery for survival has earned Agadez a reputation – 'the smuggling capital of Africa'.[4]

In addition to this unprecedented migration trend, uranium mining is another serious threat to the region. Even after the end of French colonialism, the French nuclear giant AREVA continued its mining activities in Niger. It is reported that of the hundreds of thousands who rely on the open pit mines for their livelihood, many have been exposed to radioactivity and died prematurely of undeclared diseases.[5] The alarming levels of radioactivity are found in water, soil and air samples. The contaminated aquifers have also destroyed the traditional way of life of communities.

previous page: The cultivated oasis of hope, empowerment and independence.

facing page: The Grand Paris of Niger is the outcome of efficient and poetic water management techniques and resilient landscaping strategies.

1. P Heinrigs, 'Security Implications of Climate Change in the Sahel Region: Policy considerations', SWAC, 2010, p.8

2. M Potts, E Zulu, M Wehner, F Castillo & C Henderson, 'Crisis in the Sahel: Possible solutions and the consequences of inaction', The Oasis Initiative, 2013, p.3

3. CNN, 'People for Sale: Exposing slave auctions in Libya', edition.cnn.com, 14 November 2017 [https://edition.cnn.com/specials/africa/libya-slave-auctions], retrieved 11 Nov. 2018

4. Politico, 'Welcome to Agadez, Smuggling Capital of Africa', www.politico.eu, 17 October 2016 [https://www.politico.eu/article/the-smuggling-capital-of-africa-agadez-niger/], retrieved 11 Nov. 2018

5. A Dixon, 'Left in the Dust: AREVA's radioactive legacy in the desert towns of Niger', Greenpeace International, Amsterdam, 2010

The Grand Paris promises a new hope for Niger. An oasis that is not just a shelter, a garden that is not just an agricultural field, a well that is not just a water refilling point; but the last harbour for the migrants to reconsider their treacherous journey. Whilst Paris in Europe is synonymous to love; the Grand Paris aims to cultivate the same human virtue for Africa. The love, in this context, embraces nature, climate, community lifestyles and local traditions. Furthermore, the landscape of hope seeks climate justice, and undertakes the ambitious efforts offered by the COP 21 Paris Agreement 'to combat climate change and adapt to its effects, with enhanced support to assist developing countries'.[6]

Now there is only one simple question – how do you turn sand into water and make a lush oasis in the middle of a desert? Current studies show that only one-eighth to one-third of the total rainwater is used in agriculture in arid climate development,[7] and half of the water is lost in evaporation in farming systems.[8] At the same time, the proportion of annual rainfall associated with extreme rainfall has increased over the past 50 years.[9] While the annual precipitation rate in Niger is extremely low, ranging from 50 to 100mm a year, the majority of rainfall is concentrated between June and September. In 2017, floods unleashed by the months of torrential rain killed at least 50 people and displaced nearly 120 000.[10] Therefore, the greatest challenge to cultivate resilience is to manage the unreliable distribution of water from rainfall over time, rather than the lack of it.

Practice and science have developed specific techniques for the challenge – the small-scale in-situ agricultural methods and the larger multi-purpose external water harvesting. The former, which include 'zai' pitting and half-moon trenching, are low-labour, simple and native to the region, and can significantly increase agricultural production in a good rainy season. However, as weather extremes are only to worsen, short term techniques are not enough to overcome dry spells. The use of larger multi-purpose external catchments for runoff collection immediately adds water to the balance, and they are efficient and low cost. Sand dams and sub-surface dams are curious engineering inventions to trap water behind small walls in sandy riverbeds. Sand can naturally clean water and protects it from evaporation. An integrated catchment landscape is the solution for big scale rainwater harvesting.

Poor water hygiene and lack of proper sanitation have caused infectious diseases outbreaks every year in the region. On the other hand, unmanaged human excreta, manure, and food wastes are missed opportunities for production of sustainable organic fertiliser or biogas.[11] The Grand Paris aims to completely close the water loop by differentiating the water types: blue – potable water, grey – technical water, black – sewage, and managing them accordingly. There are three main features of the sanitation loop principle.

(1) Urine and excreta are considered as valuable sources to increase soil fertility, rather than waste. Retention ponds can clean water in a natural slow manner, whilst algae and fish are by-products of the cycle. Successful precedents of biogas plants producing fuels out of human excreta and manure can be found in many developing countries.

(2) Pathogens are eliminated on site to ensure the recycling process is safe. This can be achieved by installing smaller septic tanks within housing blocks. A dispersed network is more resilient than conventional recycling plant constructions.

(3) The usage of water must be minimal and considered, an important factor in any extreme water scarcity context. As agriculture is the primary fresh water consumer, priority is given to efficient irrigation methods and digital irrigation control. Greywater should be used where possible, even substituted with sand – for cleaning, for example.

The Grand Paris combines water harvesting and treatment with habitation, and public facilities with agriculture, all in one holistic system. The new hydrological landscape follows a set of climatic principles of wind and sun, rather than a strict human-made masterplan. To implement these principles, key infrastructures of the French capital are metaphorically employed for various water management techniques and resilient landscaping strategies:

following page: Patch by patch, the desert is transformed into a resilient landscape.

- Avenues – irrigation channels and shelterbelts, which extend from north to south and block eastern winds from the Sahara.
- Railways – seasonal river sand dams, and cultivated riverine wetlands.
- Quarters – in-situ water conservation techniques.
- Fountains – public water points (laundry and bathing) and grey water and black water treatment.
- Parks – underground reservoirs.

291

The Sand Dam, located on the seasonal river, is the main source of fresh water for the community. Housing units are inserted within the downstream facing slope, structurally reinforcing the dam. At the same time, the large thermal mass of the dam ensures inner thermal comfort, while aquaponics cultures in the pools (fish and aquatic plants) purify water and provide fresh produce. In the dry season, the dam looks like an inhabited terraced hill with buried cisterns and covered pools. In the rainy season, the dam fills up and directs excess water downstream into irrigation channels to feed the surrounding blooming landscape. Runoff water is captured in reservoirs, and is protected from evaporation and contamination by canopies. The upper plate of the irrigation channel brings stormwater to communal and domestic reservoirs, while the lower plate transports sewage to reclamation ponds in settlements.

The new wealth of the Grand Paris is solar power generation, a sustainable replacement for the harmful uranium industry. The multi-purpose Wind Dam acts as a dust barrier for the solar harvesting fields and, simultaneously, as a solar chimney and cooling tower for the brick tunnel kiln. The brick factory is located on the south-facing slopes to get maximal solar gain. New bricks are baked daily and are used to extend the factory, thus constantly recreating itself. Used for local vernacular architecture since ancient times, mud bricks are undeniably the simplest available construction material for a sustainable development. The Wind Dam also captures nutritious topsoil carried by dust storms, which is used as plant fertiliser.

The Grand Paris outlines a procurement method for a positive and cost-effective hydro-environmental transformation in Niger – it is feasible for the rapidly growing population to live and prosper in a self-supporting arid landscape. With access to water, colourful triangular patterns of vegetation including shade-providing acacia trees and date gardens gradually embrace the desert, patch by patch echoing Nigerien weaving patterns. The 21st century caravanserai welcomes not only migrants en route to Europe or uranium refugees who intend to permanently settle, but also provides for the nomadic population, activating trade and exchange activities, turning the desert into a resilient landscape of hope.

6. United Nations Climate Change, 'What is the Climate Agreement', 2015 [https://unfccc.int/process-and-meetings/the-paris-agreement/what-is-the-paris-agreement], retrieved 10 November 2018

7. M Falkenmark, P Fox, G Persson & J Rockström, 'Water Harvesting for Upgrading of Rainfed Agriculture: Problem analysis and research needs', Stockholm International Water Institute, 2001, p.24

8. D Molden, 'Water for Food Water for Life', Hoboken, Taylor and Francis, 2013, p.6

9. S Nicholson, C Funk & A Fink, 'Rainfall over the African Continent from the 19th through the 21st Century', Global and Planetary Change, vol.165, Elsevier, June 2018, pp.114–127

10. Channels Television, 'Niger Flooding Kills 50, Displaces Over 100 000 in Four Months', www.channelstv.com, 11 September 2017 [https://www.channelstv.com/2017/09/14/niger-flooding-kills-50-displaces-100000-four-months], retrieved 11 Nov. 2018

11. SA Esrey, I Andersson, A Hillers & R Sawyer, 'Closing the Loop: Ecological sanitation for food security', Swedish International Development Cooperation Agency (Sida), Water Resources no.18, 2000, p.12

Sitopia – The urban future

What might cities look like a hundred years from now? Predicting the future is never easy, yet one thing is clear: if cities in the 21st century look much as they do today, only bigger, we will have failed the greatest ecological challenge of our time. Cities are expanding faster than at any time in history, and the manner of their expansion (the ad hoc arrival every week of 1.3 million rural migrants) is altering the relationship that has for millennia underpinned civilisation: that between city and country. Cities have always plundered the natural world for resources, but in the past, so few people lived in them (just three per cent in 1800) that their impact was limited. Today, with half the global population living in cities and a further three billion expected to join them by 2050, the opposite is the case. If the future is urban, we urgently need to redefine what that means.

Of all the resources needed to sustain a city, none is more important than food. In the pre-industrial world, this fact was self-evident: the sheer difficulty of feeding cities made it so. Without the benefit of farm machinery, agrichemicals, refrigeration and rapid transport (all the essentials of modern agribusiness) cities were forced to be both frugal and inventive with their food supplies. No city was ever built without first considering where were its sources of sustenance, and once established, cities kept 'food miles' to a minimum, growing perishable foods such as fruit and vegetables in the city fringes, and raising animals such as pigs and chickens within the city itself. Fresh foods, including grass-fed livestock, were consumed seasonally, with the excess preserved by salting, drying or pickling to be consumed during leaner months. The fertility of the soil was paramount, and across the centuries, various methods were employed (from blood sacrifices to crop rotation) to nurture it. No food was ever wasted: kitchen scraps were fed to pigs, human and animal waste was collected and spread as fertiliser, and leftovers from kingly feasts were handed to the needy.

In the post-industrial world, things are very different. We take it for granted that, if we walk into a restaurant or supermarket, food will be there, having arrived magically from somewhere else. Food is plentiful and cheap – so much so, that one could be forgiven for assuming that producing it was easy. Yet, however inexpensive food has become in supermarkets, its true cost is many times as great. Food and agriculture together account for one third of global greenhouse gas emissions. Nineteen million hectares of rainforest are lost every year to agriculture, while a similar quantity of arable land is lost to salinisation and erosion. Each calorie of food we consume in the West has taken an average of ten calories to produce, yet half of the food produced in the United States is thrown away. A billion people worldwide are overweight, while a further billion starve. As such statistics suggest, the global food industry is deeply flawed, yet this is the system upon which our urban life depends.

Our very concept of a city, inherited from a distant, predominantly rural past, assumes that the means of supporting urbanity can be endlessly extracted from an ever-bountiful rural hinterland. For the past two centuries, industrialisation has fuelled that assumption, both by greatly accelerating the rate at which such extraction is possible, and by extending the distance (in all senses) between city and country. The result has been an unprecedented explosion of urban development, accompanied by the creation of a dangerous illusion: that cities are somehow independent, immaculate and unstoppable. Now that the illusion is wearing off, we are in urgent need of a new urban model: one that recognises the vital role that cities play in the global ecology. But how are we to arrive at such a model?

First, we need to understand the close bond between food and cities, forged some 10 000 years ago in an area of the ancient Near East known as the Fertile Crescent. It was here that our Neolithic ancestors first began to gather wild grass seeds – experiments that were crucial to the eventual development of urban civilisation, since it was the harvesting and subsequent cultivation of grain that was to provide the first source of food capable of sustaining an urban population. As the gathering of seed gradually evolved into its deliberate cultivation and harvesting, permanent farming settlements began to be built close to the fields.

The first such settlements considered complex enough by archaeologists to be described as cities were a group of Sumerian city-states, Uruk, Ur and Kish, in southern Mesopotamia (modern Iraq), dating from around 3000 BC. These early prototypes consisted of dense urban cores surrounded by intensive farmland, made fertile through irrigation from the floodwaters of the River Euphrates. The cities were dominated by large temple complexes, which held yearly cycles of festivals mirroring the agricultural seasons. These culminated in the harvest, a convulsive moment in the city's calendar involving the entire population in complex rituals of mourning, sacrifice and rebirth. Once the grain was safely gathered in, it was offered to the gods, before being carefully stored and redistributed among the people. The temples were thus the cities' spiritual centres as well as their chief means of food distribution. Physically and spiritually, they embodied the vital bond between city and country that remains (despite appearances) fundamental to all urban life.

Throughout the pre-industrial era, feeding their citizens remained every city authority's biggest priority. Apart from the physical difficulties involved, the social and political aspects of the food supply required constant management. Most cities had laws in place to prevent various malpractices, including the formation of monopolies. The buying and selling of food was usually confined to open markets, where the trade could be most easily monitored. As a result, cities were fed by large numbers of small producers, selling from legally defined pitches at specific hours to regular customers. Since markets were the only places one could go to buy fresh food, they became powerful social and political spaces as well as commercial ones. From the Athenian agora and Roman forum to Les Halles and Covent Garden, markets were vital social hubs that linked cities to the countryside and expressed civic life in its fullest sense.

Food's role in shaping the pre-industrial city is easy to see, yet food is still shaping post-industrial ones, albeit in a far less obvious way. Modern food systems have emancipated cities from geography, disguising the effort of feeding them. But that doesn't mean the problem has gone away. On the contrary, in allowing the metropolitan carpet to roll out across field and forest, tundra and desert, industrial food systems have made the very thing they promised to make easier – feeding cities – infinitely more complex. Three billion of us now live, not merely remote from the sources of our sustenance, but utterly dependent on supply systems that are unsustainable.

Cities have always had their critics, but in the past, those wishing to escape them could do so if they chose. That is no longer true. The economic reach of cities is now such that few rural areas remain unaffected by them. Indeed, the steady flood of rural migrants to join the ranks of the urban poor is driven as much by the collapse of rural economies as by any allure of urban life. In many parts of the world, living in the countryside is no longer viable, precisely because the land has been transformed in order to feed cities.

Equally damaging is the mental transformation wrought on us by urbanity. The severance of man from nature – the essential achievement of modernity – has left us in danger of forgetting what, deep down, it means to be human. Ours is a world dominated by abstractions – credit crunches, bottom lines – that have little to do with the daily rhythms that once gave people's lives meaning. Divorced from the necessities of the everyday, we search instead for stimulation in the form of computer games and twitter, while all the while (like Dorian Gray's portrait in the attic) the true impact of our urban lifestyles wreaks havoc unseen. With all the efficiency and technology of modernity, we have neither succeeded in solving what E.F. Schumacher called 'the problem of production', nor made ourselves happy.[1] On the contrary, the further removed we are from our natural selves, the less capable we become of true contentment.

As a model for human dwelling, the city has outgrown itself. In its post-industrial form, it offers neither a good quality of life for most, nor a sustainable future. Yet it remains the dominant model for human development. Half of India's 700 000 small farms are expected to disappear over the next 20 years, as a centuries-old way of life finally succumbs to urbanity. Yet the question of what millions of displaced rural workers are going to do with themselves once their farms are gone remains unanswered. As China's recent experience has shown, the mass abandonment of the countryside for cities is no guarantee of a better life. On the contrary, 26 million Chinese migrant workers found themselves unemployed at the start of 2009, leaving the Chinese Government contemplating a 'social time-bomb'. If the credit crunch has taught us anything, it is that we urgently need to question the values by which we live, and the foundations upon which we build communities.

Our first step must be to acknowledge the essential paradox at the heart of all urban life: namely, that without a rural counterpart, it can't exist. Once we grasp that fact, we will be significantly closer to the essence of the problem we face. It is one of balance. In our rush to become civilised – to raise ourselves above the level of mere survival – we have forgotten that we remain animals, with animal needs. We may choose to live in shiny, glowing things called cities, but in a deeper sense, we still dwell on the land. What is needed is not so much a technological revolution as a mental one: a recognition that, once we lose our vital bond with nature, we too are lost. Our most urgent mission must be to regain a sense of that bond.

That is where food – so powerful in ancient cultures, so debased in ours – can play a vital part. Food is the sine qua non of our existence: the one thing none of us can live without. Its rituals of growing, buying, cooking, eating and sharing have, more than any others, shaped our civilisation. As we face our greatest ever human-made crisis, food could hold the key to our salvation. Its central role in our lives gives it a

295

1. E F Schumacher, 'Small is Beautiful', Vintage, London, 1973, p.3

unique power over us – making it the perfect vehicle for interrogating, disentangling and ultimately redesigning how we dwell on earth.

My shorthand for this approach is sitopia, meaning 'food-place' (from the ancient Greek sitos, food, and topos, place). It is a deliberate alternative to utopia, the theoretical ideal model that has for many centuries been the commonest method of addressing the dilemmas of human existence. Utopian themes – sociability, sustainability, equality, health, happiness – are unimpeachable; the problem is, because utopia aims at perfection, it can never be fulfilled. If we want to build a better world, we need a model that aims, not at perfection, but at something partial and attainable. That is where sitopia comes in. Because it uses food as a tool, sitopia already exists, albeit imperfectly. Food affects everything from the way we work, play and socialise, to the way we walk and talk, and inhabit land, sea and sky. As soon as we learn to see how it shapes our lives, we can use food in multiple ways to shape things better. At the macro scale, that will involve finding ways to reconnect ourselves to nature, and city to country. At the micro scale, it might mean anything from changing the way we design and build houses to the sorts of foods we eat for breakfast.

If the range of sitopian opportunity seems daunting, we only have to look at existing models for inspiration. Arguably, all pre-industrial cities were sitopias of sorts: societies that recognised and celebrated the primacy of food. Although nobody would suggest a return to what was undoubtedly a tough and mephitic existence in the pre-industrial city, there is much we can learn from a time justifiably named 'the golden age of urban ecology'.[2] Limited by the constraints of geography, pre-industrial cities were forced to live within their means; something that we, in the post-industrial world, must learn to do again. With the dual benefits of technology and hindsight, we must take the ancient urban model and remould it for our times; not in order to romanticise the past, but in order to seek its wisdom. For 200 years, we have suffered collective amnesia about our place in the organic order of things. Now we must re-embrace it.

Above all, sitopia is an approach that makes connections between apparently disparate aspects of our lives, and tries to establish a balance between them. Part of that process involves learning to frame the right questions. For instance, instead of asking how we can feed cities most 'efficiently' (a question which, by its very nature, can only yield one result), we should be asking what sort of communities we want to live in, and design our food systems accordingly.

Once you put the question that way round, what immediately becomes apparent is that industrial food systems are totally antithetical to the values to which we might aspire in an ideal society, summarised by the utopian themes listed above. Indeed, such systems are deliberately anti-social in nature, having been cleansed of any aspect of humanity that might interfere with their profit margins. Seen as diagrams, they are shaped like trees, with many roots (producers) channelled through a single trunk (supermarket) to feed many branches (customers).[3] They are thus structured so that the trunk exerts a stranglehold over the entire system, keeping producers and customers apart – the very thing that city authorities in the past struggled so hard to avoid.

Now imagine another system, in which city-dwellers forge direct relationships with those who grow their food. In such a scenario, customers would quickly become knowledgeable enough to influence the food network through their choices. They would effectively become collaborators in the supply process – what the founder of the Slow Food Movement Carlo Petrini calls 'co-producers'.[4] Such a food network would produce a very different society: one far more likely to foster the

296

sorts of personal connections necessary for a successful community, and one far more resilient in the face of external shocks. A society, in fact, much more like those of cities in the past.

Exercises such as this remind us that food systems exert an enormous influence, and that their essential role must therefore be not merely to feed us adequately and sustainably (no mean feat in itself) but also to nourish our quality of life. If all we are concerned with is survival, then a ruthless series of calculations aimed at maximising the ecological synergies between diet, soil, sunlight, water, energy and waste would indicate how we should arrange our lives. But if we also care about such things as joy, ethics, culture and freedom, we are faced with a far trickier – yet far more worthwhile – problem. How best to reconcile the satisfaction of our animal needs with our higher aspirations? That is the great dilemma of civilisation, one that people have long struggled to answer.

Ultimately, the dilemma boils down to a single question: what is a good life? The answer remains elusive, but surely some part of it must involve respect for food. And if that is so, then surely a life spent nurturing others through food – so long as one is respected and rewarded for doing so – must be a good one? The reduction of food to a zero-sum commodity has done more than rob us of variety, identity, taste and smell in our lives: it has taken away the most dependable source of income and sociability we are ever likely to find.

Although food has never been used explicitly as a design tool, its presence is implicit in many utopian projects. In 1902, Ebenezer Howard published a modest pamphlet entitled 'Garden Cities of To-morrow', in which he set out his idea for a 'town-country magnet': a community that would combine the benefits of town and country life, while neutralising the disadvantages of both.[5] The 'magnet' was effectively to be a city-state of 30 000 city-dwellers and 2000 farmers, consisting of a dense urban core surrounded by 5000 acres of farmland. Once its target population was reached, the Garden City would not expand; instead, a sister-city would be built some distance off, joined to the first by rail. In this way, the landscape would gradually be transformed into a network of connected, largely self-sufficient city-states. Howard received widespread interest in his idea, and even received the financial backing to build the first prototype, at Letchworth in Hertfordshire. But the project was ultimately a failure, since the radical nature of the proposal – incremental land reform – never took place.

Like all utopian projects, the Garden City was doomed to failure. However, in sitopian terms, it carries a powerful message. Whatever form human dwelling takes, the relationship between town and country will always be at its core, reconciling the two being our greatest challenge. Because Smartcities attempt to address that challenge, they are, in my terms, sitopian. Such projects give us courage to imagine the future. But you don't have to be an architect to be sitopian. How we choose to farm, shop, eat and cook is up to us, but our choices, multiplied many times over, are what will shape our future.

2. Donald Reid, 'Paris Sewers & Sewermen: Realities and Representations', Harvard University Press, Cambridge Mass., 1993, p.10

3. For a discussion on the way such systems relate to cities (or rather, don't) see the essay by Christopher Alexander, 'A City is not a Tree', 'Architectural Forum', Vol. 122, No. 1, April 1965, (Part I) and Vol. 122, No. 2, May 1965 (Part II)

4. See Carlo Petrini, 'Slow Food Nation', Rizzoli, New York, 2007, pp.164–176

5. Identified by him as unsanitary overcrowding in towns and lack of services and opportunity in the countryside

Project + Reproduction Credits

Project Credits

Smartcities, Resilient Landscapes + Eco-warriors
research team [Book]: CJ Lim, Ed Liu with Jacqueline J Barrios

Guangming Smartcity
Shenzhen, China; 2007
commissioned by: Shenzhen Municipal Planning Bureau
design team: CJ Lim/Studio 8 Architects with Pascal Bronner, Ed Liu, Daniel Wang, Lukas Wescott, Barry Cho, Nikolay Salutski, Jacqueline Chak, Anabela Chan, Dimitris Argyros, Alleen Siu, Maxwell Mutanda, Thomas Hillier, Adeline Wee, Andreas Helgesson, Tomasz Marchewka, Jonathan Hagos, Ben Masterton-Smith, Chen Chen Pang, Lei Guo, Louise Yeung
consultants: Fulcrum Consulting (environmental + sustainability engineers) Andy Ford, Brian Mark, Jules Saunderson; Techniker (structural engineers) Matthew Wells; Alan Baxters + Assoc. (transport) David Taylor; Urban Plannning + Design Institute of Shenzhen (local planners) Zhou Jin, Yang Xiaochun, Zhu Zhenlong
total area: 7.97km2

DuSable Park
Chicago, USA; 2001
commissioned by: Laurie Palmer
supported by: Illinois Arts Council, R Driehaus Foundation Graham Foundation, USA
design team: CJ Lim/Studio 8 Architects with Michael Kong
consultant: Techniker (structural engineers)
total area: 0.89km2

Tangshan Earthquake Memorial Park
Tangshan, China; 2018
commissioned by: Tangshan Municipal Planning Bureau
design team: CJ Lim/Studio 8 Architects with Daniel Wang, Martin Tang, Pascal Bronner, Dimitris Argyros, Anna Andronova
consultant: Techniker (structural engineers)
total area: 1.20km2

Remembering the Great American Plains
Fargo, USA; 2017
commissioned by: Kilbourne Group
design team: CJ Lim/Studio 8 Architects with Alex Gazetas
consultant: Techniker (structural engineers)
total area: 0.008km2

Nordhavnen Smartcity
Copenhagen, Denmark; 2008
commissioned by: CPH City + Port Development
design team: CJ Lim/Studio 8 Architects with Pascal Bronner, Kar Man Leung, Rachel Guo, Barry Cho, Thomas Hillier, Maxwell Mutanda, Yongzheng Li, Loui Lim
consultant: Fulcrum Consulting (sustainability engineers)
total area: 2.00km2

Daejeon Urban Renaissance
Daejeon, Korea; 2007
commissioned by: Daejeon Metropolitan City
design team: CJ Lim/Studio 8 Architects with Pascal Bronner, Barry Cho, Maxwell Mutanda, Frank Fan
consultant: Techniker (structural engineers)
total area: 0.89km2

The Tomato Exchange
London, UK; 2009
design team: CJ Lim/Studio 8 Architects with Jen Wang, Yongzheng Li, Frank Fan, Barry Cho
consultant: Techniker (structural engineers)
total area: 0.008km2

Central Open Space: MAC
Yeongi-gun, Korea; 2007
commissioned by: Government Administrative City Agency + Korean Land Corporation
design team: CJ Lim/Studio 8 Architects with Pascal Bronner, Dimitris Argyros, Daniel Wang, Alleen Siu, Thomas Hillier, Martin Tang
consultants: Techniker (structural engineers) Matthew Wells; Fulcrum Consulting (environmental + sustainability engineers) Brian Mark; KMCS (quantity surveyors) Colin Hayward
total area: 6.82km2

The Linear Park
Shunde, China; 2018
commissioned by: Dongseng Real Estate Development
design team: CJ Lim/Studio 8 Architects with Rachel Guo, Pascal Bronner, Jen Wang, Jason Ho
local architect: OS Partnership China
consultant: Techniker (structural engineers); Fulcrum Consulting (environmental + sustainability engineers)
total area: 0.5km2

A Workplace in a Garden
Wexford, Ireland; 2018
design team: CJ Lim/Studio 8 Architects with Pascal Bronner
consultant: Techniker (structural engineers), Fulcrum Consulting (environmental + sustainability engineers)
total area: 0.01km2

Guangming Energy Park
Shenzhen, China; 2008
commissioned by: Shenzhen Municipal Planning Bureau
design team: CJ Lim/Studio 8 Architects with Dimitris Argyros, Barry Cho, Kelly Chan, Louise Yeung
consultants: Fulcrum Consulting (sustainability engineers)
total area: 2.37km2

Newark Gateway Project
Newark, USA; 2009
commissioned by: AIA Newark + Suburban
design team: CJ Lim/Studio 8 Architects with Jen Wang, Barry Cho, Pascal Bronner, Daniel Wang
consultant: Techniker (structural engineers)
total area: 0.25km2

The City of a Thousand Lakes
Nanjing, China; 2015
commissioned by: Gaochun District Government, Gaochun Municipal Planning Bureau, China
design team [London]: CJ Lim/Studio 8 Architects with Steve McCloy, Ryan Hakiaman, Jason Lamb, Samson Lau, Woojong Kim, Eric Wong, Yu-Wei Chang, Nick Elias
design team [Nanjing]: Hongyang Wang with Dongfeng Zhu, Yi Chen, Jingjing Li, Wen Yuan, Yan Zhang
consultants: Fulcrum Consulting (environmental + sustainability engineers); Techniker (structural engineers)
total area: 80.20km2

Rifle Range Regeneration
Penang, Malaysia; 2010
commissioned by: Penang State Government, Malaysia
design team: CJ Lim/Studio 8 Architects with Julia Chen, Frank Fan, Barry Cho, Martin Tang
consultant: Fulcrum Consulting (environmental + sustainability engineers); Techniker (structural engineers)
total area: 0.08km2

Dongyi Wan East Waterfront
Shunde, China; 2009
commissioned by: Dongseng Real Estate Development
design team: CJ Lim/Studio 8 Architects with Rachel Guo, Jen Wang, Pascal Bronner, Kar Man Leung, Julia Chen
local architect: OS Partnership China
consultant: Techniker (structural engineers); Fulcrum Consulting (environmental + sustainability engineers)
total area: 0.22km2

Brockholes Wetland + Woodland Reserve
Preston, UK; 2001
commissioned by: Lancashire Wildlife Trust, Northwest Regional Development Agency, The Tubney Charitable Trust
design team: CJ Lim/Studio 8 Architects with Thomas Hillier
consultant: Techniker (structural engineers)
total area: 1.06km2

The Green Pension Plan
London, UK; 2018
supported by: Architecture Research Fund [The Bartlett, UCL]
design team: CJ Lim/Studio 8 Architects with Ivan Chan, Dean Walker, YoonJin Kim, Alex Gazetas, Martin Tang
consultant: KaMan Lai [Health Infrastructure Research Centre], Francesca Medda [Quantitative and Applied Spatial Economics Research Laboratory UCL], Techniker (structural engineers)
total area: –km2

WanMu Orchard Wetland
Guangzhou, China; 2012
commissioned by: Guangzhou Haizhu District Goverrnment + Guangzhou Municipal Planning Bureau
design team: CJ Lim/Studio 8 Architects with Lik san Chan, Alex Gazetas, Pascal Bronner, Martin Tang, Barry Cho, Yu-Wei Chang, Savan Patel, Ned Scott, Franky Chan, Dean Walker, Xiaoqing Qian, Tingting Wu, Tao Yang
consultants: Mott MacDonald Hong Kong (ecology); Mott MacDonald London (sustainability); SCUT Guangzhou (heritage development); Techniker (land engineering); Space Syntax Lab
total area: 25.96km2

Imagining Recovery
USA; 2009
design team: CJ Lim/Studio 8 Architects with Ed Liu, Rachel Guo, Jen Wang
total area: –km2
Note: Illustrated in the chapter 'Urban Utopias and the Smartcity'

299

Reproduction Credits

8 NASA Earth Observatory

42–43 Pascal Bronner

44–45 Martin Tang

192–193 Alexandra Dumitras

287, 288, 292 Anna Andronova

all other images CJ Lim/Studio 8 Architects

Special thanks to Ivan Chan, Jason Ho + Alanna Donaldson

Index

301

304

305

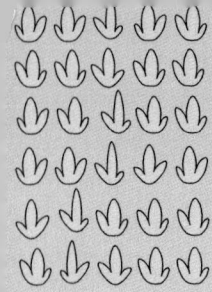

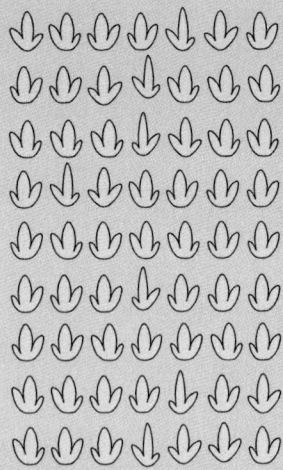

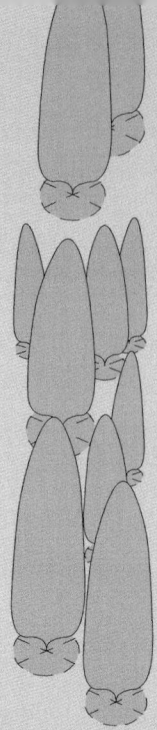

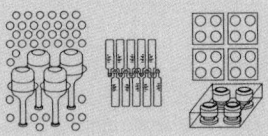

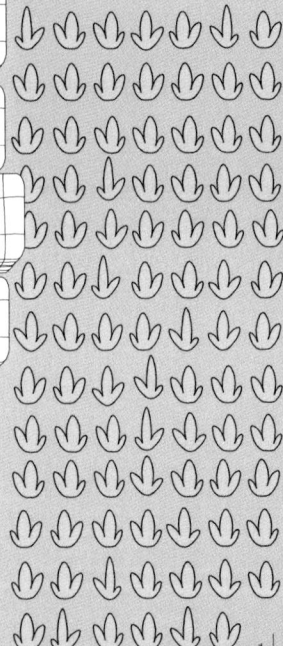

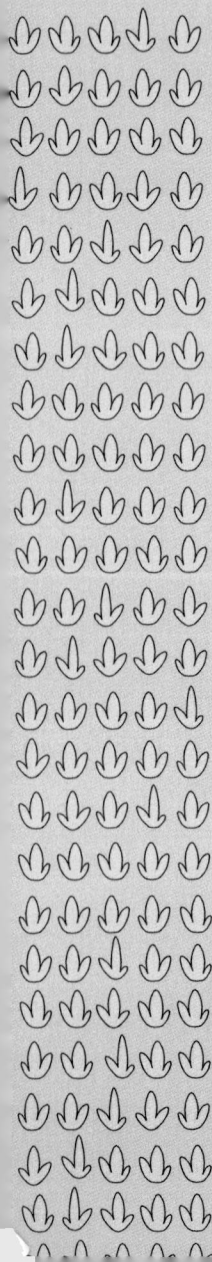

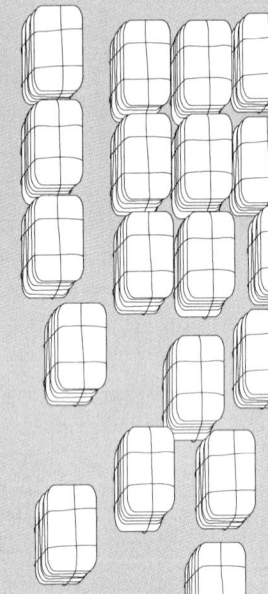

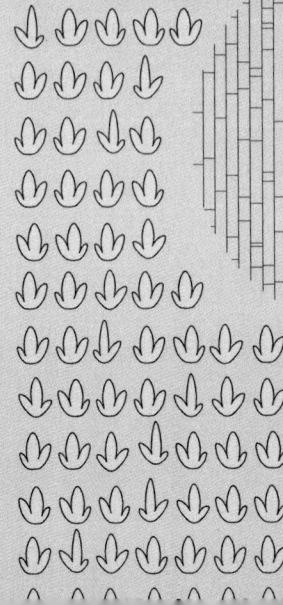

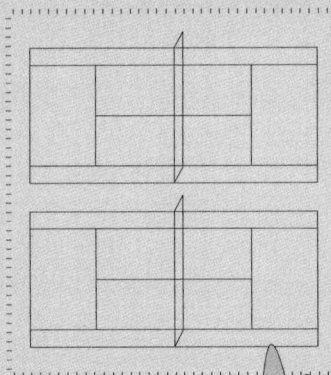

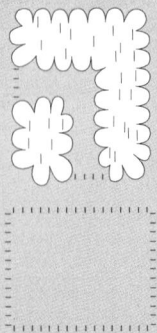

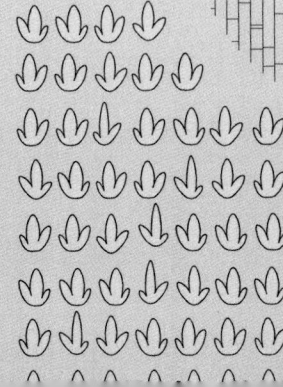

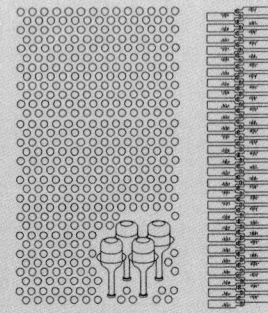